Charles Emmerson was born in Australia in 1976 and grew up in London. He studied modern history at the University of Oxford and international relations at the Institute of Political Science in Paris. Since then, he has worked on a broad range of global policy issues at the European Commission, the International Crisis Group conflict prevention organisation, and as an Associate Director of the World Economic Forum, responsible for their work on global risks. He currently lives in London.

CHARLES EMMERSON

The Future History of the Arctic

How Climate, Resources and Geopolitics are Reshaping the North, and Why it Matters to the World

VINTAGE BOOKS
London

Published by Vintage 2011

2 4 6 8 10 9 7 5 3 1

First published in Great Britain by The Bodley Head in 2010

Vintage
Random House, 20 Vauxhall Bridge Road,
London SW1V 2SA

www.vintage-books.co.uk

Addresses for companies within The Random House Group Limited can be found at: www.randomhouse.co.uk/offices.htm

The Random House Group Limited Reg. No. 954009

A CIP catalogue record for this book is available from the British Library

ISBN 9780099523536

The Random House Group Limited supports The Forest Stewardship Council (FSC), the leading international forest certification organisation. All our titles that are printed on Greenpeace approved FSC certified paper carry the FSC logo. Our paper procurement policy can be found at www.rbooks.co.uk/environment

Printed and bound in Great Britain by
CPI Cox & Wyman, Reading, RG1 8EX

To my mother and father,
who are responsible for much more than they know.

Contents

Maps

Illustrations

Photo section one

P. 1. Greenland. Photograph by Charles Emmerson. P. 2. Illustrations by Édouard Riou in *Voyages et aventures du capitaine Hatteras: Les Anglais au pôle Nord* (*The Voyages and Adventures of Captain Hatteras: The English at the North Pole*), Bibliothèque d'éducation et de récréation, Paris, 1866; Fridtjof Nansen. Photograph published by Bain News Service, 29 April 1915. George Grantham Bain Collection, courtesy Library of Congress. P. 3. Vilhjalmur Stefansson. Photograph published by Bain News Service. George Grantham Bain Collection, courtesy Library of Congress; 'The Path of Supremacy' from *The Northward Course of Empire*, Vilhjalmur Stefansson, George G. Harrap & Co. Ltd., London, 1922. P. 4. Josef Djugashvili, aka Stalin, 1913, photographer unknown. David King/David King Collection; Stalin greeting Valerii Chkalov, 1936, photographer unknown. David King/David King Collection; *SSSR na stroike*, May 1941. David King/David King Collection. P. 5. *Vstrecha* (return), photo-montage celebrating the 1934 airlift of survivors from the *Cheliuskin*. David King/David King Collection; Construction of the Belomorkanal (White Sea Canal), 1931–1933, photographer unknown. David King/David King Collection.

P. 6. Solovetsky monastery, 1915. Photograph by Sergei Prokudin-Gorskii, courtesy Library of Congress; Book jacket, *Arkhipelag Gulag, 1918–1956: Opyt knudozhestvennego issledovaniia*. Paris: YMCA-Press, 1973–1975. David King/David King Collection. P. 7. Solvetsky monastery today. Photograph by Charles Emmerson. P. 8. Painting depicting the Alaska purchase, courtesy the Alaska State Library, Alaska Purchase Centennial Commission Photograph Collection, P. 20–181.

Photo section two

P. 1. The flag raised by Robert E. Peary at the North Pole in 1909, National Photo Company Collection, courtesy Library of Congress; Arthur Chilingarov, August 2007, Reuters/Alexander Natruskin. P. 2. *USS Skate*, 1959, photograph courtesy The Ohio State University Archives, Papers of Sir George Hubert Wilkins, RG 56.6, image #35_5_4; Canadian Prime Minister Stephen Harper arriving on *HMCS Toronto*, 2008, Reuters/Andy Clark. P. 3. Murmansk. Photograph by Evgenii Khaldei, 1942, courtesy Library of Congress; Murmansk today. Photograph by Charles Emmerson. P. 4. Satellite images showing Arctic ice coverage. NASA, courtesy nasaimages.org; Ny-Ålesund, Spitsbergen, Photograph by Charles Emmerson. P. 5. ". . . But first, let's hear your position on the Alaska pipeline . . ." Cartoon by Pat Oliphant © 1998, Universal Uclick. Used by permission. All rights reserved; Melkøya Liquefied Natural Gas plant, Norway. Photograph by Charles Emmerson; Alexei Miller, CEO, Gazprom. Reuters/ Sergei Karpukhin. P. 6. "Will you take care of him for the duration?". Cartoon by C.K. Berryman, published 12 April 1941, probably in the *Evening Star* (Washington, D.C.), courtesy Library of Congress; View of Nuuk, Greenland. Photograph by Charles Emmerson; Denmark's Queen Margarethe II, Greenland, 21 June 2009. Reuters/Scanpix Denmark. P. 7. End of the road in the Icelandic highlands. Photograph by Charles Emmerson; Icelandic riot police protect the country's parliament from protestors, 2008 Reuters/STR New. P. 8. Alaska's North Slope, August 2008. Photograph by Charles Emmerson.

Foreword

Since the hardback edition of *The Future History of the Arctic* was published in early 2010, some details in the portrait of the Arctic offered here have inevitably been updated by events. Some disputes have moved closer to resolution; others have remained stubbornly unsolved. Some commercial developments in the Arctic have slowed; others have entered a higher gear. The environment, geopolitics and geo-economics of the Arctic remain very much in flux.

Norway and Russia have come to agreement over ownership of a large section of the Barents Sea, opening the way for potential oil and gas development. Canada has restated its foreign policy objectives in the Arctic, citing the United States as its 'premier partner', and stating the resolution of ownership disputes as a major objective. An American Secretary of State has, for the first time, attended a meeting of Arctic coastal states. China has pushed ahead with plans for its first icebreaker, to be operational by 2013. Gas has been struck by a British company off the coast of Greenland. A shipping route across the Arctic coast of Russia, the so-called Northeast Passage, has attracted an increasing (though still low) number of commercial shipments. Joint military exercises involving Canadian, American and Danish forces have taken place in the Canadian Arctic.

These events, and others, have tended to reconfirm the central arguments of this book: the Arctic is increasingly central to global affairs, many of our popular preconceptions of the region will have to adjust to the more complex reality now emerging, while challenges to the Arctic environment have intensified. The basic facts of climate change have been sharpened by Arctic science, and their direct consequences have been experienced by communities throughout the north. The longer-term political, strategic and commercial consequences of climate change – both at the Arctic and at the global level – are beginning to enter the mainstream of public discussion.

There have been false leads, too. The tragic explosion of the Deepwater Horizon rig in the Gulf of Mexico with the loss of eleven lives in April 2010 – and the appalling oil spill which followed – has turned public attention towards the activities of oil companies in difficult and sensitive environments across the world, including the Arctic. For many, the Gulf of Mexico spill marks a turning point in global attitudes to oil – or, at the very least, it should.

Whether things will turn out that way is far from clear. At the very least, those who argue that Arctic oil and gas development carries immense environmental risks have been strengthened. Oil companies will, rightly, be subject to far more formal and informal monitoring. Some development of oil and gas resources in the American Arctic will be delayed. However, in other parts of the Arctic – in Norway and Greenland – development will probably continue if justified by the scale of resources, the economics of their production, the strength of political support and the frameworks of environmental regulation. In Russia, exploitation of Arctic resources is considered a strategic imperative – and is therefore highly likely to continue if the economics and politics stack up.

What is absolutely clear is that the debate on whether resource development should continue in the Arctic – and on what terms – is not going away. Indeed, it is likely to become more intense. The future history of the Arctic – requiring us to reshape our image of the region – has only just begun.

London, Autumn 2010

Introduction

The first time I crossed the Arctic Circle, I missed it entirely.

I think my mother's original plans for a family holiday to Sweden had not involved engaging in what my ten-year-old mind imagined as an adventure. She had more sensible destinations in mind – Stockholm, Uppsala, Gothenburg: the Sweden of lakes, of pretty painted houses, fluttering blue-and-yellow flags and girls with flowers in their hair. But I insisted.

In the mind of a ten-year-old almost any line on a map is worth crossing for the sake of it. I had crossed the International Date Line before I was seven years old which, at the time, I accounted as a great personal achievement, though really I was just a passenger with three hundred others on a trans-Pacific flight. I had been over the Equator. I had even crossed – and, more to the point, been aware of crossing – the tropics of Capricorn and Cancer on the way to Australia from Europe. The Arctic Circle, I felt, was the only great line remaining. There had been a television drama a few years before which stuck in my mind, about Scott and Amundsen's race to the Pole. The objective of that explorers' duel had been the South Pole of course, not the Arctic North Pole, but what did that matter? Some children collected football stickers; I collected destinations.

I determined that our adventure, should we be successful – remembering Scott, I thought it irresponsible to guarantee success given the obstacles ahead – would take us to the Swedish city of Kiruna. If we were lucky, and weather permitted, perhaps we could press on to Narvik, on the coast of Norway, which I had imagined as a kind of Arctic Shangri-La. Though it was mid-summer, I pictured snowstorms and train carriages filled with the musty smell of fear, Aquavit and pipe tobacco. What I got was the clean efficiency of the national Swedish rail system.

My mother's calmness perplexed me. She did not seem too alarmed by the potential dangers of our adventure to the outer reaches of human habitation. Perhaps, I told myself, she was merely hiding her nerves.

The train pulled out of Stockholm's Central Station and started north-west. First, the suburbs raced past, with the mix of office blocks, industrial depots and apartment blocks steadily giving way to houses, and houses giving way to fields. Then the commuter villages, strung out along the railway line, the last outposts of the city. Then, as our route began to right itself towards due north, the rhythm of human settlements slowed. Ten minutes of empty landscape, then a blur of houses, then twenty minutes, then thirty. As each gap got longer I sank further into my highly sprung dark-blue seat.

Initially, I had confidently expected to see a troupe of elk, at the very least, roaming just beyond the city limits. Later, with the weak northern sun still streaming into our train carriage almost horizontally I would have settled for a pack of foxes, a lone wolf, an Arctic hare, anything. I looked at the map of Sweden again. True, we were still hundreds of miles from the magic curved line, the Arctic Circle. I had to be patient, I reminded myself, and fell asleep to the slow rocking of the train.

When I woke the next morning, my mother told me to start getting ready – we'd be in Kiruna in half an hour. I got dressed quickly and, my excitement returning, glued my face to the window.

'What on earth are you looking for?' my mother asked, packing the last of our things into an overnight bag.

What a question. 'The Arctic Circle, of course,' I replied.

'Oh, we crossed that ages ago, not that there's much to see,' my mother informed me, matter-of-factly.

I could barely contain my disappointment. We had come all this way, and, at the end of it all, I had missed the moment of triumph. I had read in a book somewhere that the Arctic Circle was marked by a line of white stones, and I had imagined great rune-inscribed menhirs towering over the landscape around, the silent guardians of the Arctic's secrets. And I had missed it.

What was it, exactly, that fascinated me about the Arctic as a child?

Part of the answer lies in the way in which it seemed to break down accepted categories. Who can set a clock by a sky which seems to extend days and nights into whole seasons? Who can define direction in a place where all the lines of longitude slowly converge on a single point? The Arctic was a permanent trick played on teachers who would have you believe in straight lines and being on time.

More enthralling even than the overturning of accepted categories, however, was the space the Arctic allowed for imagination to take its place. The Arctic became a place where extraordinary events and exceptional individuals could be readily accepted as true, and where fiction looked as real as a textbook. The Arctic was a place where disbelief could be permanently suspended. What I knew to be impossible in my own world might conceivably exist up there. And though this reversed the adult world's hierarchy of reality and fantasy, that upending of the rules has resonance for all of our imaginations. The secret of the longevity of our half-imagined Arctic is that, above all, we want it to exist.

But my fascination predated even that first, failed, journey across the Arctic Circle. Maybe it started with an article on the Svalbard archipelago – 'Norway's Strategic Arctic Islands' – in a 1978 *National Geographic* left lying around the house. Among grainy colour-saturated photographs I discovered a place where street signs warned of polar bears; a place where one might encounter a citizen of that fantastical dark empire called the Soviet Union. The Arctic seemed filled with hidden importance.

Or perhaps my yearning for the land beyond began with a map covering one of my bedroom walls which had a huge off-white Greenland dominating the North, and the unfamiliar names of islands, inlets and towns beyond the dotted line: Novaya Zemlya, Disko Bay, Thule.

As with so many childhood obsessions, my fascination for the Arctic has retained its mystery and its strength over the years. Part of what draws me back to the Arctic today is that childish dream of the Far North – with its pristine isolation, extraordinary nature, oversized characters and magical connotations. But there is something else which pulls my mind inexorably north, an adult realisation that the Arctic is not just a theatre for my imagination, but a stage on which history is being played out.

What is the Arctic? For the geographer, it starts with that line on a map, the Arctic Circle, endlessly running round the earth at a latitude of 66° 33′ 39′′ North. For the climatologist, it is an area defined by average temperature, with the Arctic's border extending far south or north of the Arctic Circle depending on which temperatures from which seasons are considered most important. For the biologist, the Arctic is a way of describing a set of conditions in which a certain ecology may or may not flourish. For the political scientist, the Arctic extends far to the south of the Arctic Circle, limited only by the ambitions of states to claim an Arctic role, as do China, France and the United Kingdom, to name a few. While the geographer's Arctic is fixed, the political scientist's Arctic is in a constant state of flux. And for the historian? For that unlucky soul, the Arctic is all these things and more.

But for most of us, the Arctic is above all an idea. It cannot be mapped, it can only be described. Cold, isolated, empty, white, pristine – these are the terms which the word 'Arctic' calls to the mind. This is the mental framework inside which our personal picture of the Arctic develops. These are the preconceptions through which we filter all subsequent information about it. Yet many of those preconceptions – the same ones which, as a child, drew me towards the Arctic – are wrong. The Arctic is not empty, it is populated. Though much of the Arctic is isolated, parts of it are easily accessible. And while the Arctic is in many places a beautiful and unspoiled wilderness, in others it is heavily polluted.

There is perhaps one other word which the Arctic evokes: unchanging. The thought of a place which lies outside time – a source of stability, even refuge – is appealing to the modern mind. There is comfort in the idea that, beyond the clamour, complexity and rush of urban life there is the possibility of escape. But, again, the notion of the Arctic as unchanging is wrong. It always has been.

Long before planes, factories and cars brought prosperity and pollution to our world, climate change was a fact of Arctic existence. Thousands of years ago, human life in the Arctic beat to the rhythm of climatological time – a timescale which man-made global warming has bent out of shape. In the deep history of the Arctic, environmental factors – whether climate change or different approaches to environmental management – were key.

Take Greenland. The first waves of population on the island – the early Dorset culture and their Inuit successors – advanced and retreated over time, often in close synchronisation with global climatic changes. The Dorset culture arrived on Greenland in around 800 BC travelling across the frozen ice between Canada and Greenland.[1] Two thousand years later, around AD 1200, after the freeing up of the waterway between Greenland and Canada as a result of a period of global warming, the Inuit followed the migrations of their staple food – the whale – and set up hunting colonies in northern Greenland. To the south, a Norse population, mostly subsisting on onshore agriculture, flourished at around the same time. The arrival of the Norse on Greenland – the first settlements were established at the end of the tenth century – was partly related to environmental factors, namely the shortage of agricultural land in Iceland.

The growth of the Norse colonies of the south in the first half of the last millennium was linked to the same warm period in global history that had prompted the Inuit colonisation of the north. When the original Norse colony died out, five centuries or so after it had been established, it was a combination of a number of factors, principally environmental mismanagement and climate change, which led a marginal community on the fringes of European settlement to decline, wither and die.[2] Following the return of a permanent European population in Greenland after 1721, the colony's survival became linked not to the land but to the sea.

The interaction of climate change with human society on Greenland cannot just be consigned to the distant past. Much more recently, in the 1930s and 1940s, Greenland's fishing industry boomed as a result of the island's expanding cod fishery, itself the consequence of, amongst other things, a spike in temperature. Now, in the twenty-first century, the warming of the Arctic has, once again, unleashed a complex set of challenges and opportunities for Greenland's population.

The idea of an unchanging Arctic – a place on earth without history, where things have always been as they are now – is categorically wrong. True, much of what has happened in Arctic history has happened on timescales far longer than those we are used to thinking in: generations rather than lifetimes, decades rather than years. Nonetheless, changed it has.

To say this is not to deny that much of the Arctic is special, or that it is beautiful, or that it is inspiring. No one who has stood atop a mountain on Spitsbergen, or peered through the morning mists at an iceberg off Greenland, or looked down from a plane as it soars over the polar ice cap, can fail to be struck by the Arctic's inherent majesty and wish for its protection. Moreover, to say that the Arctic has always changed is not to deny that it is changing now more quickly than ever, and largely as a result of human activity.

But this book is not about the Arctic as we tend to think of it: picturesque, pristine and unchanging. This book is not about the Arctic of my childhood. There are other books that describe the human fascination with the Arctic, or which expound the poetic qualities of Arctic nature far better than I ever could.[3] And there are other books, though relatively few, which more classically describe the Arctic's past, taking into account both the perspective of its indigenous population and that of the Arctic's later inhabitants.[4] Finally, exploration of the Arctic and the heroic deeds of those racing for the North Pole, fascinating though they are, are not the main focus of this book.

Instead, this book looks at an Arctic that has come of age. In the twenty-first century, the Arctic's historical isolation is untenable. Its role in the major issues of our global future – energy security, climate change, globalisation, the balance between economic development and environmental protection – is unavoidable. The Arctic has become a lens through which to view the world. And this, ultimately, is why the Arctic matters. It is why we should care about the Arctic's history and why the Arctic's future, uncertain as it is, will affect all of us.

What is a 'future history'? The two ideas – history and the future – seem at first glance opposed. History is an account of events which have occurred and about which the facts are, for the most part, known or knowable. The future, on the other hand, is necessarily a matter of conjecture, shorn of the certainty of facts.[5]

But there are, in fact, multiple ways in which history and the future interrelate. The first is that there is a history to ideas about the future. After all, ideas about the future have changed over time. This is particularly true of the Arctic, where the very blankness of its history has made it an attractive screen for a broad range of projections of the future, both those projections intended as entertaining speculation and those intended as more serious prediction. The future history of the Arctic is, in part, the history of how ideas of an Arctic future emerged, what those ideas have meant and how they inform our current ideas of the Arctic's future.

The second way in which history and the future interrelate is more classical and perhaps more familiar: the past does not determine the future, but it moulds it. Grasping the history of the Arctic is a precondition to understanding the motivations, hopes, fears and ambitions of those who will shape its future. Understanding the history of the Arctic will help to narrow the gap between pure speculation about the Arctic's future and considered supposition. Just as the future of a country depends, to some degree, on its history, so it is with the Arctic.[6]

The journey of this book, therefore, is one which often takes us back into Arctic history. But it does so with a purpose: to help us think about the Arctic as it is now and as it may be. After all, many of the forces which have shaped the Arctic's history will also, inevitably, shape its future. It is around those forces – visions, power, nature, riches and freedom – that this book is structured.

Visions takes us back to the origins of the idea of the Arctic as outside the current of human history, and how that notion of the Arctic began to be overturned in the twentieth century. *Power* takes us inside the articulation of those visions: the historical struggles which shaped the borders of the modern Arctic, the unfinished process of defining national frontiers in the twenty-first century, and the emergence and future of the Arctic as an arena of military competition. *Nature* examines the importance of the Arctic for our understanding of climate change and how the Arctic and the economic geography of the world are being shaped by the consequences of Arctic warming. *Riches* explores the role of natural resources in Arctic history, the key question of where and how those resources are likely to be exploited in the future, and what the environmental and geopolitical consequences of exploitation will be. *Freedom*, most fundamentally of all, takes us inside the choices facing smaller Arctic nations as they seek to choose their own future.

As this book strives to emphasise, the Arctic is not a single place, fenced off from the world. It is a fractured region, increasingly tied to economic and political interests outside it, in Asia and Europe as well as in the Arctic countries themselves. The views from Moscow, Helsinki, Reykjavik or Washington are very different. While in the United States the Arctic is an Alaskan afterthought associated principally with energy and polar bears, in Russia it is considered a fundamental part of national identity and a repository of national power. In Canada the Arctic is a hinterland fought over through domestic politics, and it is a symbol of Canadian distinctiveness. In Norway the Arctic is the country's source of future wealth and claim to historical greatness. Finally, in Greenland and Iceland, the Arctic is simply what is all around.

The research for this book has taken me to many of the places I dreamed of as a child, looking up at the empty spaces of the map and wondering at their mystery. But it has also taken me inside the government ministries, oil companies, environmental organisations and research institutes which are shaping the Arctic's future, sometimes in cooperation with one another, often in competition. And so in writing this book I have spoken to politicians and diplomats, environmentalists and corporate executives, scientists and sailors, businessmen and engineers, spies and soldiers, and to the local populations whose futures are so closely bound up with that of the Arctic. I have looked, listened and, I hope, learned.

PART ONE
Visions

'The history of Arctic discovery shows how the development of the human race has always been borne along by great illusions'

Fridtjof Nansen, 1911

1

Oracles and Prophets: Rethinking the North

'What do we, in reality, see in the polar and frozen countries of the North . . . ? During the greatest part of the year life seems almost extinguished by the rigorous cold of a perpetual winter . . . The absence or scarcity of arborescent vegetation gives it that character of poverty and uniformity which strikes us in these desolated lands'

Arnold Guyot, French geographer, 1850

'Behold! Far north along the shores of the Arctic a quiver of upspringing settlements fringes the coast. Boats swarm around canning factories, smoke flutters above smelters, herds of reindeer dot the prairies. Motor drawn sledges glide over the snow. Aeroplanes whiz across the sky . . . And here or there, on every street-corner, glimmer out the lights of theatres where moving-pictures entertain white people through the sunless weeks of the midwinter dancing-time, the singing-time, the laughing-time of Eskimo Land'

Northward Ho! An Account of the Far North and Its People, 1927

For the historian–geographers of the ancient world, and for most of their successors down to the twentieth century, the North mattered little. Their focus was on the 'temperate zone' of the northern hemisphere, an area bounded by the Tropic of Cancer to the south and the

Arctic Circle to the north. Everything that mattered in the grand narrative of human history – the birth of civilisation, the invention of commerce, the construction of cities, states and empires – had taken place between those lines. What lay beyond them was unimportant.

Combining observation with rudimentary geography, the philosophers of Greece and Rome viewed civilisation and latitude as inseparable. Informed by the disposition of human population in their own world, with its towns and cities clustered around the shores of the Mediterranean and the fertile Middle East, they concluded that only a temperate climate could produce history. It was they who first expressed a remarkably enduring prejudice: that both the Far South and the Far North were unfit for human civilisation.

In Aristotle's view the 'good life' could only be pursued in the temperate area that lay between the 'torrid' South and the 'frigid' North. Later, Ptolemy described the world as divided between twenty-one parallel zones, corresponding to different climates with differing levels of suitability for human life. The northernmost extent of his map was a line of latitude of 63° North, beyond which he considered human life to be impossible. Drawn today, such a northern border would exclude half of Scandinavia, much of Russia, all of Iceland, and the vast bulk of Greenland and Alaska.

The vision of a world divided by latitude was not just a Greek or Roman conceit. At the height of Islam's intellectual influence in the Early Middle Ages, Arabic geographers drew maps separating the world into seven climes of habitation – the most inhabited being the fourth and fifth, with little habitation in the seventh. Remarkably, there was some Arab acquaintance with the North in the Early Middle Ages. Several writers described the Northern Lights; one described the use of sleighs. But knowledge was fragmentary, and tended to reinforce the idea of the North as obscure, exotic and outside the main current of historical development.

Elsewhere, in the lands at the north-western corner of the known world, the arguments of latitude were employed as a tool to convert heathens to Christianity. Writing to St Boniface in the early eighth century, Daniel, Bishop of Winchester, advised that a comparison between the geographies of the South and the North would offer conclusive evidence of God's favour to Christians:

And while they, that is, the Christians, possess fertile lands, and provinces fruitful in wine and oil and abounding in other riches, they [the gods] have left to them, the pagans that is, with their gods, lands always frozen with cold . . .[1]

In Scandinavia, still further to the north, Aristotelian ideas linking latitude, climate and civilisation were accepted without question. The Norse viewed themselves as living at the edge of the liveable world, 'nearer the frigid zones, where the frost is able to use its chilling powers'. Beyond them, wrote an unknown Norse writer in the thirteenth century, there was nothing:

It has been stated as a fact that Greenland lies on the outermost edge of the earth toward the north; and I do not believe there is any land in the home-circle beyond Greenland, only the great ocean that runs around the earth.[2]

The equation of latitude with civilisation has persisted through time. In *The Character of Races*, the most famous American geographer of the early twentieth century, Ellsworth Huntington, wrote that:

. . . the strain of life in the far north tends to eliminate the very type which is most likely to start some new idea and thereby cause progress . . . The man of action, the one who has ideas, generally leaves Alaska because he is the one to whom the strain of inaction is least endurable during the unduly long winter.[3]

Taking this argument one step further, Huntington saw climate as not only a determinant of prosperity but also as an explanation for perceived racial characteristics:

We realise that a dense and progressive population cannot live in the far North . . . simply because the difficulty of getting a living grinds men down and keeps them isolated. We know that the denizens of the torrid zone [the South] are slow and backward, and we almost universally agree that this is connected with the damp and steady heat.[4]

Such categorical assertions were not controversial in Huntington's day. Popular history, economics and the burgeoning pseudo-science of race seemed to confirm them.

To all of these historians and geographers, from Aristotle to Huntington, the idea of civilisation and prosperity as a function of latitude and climate appeared to accurately describe the experience and direction of global history.[5] Most events of perceived global importance had, after all, taken place in the 'temperate zone'. Were not all the great civilisations – the Egyptians, the Greeks, the Romans, the Arabs and the Chinese – products of these latitudes?

In this view of the world, history spun along an axis of East–West development, not Pole to Pole. It was the project of connecting East and West which occupied the minds of explorers, geographers and traders.[6] In the sixteenth and seventeenth centuries, when European navigators began to map some parts of the Arctic shoreline, their intention was to chart a route through the Arctic to the Orient, rather than to settle or develop the Arctic region for itself.[7]

In effect, the territory of the Arctic remained the possession of those who had lived there the longest – the indigenous communities of the North. European settlement of the Arctic – except for a few small Norse communities in Greenland, Iceland and Scandinavia, and Russian settlements beyond Novgorod – was out of the question. Throughout the 'Age of Exploration', running from the fifteenth century to the seventeenth, the Arctic was not so much an objective as an obstacle in the way of grand designs to girdle the earth. The Arctic hinterland, meanwhile, was left untouched and unknown.

Without a firm understanding of the Arctic, either on land or at sea, speculation flourished. Maps would either leave the Far North as a blank space on the page or substitute conjecture for knowledge, as with Mercator's 1595 map of the Arctic, featuring four entirely imaginary islands surounding an open polar sea.

Much of what passed for information about the Arctic was false or exaggerated. Laurent Ferrer Maldonado, a Spanish navigator who claimed to have found the Northwest Passage at the end of the sixteenth century, wrote a widely translated account of his journey, now discredited, describing an easy voyage across the Arctic coast of North America.[8] Having passed through what he called the 'Anian Strait', he claimed, in 1588, to have come across Russian ships laden with Chinese goods. His story was almost certainly false: the Russians themselves would not discover what is now called the Bering Strait for another sixty years.[9] Perhaps the true purpose of his account, with its many nods to the advantages the King of Spain might glean from the control of the 'Anian Strait', and his detailed description of what would be necessary to furnish a second voyage, was to raise money.

A popular seventeenth-century account of travels in the Arctic by a French traveller, Pierre Martin de la Martinière, merged genuine experience with a more fictionalised taste for the obscure and the mysterious, writing that 'nearly all those who live above the [Arctic] Circle . . . are sorcerers, who control the winds at will'.[10] He described the presence, in the house of every 'Lappon' family, of large black animals each of which, 'although, by its terrifying aspect, looks like a cat, I believed and still believe to be family demons'.[11] De la Martinière's account is full of semi-accurate observations, patent exaggerations and pure fictions. Despite – or because of – this mixing of fact and fiction his book was translated into six languages and went through sixteen editions.[12]

Given such unreliable information, it is perhaps not surprising that when, at the end of the eighteenth century, de la Martinière's countryman, Denis Diderot, was compiling what he hoped would be the first comprehensive view of the state of accurate human knowledge, he afforded a scant few lines to the 'Arctique' in the multiple volumes of his *Encyclopédie.*[13] A few decades later, in a world increasingly obsessed with lists, classifications, encyclopedias and almanacs, John Macgregor's *Resources and Statistics of Nations* had little more – and nothing positive – to say about the Arctic. While commenting on Russia that 'no other nation, is, in reality, *so little known*, nor any whose power has been *so greatly exaggerated*', Macgregor was even more scathing about Greenland and Iceland:

> Iceland and Greenland are scarcely valuable for any purpose, but the fisheries, and some traffic in furs; the former yields a little barley, hemp, and flax. Its wild animals are white bears and foxes.[14]

It was only in the mid-nineteenth century that the Arctic itself became a major goal of exploration. Only then were regular national surveys sent out to assess the value of the lands of the Arctic, rather than simply to map its edges for East–West navigation. In the United States, surveying of Arctic territory started only with the acquisition of Alaska in 1867. Mapping the Canadian Arctic began in earnest at the turn of the nineteenth and twentieth centuries.

While such geographic and statistical surveys brought an increase in accurate information about the region, it was the literary Arctic – terrifying, cruel and mystical – which dominated public imagination. In *Frankenstein* (1818) it is while searching for the North Pole that Captain Robert Walton, locked in the ice pack, meets Dr Frankenstein – described as the 'divine wanderer' – and hears his story of scientific endeavour and its tragic consequences.[15] In the English translation of Jules Verne's *Voyages et aventures du capitaine Hatteras* (1866), the eponymous hero of the book is finally driven insane by his obsession with the North, succumbing to what his doctor notes as 'polar madness':

> For some time Captain Hatteras, followed by his faithful dog, that used to gaze at him sadly, would walk for hours every day but he always walked in one way, in the direction of a certain path. When he had reached the end, he would return, walking backwards. If any one stopped him, he would point his finger at a portion of the sky. If any one tried to make him turn round, he grew angry, and Duke would show his anger and bark furiously.
>
> The doctor observed carefully this odd mania; he understood the motive of this strange obstinacy; he guessed the reason of this walk always in the same direction, and, so to speak, under the influence of a magnetic force.
>
> Captain John Hatteras was always walking towards the north.[16]

Often, the boundaries between fact and fiction blurred. The true saga of the search for the lost Franklin expedition – a mid-nineteenth-century attempt by the British Royal Navy to discover the Northwest Passage – generated immense newspaper coverage, at least as sensational and colourful as that of any fictional account of the Arctic.[17] Later, in the 1890s, when the great Norwegian explorer Fridtjof Nansen set off on his famous voyage into the Arctic ice pack, he named his ship *Fram*, a Norwegian translation of the name of the ship which carried the fictional Captain Hatteras in Jules Verne's novel. By design or by accident, fact and fiction merged.

By the turn of the twentieth century, however, thinking about the Arctic was developing beyond the heroic, the uncanny and the fantastic. Some were willing to question the Aristotelian link between latitude and civilisation with a far more positive view of the North, its importance and its future. The old idea of the Arctic as at the margins of human history was being superseded by the idea of the Arctic as frontier. Two men in particular – Fridtjof Nansen and Vilhjalmur Stefansson – challenged the traditional view of Arctic marginality, placing the

North at the core of their visions of human development and progress. At last, the Arctic was viewed as something more than an accessory to the course of history.

Fridtjof Nansen (1861–1930) was the very symbol of the Arctic at the turn of the nineteenth and twentieth centuries. An overachiever in every field to which he directed his mind and his energy, he was considered the father of modern polar exploration, treated as an oracle by the Arctic and Antarctic explorers of the early twentieth century. He proved his theory that Arctic pack ice was in constant movement – the theory of 'polar drift' – and thereby undercut centuries of speculation about an open ice-free polar sea (or even a polar continent) without actually ever reaching the North Pole. He redefined the popular vision of the Arctic in his image, straddling the romantic and the modern. At once mystical and scientific, poetic and practical, patriotic and internationalist, Nansen personified the paradoxes of turn-of-the-century attitudes to the Arctic.

Vilhjalmur Stefansson (1879–1962), the Icelandic–Canadian–American explorer, was more controversial than Nansen and his claims for the future of the Arctic were more strident. The North, in his view, had received a bad press. He devoted his life to correcting what he considered to be the romantic and literary misappropriation of the Arctic, and to overturning a set of prejudices about its possibilities. In so doing, he earned himself a remarkable epithet: 'prophet of the North'.[18] Through books, speeches, articles and pamphlets Stefansson criticised the notion of an unbreakable link between latitude and civilisation as unsubstantiated, unscientific and unimaginative. The view of the Arctic as uninhabitable and unprofitable was, he insisted, quite wrong. Turning received wisdom on its head, Stefansson outlined a northward turn in the course of human history. The future, as he saw it, lay north.

Fridtjof Nansen's attitude towards the Arctic was ambivalent. On the one hand, Nansen put Arctic exploration at the very core of his vision of human progress – viewing the scientist's experiment and the explorer's expedition as two sides of the same coin. Envisaging exploration as a common endeavour for the future of mankind, Nansen wrote to the famous journalist W. T. Stead, 'true civilization will not have been reached until all nations see that it is nobler to conquer nature than to conquer each other'.[19] He identified the resources of the Arctic – particularly the Russian Arctic – as a source of future prosperity.[20]

But Nansen's view of the Arctic was not confined to modernising fantasies of exploration, conquest and exploitation. On occasion, Nansen described the Arctic in more romantic terms, as a place with the spiritual power to redeem the ills of the modern world. Just as Nansen's moods swung between euphoria and despair, his confidence in human progress swung between boundless optimism and a belief that only spiritual renewal could pave the way for human development. Vaunting the importance of character, integrity and independence, he argued that the Arctic was one of the few places on earth where humanity could reconnect with those values. In 1926, towards the end of his life, Nansen told the assembled students of St Andrew's University – of which he was Rector – that 'deliverance will not come from the rushing, noisy centres of civilization; it will come from the lonely places'.[21] As a man who had built his own life around the North – but who tragically never attained the holy grail of the Pole – it was perhaps understandable that he viewed the Arctic as imbued with spiritual resonance.

If Nansen's life could be encapsulated in a single word it would be 'forwards' or, in Norwegian, 'fram'.[22] While it was Nansen's crossing of the Greenland ice cap, in 1888, which made him a national hero in Norway, it was the voyage of the *Fram* between 1893 and 1896 that gave him international repute. Fittingly, when Nansen died, a public campaign was launched to save the *Fram* in an expression of reverence for the great Norwegian. The man might be gone, but the object most closely associated with his life and achievements could still be rescued.

But 'fram' was more than this. 'Fram' was the philosophy by which Nansen lived his life and the core of his vision of human history. A man of extraordinary energy and drive, Nansen was constantly pushing forward when others would not. Extending this idea of forward movement to mankind as a whole, he explained historical progress in terms of the ceaseless human impulse to extend the boundaries of knowledge. For him, Arctic exploration and scientific advance were, in fact, a 'single, mighty manifestation of the power of the unknown' and of man's desire to overcome it.[23] Through his hugely popular writings and lectures – not to mention his very public persona – Nansen projected an image of the Arctic as intimately connected with the story of human progress.

Nansen's iconic status – combining the traditional romance of adventure in far-off places with a thoroughly modern attitude toward science and progress – arose initially from his innovative approach to Arctic exploration.

In 1888, Nansen set off to cross the Greenland ice cap on skis. It had never been done before. Whereas all previous attempts were relatively large expeditions starting from the populated west coast of Greenland, thus allowing for a line of retreat, Nansen travelled light, starting from the unpopulated east coast and moving west. To many, this seemed an extraordinary – and potentially suicidal – decision. But Nansen thought his approach coldly logical, justified by the psychological incentives the route would provide to press forwards. By travelling on skis and by placing his faith in a small expedition, Nansen thought that he could move quickly and lightly. The strategy worked. In 1889, Nansen returned to Christiania (Oslo) a conquering hero, having confounded his critics and proven the possibility of applying new methods to Arctic exploration. Instead of heroic failure, Nansen had given the Norwegian people – still conjoined with Sweden under the Swedish monarchy – a focus for national pride.

In the 1890s, Nansen undertook an even more extraordinary feat – one that confirmed his status as an Arctic icon and dramatically advanced the cause of the polar science. Whereas the traditional rule of Arctic navigation had been to avoid being locked into the ice pack at all costs, Nansen announced his deliberate plan to do just that. Rather than pitting himself against the power of the Arctic, Nansen intended to harness it. Nansen reckoned that he and his men would be carried far closer to the North Pole – and in far greater comfort – if they were to allow themselves to be drawn northwards by the drift of polar ice.

Many viewed the project of harnessing polar drift as foolhardy. For a start, polar drift remained a hypothesis rather than a proven fact. Some doubted that a boat could be built which could withstand the enormous pressure of the ice pack. If the hull of the boat were to crack, the crew would face the task of trekking to safety over several hundred miles of Arctic ice.

But Nansen, having taken a keen interest in the design of the *Fram* (intended to ride up above the ice rather than relying on withstanding its lateral pressure) believed that the risks he and his crew would take had been adequately identified and minimised. If the expedition were to succeed, it would not only reflect positively on the spirit of Norwegian endeavour, but demonstrate the ability of man to harness the natural power of the Arctic and also would prove a number of theories about the physical nature of the Arctic.

The *Fram* expedition was a success in all respects but one: Nansen did not reach the North Pole. Having been carried as far north by the ice pack as he thought possible, Nansen attempted a 'dash for the north' on foot, leaving the *Fram* behind under the command of Otto Sverdrup.[24] This time, however, the drifting ice worked against Nansen. Along with his long-suffering companion Lieutenant Johansen, he only got as far north as 86° 14′ before he was forced to turn back and begin the long trek home.

The failure to reach the North Pole did not diminish the enormity of Nansen's feat, nor dampen the enthusiasm with which he was greeted when he returned south. The first 40,000-copy print run of *Farthest North*, his two-volume account of the voyage of the *Fram*, quickly sold out. In 1897, Nansen pursued a lucrative lecture tour in the United Kingdom.[25] Later in the year, he travelled to the United States, where he met both President McKinley and his successor, Theodore 'Teddy' Roosevelt, a fellow advocate of the idea that frontier development could forge a nation. (When Roosevelt became President, actively sponsoring Robert Peary's attempts on the North Pole, he wrote that it was Nansen himself who had confirmed Peary's suitability for the task.)[26] By his mid-thirties Nansen had firmly established himself as a kind of polar oracle – a man sought out by Arctic and Antarctic explorers for advice, support and benediction.

Despite his scientific training – some viewed him as the father of neuro-science – Nansen was fully aware of the almost magical quality of the pull which the Arctic exerted on the human mind. Rather than dismissing the myths and legends of the North out of hand, Nansen underlined the useful purpose which they had played in the past. In his great argument for Arctic exploration as the vehicle and symbol of questing human progress – a book bearing the somewhat ambiguous title *In Northern Mists* – Nansen pointed out that it was dreams, myths and illusions about the world which had pushed explorers forward when, by any reasonable standard, they should have stopped and turned back for the warmth of home, be it Columbus's Genoa, Cabot's Venice or Nansen's Oslo.[27] If dreams and illusions of the Arctic had provided a forward impulse for human progress, Nansen argued, so be it.

In Northern Mists was published in 1911. By then, of course, Nansen was no longer the man to have travelled *Farthest North*. Such was the pace of Arctic exploration at the end of the nineteenth century that his record had lasted only five years. In 1900, the Italian explorer, Luigi Amadeo, had reached 86° 33′.[28] In 1908–09, the North Pole itself had been attained, though a bitter dispute remained between the claims of Dr Frederick A. Cook and Robert E. Peary – a man who resented Nansen for having beaten him across the Greenland ice cap.[29]

There was a distinct element of national self-aggrandisement in *In Northern Mists*, which placed Norway as the key country in mankind's Arctic endeavours. The Norwegians, according to Nansen, had been 'the first people in history who definitely abandoned the coast-sailing navigation universally practised before their time, and who took navigation away from the coasts and out on to the ocean'.[30] It was the Norse who first discovered North America, Nansen continued, centuries before Cabot and Columbus. Ancient references to the land of Thule were, in Nansen's argument, really references to Norway.

The historical case which Nansen made for Norway's unique role in Arctic history fitted with a wider agenda of Norwegian nation-building. Nansen had actively supported Norway's final separation from Sweden in 1905, writing a book (in English) on the subject and serving as his country's minister in London between 1906 and 1908.[31] There was speculation that Nansen had been approached about the possibility of becoming Norway's king.

In 1913, Nansen undertook another, less arduous, Arctic trip – this time to Siberia. His account of that trip was published in 1914, just as Europe was entering the First World War.[32] In it, Nansen described the potential of a trade route from the Arctic Kara Sea to Western Europe to open up the almost 'inexhaustible wealth' of Siberia.[33] Siberia's main future importance, according to Nansen, was as a major source of food supply. Meanwhile, in sub-Arctic Sakhalin – which would become the site of major oil and gas developments in the twenty-first century – Nansen reported that 'oil wells have been discovered, which, it is asserted, might supply the whole of Siberia with petroleum, and, indeed, Australia as well'.[34] Elsewhere, Nansen wrote of the rich coal seams around Dudinka, where 'the coal is as good as the best that comes from Cardiff'.[35] On the Yenisei River, he reported graphite deposits.

The Arctic dominated Nansen's life. Even when he was the Norwegian government's minister in London, from 1906 to 1908, he still found time to lecture at the Royal Geographical Society, showing himself quite the master of the latest advances in Arctic oceanography and providing his own personal view of 'the north polar problems of the future'.[36] Twenty years later, in 1927, he penned a long article combining two of his greatest passions – history and science – explaining the role of climate change on the history of the North.[37] He never quite gave up the dream of reaching the North Pole.[38] In the last decade or so of his life, however, from the last years of the First World War to his death, aged sixty-eight, in 1930, the Arctic shared its place in Nansen's affections with two other causes: humanitarian relief and Russia.

In the 1920s, as a firm supporter of the newly established League of Nations, he became the High Commissioner for Refugees, dealing with the complex problems of repatriation, refugees and nationalities as the borders of Europe were redrawn. Under the auspices of the Red Cross, Nansen was given the job of organising relief to those struck by famine in Russia in 1921–22, work for which he was awarded the Nobel Peace Prize in 1922. Proud Norwegian patriot, one of the fathers of his country, Nansen proved himself a keen internationalist, strongly supporting the League of Nations and warning, after the First World War, that nationalist backsliding would only lead to another. 'The only policy which can save Europe,' he wrote, 'is that which resolutely regards all problems from an international economic point of view.'[39] Could not the North act as a model of international cooperation, just as it had provided, in Nansen's view, an impulse for the advance of human knowledge?

In Norway, Nansen was the country's leading national hero: a living legend, a strong patriot and a chronicler of his country's ancient northern heritage.[40] When he died in 1930, the flags flew at half-mast over Norway and news of his death was carried on the front page of the *New York Times*, with testimonials from London, Geneva and

Russia.[41] A speaker at a meeting in London held in his honour concluded that 'Nansen was one of three people who could have been President of our [European] United States . . . One of the best men of our time is dead.'[42]

With the death of Nansen, the world had lost a figure of immense stature. Nansen's amalgamation of nineteenth-century heroism with twentieth-century science had struck a lasting chord – one that would persist in popular thinking about the Arctic for decades to come.

Vilhjalmur Stefansson, on the other hand, tried to break down the image of the heroic Arctic. In its place, he proposed a more modern, more commonplace Arctic, humming with human activity and economic development.

Stefansson would never be comfortable with the mantle of Arctic hero. The drawn-out failure of one of his most important expeditions and the messy dispute which resulted, trailed him throughout his career. To his critics, he was a 'mere populariser and part-time charlatan' who was responsible for the deaths of the crew members of the *Karluk* in 1914, and for the fiasco of Wrangel Island in the 1920s.[43] Speaking to a Canadian newspaperman about Stefansson's assertions of the Arctic's liveability, Roald Amundsen blurted out, 'Of all the fantastic rot I have ever heard of, this comes close to the top.'[44] But, for many, Stefansson was a prophet of an Arctic future, who for over half a century stressed the liveability of the Arctic, its strategic importance and the need for its economic development. Engraved on Stefansson's tombstone in Hanover, New Hampshire, is the simple epitaph, 'Prophet of the North'.

In Stefansson's mind, the North had been misunderstood, misrepresented and undervalued. As he saw it, the Arctic had been hijacked by grandiose notions of heroism and worse, unscientific notions about the Arctic's natural conditions that prevented the region's development. The problem was one of perception, he argued: 'It is chiefly our unwillingness to change our minds which prevents the North from changing into a country to be used and lived in just like the rest of the world.'[45] The only question in Stefansson's mind was who the beneficiaries of the Arctic future would be: the country of his birth (Canada), the country of his citizenship (United States), or some other nation.

Stefansson was even more prolific a writer on the Arctic than Nansen, writing sixteen books and over fifty articles. But his advocacy of Arctic development went far beyond the use of the pen. He lobbied the Canadian and American governments to take a far more proactive approach to establishing themselves in the Arctic. In 1921, with what Stefansson understood as tacit support from Ottawa, he took matters into his own hands, launching a mission to establish Canadian or British sovereignty over Wrangel Island, north of Siberia.

As it turned out, Stefansson had overstretched himself. Without firm Canadian or British support, the Wrangel Island mission ended in fiasco. But decades later, Stefansson's concern with securing Arctic territory seemed prescient. As the geopolitical significance of the Arctic came into sharper focus in the 1940s, with improved accessibility and the emergence of the Soviet Union as an increasingly threatening power, Stefansson was called upon to help the US military to work out how to fight and survive in what became the newest arena of the Cold War. In the late 1950s, when the US Army set up the Cold Regions Research and Engineering Laboratory, they decided on Hanover, New Hampshire, in part because of interest and expertise at Dartmouth College, where Stefansson, then in his seventies, had a long-standing affiliation.

Unlike Nansen, whose exploration focused on the ice and the sea, Stefansson's exploration focused on the land of the Canadian and American Arctic. It was there that Stefansson lived for much of the period from 1906 to 1918, largely with the native population of the North. His first trip was the shortest, eighteen months over 1906 to 1907. His second, from 1907 to 1912, was his most groundbreaking from the point of view of learning the local Inuit language and studying its culture.[46] His third, from 1913 to 1918, was the most organised and the most successful in terms of its stated aims: to discover and claim Arctic land for Canada.

It was the 1907–12 expedition which provided the material for Stefansson's first book, *My Life with the Eskimo*. The book was, in essence, a work of anthropology, which described the language, culture and religious beliefs of the native population. He had arrived in the Arctic, according to his own description, carrying little more than might be required for a day trip:

I found myself, in accord with my own plan, set down two hundred miles north of the polar circle, with a summer suit of clothing, a camera, some notebooks, a rifle, and about two hundred rounds of ammunition, facing an Arctic winter, where my only shelter would have to be the roof of some hospitable Eskimo house.[47]

For the next five years, Stefansson would live with the Inuit, gaining understanding and respect for their way of life. In some places, *My Life with the Eskimo* took on the character of a travel narrative, with Stefansson describing the 'all-pervading odor' of tar in the Alberta tar sands around Fort McMurray, viewed then, as now, as a major source of 'unconventional' petroleum. A flame from a natural gas well in the area was interpreted, with true Stefanssonian flair, as 'the torch of Science lighting the way of civilization and economic development to the realms of the unknown North'.[48] Elsewhere, Stefansson struck a more pessimistic tone. With the decline of commercial whaling, Inuit society was in flux. 'The Eskimo are facing a new era and the change will be hard on them,' he wrote.[49]

Overall, however, *My Life with the Eskimo* was a descriptive book. In contrast, the book which Stefansson published in 1921, *The Friendly Arctic*, was a far more political book. Sir Robert Laird Borden, Canadian Prime Minister between 1911 and 1920, provided its introduction. Borden had been the political benefactor of Stefansson's third expedition to the Arctic between 1913 and 1918, when Stefansson had essentially acted as an agent of the Canadian government in the North, discovering and claiming land for Ottawa.[50] Stefansson had added 'many thousands of square miles' to the territory of Canada over those years, Borden wrote.[51] But, as its title suggested, the book was far more than an account and proof of those travels and claims. Stefansson wished to make a far broader point – that the Arctic was a place that could and should be settled, developed and exploited. It was, in his argument, far more liveable – or 'friendly' – than most people thought.[52]

The Friendly Arctic laid out the lines of an argument that Stefansson would continue to develop for the rest of his career. Above all, he thoroughly rejected the notion that the Arctic was barren or inhospitable. Much depended, according to Stefansson, on one's perspective. A Greenlander, he wrote, 'would find it no more remarkable that he could survive the cold of Greenland than a zulu finds it that his neighbours can survive the heat of Africa'.[53] The purpose of *The Friendly Arctic*, Stefansson said, was to prove that men could easily subsist in the Arctic and, indeed, to show that 'when the polar regions are once understood to be friendly and fruitful, men will quickly and easily penetrate their deepest recesses'.[54] At heart, it was a manifesto for demystifying the Arctic and urging its development.

One chapter – 'The North that Never Was' – was devoted to debunking what Stefansson saw as the unhelpful myths which imprisoned the Arctic within a literary and romantic trope that his own experience of Arctic conditions had demonstrated as false. The common view of the Arctic, Stefansson wrote, was that:

> The land up there is all covered with eternal ice; there is everlasting winter with intense cold; and the corollary of the everlastingness of the winter is the absence of summer and the lack of vegetation. The country, whether land or sea, is a lifeless waste of eternal silence. The stars look down with a cruel glitter, and the depressing effect of the winter darkness upon the spirit of man is heavy beyond words. On the fringes of this desolation live the Eskimos, the filthiest and most benighted people on earth, pushed there by more powerful nations further out, and eking out a miserable existence midst hardship.[55]

In exploding these myths, Stefansson was well aware that he might also destroy the romantic notion of the hero–explorer. It is hard to imagine that Stefansson did not have Fridtjof Nansen in mind when he described the grand self-image of the traditional hero–explorer combining an 'insatiable desire, mysteriously implanted in our race to throw ourselves against obstacles' with 'a thirst for knowledge, who struggles through the arctic night with the same spirit that keeps the astronomer at his telescope'.[56] 'It may be a pity to destroy the illusion, for the world is getting daily poorer in romance,' Stefansson opined. After all, 'elves and fairies no longer dance in the woods and it appears a sort of vandalism to destroy the glamorous and heroic North by too intimate knowledge'. But these were rather crocodile tears.

Stefansson took the romantic view of the Arctic to task on five specific points. First, he wrote, the Arctic was not necessarily the coldest place on earth. He argued that the traditional link between latitude and liveability was, itself, scientifically and statistically unfounded. The January temperature of Reykjavik was no lower than that of Milan in Italy. The coldest places in North America were not in the Arctic at all, but in Montana or North Dakota, both inhabited parts of the continent.

Second, in a dig at many of his fellow explorers who had devoted their lives to the pursuit of the North Pole, Stefansson pointed out that while the North Pole certainly had 'glamour' it was not, in fact, the most inaccessible place on the globe. 'The world in general has imagined the North Pole to be to the Arctic what the mountain top is to the mountain,' he wrote. But this, he argued, was a false analogy. The northern point on the globe most distant from land was not the North Pole.

Third, 'in the process of removing the imaginary Arctic from our minds we come to the proposition that all land in the far north is covered with eternal ice'.[57] This, Stefansson suggested, was a misconception based on the erroneous notion that Greenland – nearly wholly covered in an ice cap – was typical of the Arctic as a whole. Snowfall, Stefansson pointed out, was actually considerably lower in some parts of the Arctic than in areas further south. To describe the Arctic as 'barren ground', in Stefansson's mind, was viewed as practically 'libellous' given the wealth of its natural life and the prospects for agricultural development. The intention of such descriptions, he concluded, was simply to convey 'the impression that those who travel in the North are intrepid adventurers'.[58]

Fourth, Stefansson took issue with the view that lack of light and fear of darkness provoked depression. This might be true with 'the typical sailor or Alaska miner' he wrote, scathingly, but 'college graduates' could be persuaded otherwise. And those who had no preconception of the depressive effects of an Arctic winter – 'Hawaii Islanders, Cape Verde Islanders, or southern negroes' – suffered no ill effects.[59]

Finally, he criticised the tendency of those from the South to pity the North and to view the native inhabitants of the North by their own, imperfect, southern standards. This approach, Stefansson contended, was misguided, and as fundamentally condescending towards the North as 'Orientalism' was to the East:[60]

Point Hope is just beyond the reach of tourists and of the journalists who write fascinating magazine articles about 'primitive people untouched by civilization'. It lies in that tame intermediate zone where missionaries, equipped with victrolas and supplied by yearly shipments of canned goods, labor heroically for the betterment of the natives, who realise that they are badly off just as soon as they are told about it. It is one of the anomalies of our world that it should take the efforts of so many self-denying people to awaken the wretched to a consciousness of their wretchedness.[61]

In his conclusion to the book, Stefansson expressed his hope that 'we have brought the North a good deal closer and have made it look more than it used to like Michigan or Switzerland'.

A year later, Stefansson took the argument one step further. The Arctic had been misunderstood and misrepresented but, in reality, it offered all the potential for a bright future. *The Northward Course of Empire* (originally an article in the *National Geographic* magazine) argued that, through history, the North had gradually become more important. Global power had shifted steadily northward – from Babylon and Alexandria to Athens and from Rome to Paris, London and New York. Why should this historical process not continue?

Stefansson didn't deny that sound agricultural and environmental reasons lay behind man's origins in the tropics, and civilisation's growth in the temperate zones. But Stefansson argued that what constituted a 'good' climate was not set in stone. He pointed out that, historically, relatively poor climates often ended up producing better long-term results. Britons who emigrated to the West Indies in the seventeenth century, Stefansson wrote, ended up being part of a community of 'poor white trash', forgotten by history. Those who went to the far less hospitable climate of Plymouth, Massachusetts, founded one of the most powerful nations on earth.[62] What was more, he continued, the resources of the Arctic were available and untapped. Only short-sightedness prevented the future strategic importance of the North from being more widely recognised.

Stefansson thought it a great mistake that Britain had allowed Spitsbergen to pass under Norwegian sovereignty in 1920 – albeit with an international treaty mandating equal access to its resources. He argued that Spitsbergen was one of the few places where good coal and good iron – both vital for the British navy – existed side by side.[63]

Perhaps, Stefansson suggested, Britain's acquiescence on Norwegian sovereignty was the result of some dark 'secret political bargain'.[64]

In terms of global food production, Stefansson contended that the Arctic represented 'the next largest grazing area in the world [after Europe], one and a half or two million square miles of prairie land, equal to half the area of the United States'.[65] In 1919, under the auspices of Canadian interior minister Arthur Meighen, a Royal Commission had been set up to investigate the possibilities for the commercial production of musk ox and reindeer in northern Canada. Its conclusions were cautious, though broadly supportive of Stefansson's faith in the North as 'the greatest meat-producing area of the world'.[66] He anticipated a world where the North would be given over to meat production – particularly musk ox and reindeer – while the South would become more exclusively used for growing cereal crops.

Could not the reindeer form part of a new chain in the global food trade? 'Companies have already been formed to take up, as soon as economic conditions become stable, the erection of packing plants and the installment of refrigerator steamers,' he wrote. 'The route is not as long as that which brings Argentine beef to England, only half as long as that which brings Australian mutton,' he pointed out. Refrigeration, which had allowed Argentina and Australia to become major suppliers of fresh meat to Europe, would be easier on shorter voyages within the northern hemisphere than over longer voyages passing through the Equator. And there were the prospects of the growing markets of Asia and the Pacific: 'Should the Japanese and Chinese or the people of Hawaii and San Francisco develop a taste for reindeer meat, the herds of north-eastern Siberia are readily available to ships plying the north Pacific and the north-east arctic portions of Asia.'[67]

Stefansson argued that, in the face of growing global population, northward expansion and development was a matter of necessity as much as a matter of profit. 'With reference to the world of a hundred years from now,' he argued, 'if we avoid destructive wars and do not adopt birth control, this supply [of meat from the North], vast in itself, will be insignificant.'[68] But, without it, the coming food security crisis would be even worse. Dismissing as temporary the fall in meat prices in the early 1920s, he insisted that supporting agricultural production in the North was simply good forward planning:

There are some who say that long before the year 2000 we shall have released the energy of the atom and shall stop using petroleum, and that long before then we shall learn to make food directly out of the air thus doing away with pig-stys and wheat fields. That may prove so, but it is well to have two strings to our bow and to plan to conserve fuel and produce food so that we shall have something to fall back on in case the dreams of our chemists are not realized fast enough to keep step with the increase of population.[69]

Further, Stefansson cautioned that new means of transport would allow for the overflight or underpassage of the Arctic, vastly increasing the area's strategic and economic importance. 'The airplane, the dirigible, and the submarine are about to turn the polar ocean into a Mediterranean and about to make England and Japan, Norway and Alaska, neighbors across the northern sea,' he wrote.[70] This was four years before the first balloon overflight of the North Pole, more than a decade before the first aeroplane landing at the North Pole, and thirty-six years before the first submarine passage under the North Pole by the *USS Nautilus*, in 1958.[71]

According to Stefansson, what was needed, above all, was visionary leadership. The Arctic was a call to 'men of enterprise' and 'captains of industry' with the power and initiative to invest for the long term. Canada's railway network remained undeveloped in the North, Stefansson complained. Australians did not think twice about building a railway across the continent from New South Wales to Perth despite the fact that the country was sparsely populated. Why did Canadians lack a similarly grand vision of their future?

The central irony of Stefansson's *The Northward Course of Empire*, which was at heart a call to the Canadian government and to Canadian authorities to take advantage of their huge Arctic hinterland, was the success of the book outside Canada, and its poor reception within the country. There, Stefansson's reputation had been irretrievably damaged by controversial claims of 'blond Eskimos' and by talk of mutiny on his third expedition into the Canadian Arctic. His standing in Canada was further undermined by what Stefansson called 'the adventure of Wrangel Island'.[72]

Stefansson's association with Wrangel Island – due north of the Siberian mainland and straddling 75° North – went back some years.[73] For many, the island was a symbol of the tragedy which marred

Stefansson's 1913–16 Arctic expedition. On that occasion, eleven crew members of one of Stefansson's ships, the *Karluk*, had perished as the result of events that some put down to Stefansson's lack of preparation and leadership. Those crew members who survived had abandoned their sinking ship and made it over the ice to Wrangel Island.[74]

In early 1921, Stefansson hatched a plan which he believed might not only wipe away that negative association, but might also extend Canadian Arctic sovereignty to the west and offer up a possible commercial opportunity: he would set up a colony on Wrangel Island. Russia, the United States and the United Kingdom all had potential claims on Wrangel. Though a Canadian claim had not officially been lodged, Stefansson thought that one would be forthcoming once he had demonstrated that it was possible to establish a colony there. Perhaps the island could become a landing place for trans-polar flights, Stefansson thought, or perhaps a self-contained ranch for Arctic reindeer.

The first attempt to establish a colony was a fiasco, as settlers struggled to survive the brutal Arctic winter. Stefansson's attempt to bounce Canada into claiming sovereignty – Ottawa had never officially supported Stefansson's designs – backfired disastrously. The failure of the Wrangel colony persuaded Canada to disavow any relationship with him. (A relief mission to the settlers was organised with private money from a number of sources. Sir Arthur Conan Doyle was one of the donors.)

In 1923, a second group of colonists was organised to replace the first group. This second group differed from the first in that it was mostly made up of American Inuit from Nome, Alaska – with far better prospects of surviving on Wrangel and establishing a workable colony. By now, however, the politics of Wrangel Island had shifted. Ottawa rejected the notion that it would make a claim, and London decided against it. Stefansson turned to the United States. In May 1924, in an attempt to further Americanise the expedition, Stefansson sold his 'rights' to Carl Lomen, reported to have a herd of some 40,000 reindeer in Alaska.[75] But it was too late.

A Soviet ice-breaker, the *Krasny Oktyabr* (*Red October*), made it to the island in late summer 1924. Stefansson's colonists were removed and taken to Vladivostok, in Russia's Far East. In 1925, they turned up in Chinese Manchuria. One, Charles Wells, died there. The remaining eleven were picked up by a Japanese steamer in Harbin and arrived back in the United States in spring 1925.

The disastrous saga of Wrangel Island did not support Stefansson's claims of a 'friendly Arctic' and made him *persona non grata* in Ottawa. By discrediting Stefansson's approach to Arctic development, it affected the course of Canadian history. The irony of the affair was that it may have contributed materially to awakening the Soviet Union to the possibilities and challenges of the Arctic. In one Canadian newspaper of the time, the Soviet commander who removed the colonists from Wrangel Island was reported as saying that 'raising the red flag on Wrangel Island is of the greatest importance to Soviet Russia. It terminates the efforts of imperialists to seize the island, on which they indeed planned to organise a base for airplanes flying from the United States to Europe.'[76] And, he added, it offered the possibility that the island would now be developed by the Soviet Union instead.

By the late 1930s, after three decades lobbying for the development of the American Arctic – in Canada and the United States – Stefansson felt that both his country of birth and his country of citizenship had failed to live up to his vision of a booming North. Instead, the Arctic future which Stefansson had envisaged for the Canadian Arctic and for Alaska was being built in the Soviet Union:

> During the twenty years or so between the two wars no Alaska town north of the Arctic circle grew materially. In that period the largest Arctic town of Canada's largest river, Aklavik on the Mackenzie, grew from a few dozen to something between 200 and 300. In the same period the Arctic village of Igarka, on Siberia's comparable river the Yenisei, grew from less than 100 to more than 20,000. Just before 1940 more than 100 steamers were plying the Arctic coasts of the Soviet Union – more than 20 of them going all the way through the Northeast Passage between Atlantic and Pacific. On the Arctic coast of North America there were a half-dozen steamers, and none went through the full Northwest Passage. In a single year a dozen or so different Canadian airplanes crossed the Arctic Circle and flew there a total of a few hundred miles, and the like was true for Alaska; more than a thousand airplanes were then operating in the Soviet Arctic, flying several hundred thousands of miles.[77]

In the 1930s, Stefansson was still treated with scepticism in Ottawa, while Washington was only beginning to think about its own role as a future Arctic power. At the same time, however, many of Stefansson's books were available, translated into Russian, in Moscow and Leningrad. Nowhere were his ideas more enthusiastically taken up than in the Soviet Union.

In 1914, as Stefansson was sledging through the Canadian and American Arctic on his third expedition to the North, Josef Djugashvili, a dangerous criminal, political agitator and editor of the Bolshevik newspaper, *Pravda*, was familiarising himself with his new home in Russia's Arctic. Djugashvili was originally from the Russian imperial province of Georgia, in the south Caucasus. But in 1914, he had been arrested by the Tsarist secret police, and exiled to the area around Kureika, high on the Yenisei River. He would stay there until his release in March 1917.

Nansen knew that part of the Russian Arctic from his 1913 trip to Siberia. Had Djugashvili been exiled a year earlier, the paths of the Georgian criminal and the great Arctic explorer might have crossed. Indeed, Nansen did meet one Georgian exile, 'a remarkably handsome type, dark, with a short black beard and melancholy eyes, as soft as velvet'. His book on Siberia described the area to which Djugashvili had been exiled – 'the same flat, endless forest country' – and taking tea with the local police chief at Turukhansk, an Ossetian named Ivan Kibirov, who was later to prove instrumental in an attempt by Djugashvili to escape from his northern exile.[78]

In later years, Djugashvili – then known as Josef Stalin – would claim to have become a great Arctic hunter in his years of exile at Kureika. On one occasion, he claimed, he had skied forty-eight versts (about thirty-two miles) on a single winter's day, a distance of which Stefansson or Nansen would have been proud. It was, no doubt, an exaggeration. But no one would have been likely to point this out to the man who, by then, had become the Soviet Union's greatly feared supreme leader.

Whatever the true extent of Stalin's activities in Kureika, the Arctic had left its mark on the man. One of his colleagues wrote that 'a little bit of Siberia remained lodged in [Stalin] for the rest of his life'.[79] With hindsight, however, he might have described the relationship in a different way. For whatever influence the Arctic might have had on the young Stalin – his biographer speculated that perhaps 'Siberia froze some of the Georgian exoticism out of him' – it was surely surpassed by the influence on the Arctic that Stalin would have as Soviet dictator.[80] By imposing on the Arctic his own vision of its future – as a source of wealth, power and strategic advantage for the Soviet Union – a little bit of Josef Stalin would remain lodged in Siberia, indefinitely.

2 Through a Glass Darkly: The Soviet Arctic

'The Arctic and our northern regions contain colossal wealth. We must create a Soviet organization which can, in the shortest period possible, include this wealth in the general resources of our socialist economic structure'

Josef Stalin, 1932

'Kolyma means death'

Gulag saying

SOLOVETSKY ISLANDS, RUSSIA, 65° North: The Solovetsky monastery complex is no evidence of a single creative genius. There's no attempt at symmetry, no layout. The stones of the outer walls are rough and misshapen, looking more flung together in an Old Testament rage than placed with an architect's deliberation.

Within the complex, each building differs from its neighbour. The refectory is massive and workmanlike – the first stone building on the Solovetsky Islands, made up of boulders lugged from the surrounding island of Solovki on crude wooden sledges, drawn across ice and snow. Next to it stands the eighteenth-century bell tower, as light and as fanciful as the refectory is heavy and functional. Built at the height of the reign of Catherine the Great, the bell tower seems more German than Russian. And finally, dominating the complex, is the absurd Russian decorativeness of the cathedral itself, crowned with five perfect wooden onion domes.

The monastery is a physical reflection of the ebb and flow of its fortunes: the falling in and out of favour of its abbots, times of Russian introspection followed by times of Westernisation, periods of prosperity when the Orthodox church was ascendant, followed by periods when church and state were locked in conflict over the direction of the Russian nation.

But the bloodiest phase of the monastery's history – as one of the Soviet Union's first labour camps from 1923 to its closure sixteen years later – is almost totally effaced. A one-room exhibition from the late 1980s, a plaque on the local supermarket – formerly barracks for the prison guards – and a huddle of crosses on a nearby hillside are all the evidence for 100,000 prisoners who arrived on the Solovetsky Islands, shipped from the railhead across the water at Kem. Over 40,000 never left this place.

Perhaps, buried deep in the stones of these three-foot-thick walls there is the faintest tremor, the final pulse of a sound wave from the cries of the thousands of men, women and children who were incarcerated and died here. But these days the gulag is only to be whispered within the walls of the Solovetsky monastery.

The history of Solovetsky is, in a sense, the three-times-distilled history of the Russian Arctic. A place of retreat, a place of veneration, a source of national identity, a strategic bastion, a prison, a labour camp – Solovetsky, like the Russian Arctic, has been all these things. Russians' conflicted associations with the Arctic today are the product of that fragmented history.

Why put a prison camp here? Above all, isolation – the same Arctic isolation which had attracted Savvaty and Herman, the founders of the monastery in 1429. The same Arctic isolation which, from the sixteenth to the eighteenth centuries, has made the north and east of Russia the natural place to send the unwanted, the dangerous and the criminal, from the court opponents of the Tsar to the political prisoners of Stalin. The Solovetsky Islands are, in any case, a natural prison: a few degrees south of the Arctic circle but arctic in climate. In summer they are inaccessible except by boat across the White Sea; in winter the sea around them freezes.

The main complex of buildings on Solovki was more or less empty by 1923. The monastery itself had been shut down by the Soviet authorities in 1920. In the chaos of the civil war which accompanied the birth of the Soviet Union, a production office remained to manage the monastic estate – including sixty monks – but that burned down in the summer of 1923. Some say the fire was arson – to cover up bureaucratic theft from the estate.

Whatever the cause of the fire, the demise of the Soviet production office opened the way for a new enterprise for Solovetsky: the 'northern camps of special significance', *severnye lagerya osobogo naznacheniya*, or, simply, SLON. On 13 October 1923, Solovetsky was officially handed over to the GPU secret police – the successor to the Cheka, and the forerunner of the OGPU. Nearly five hundred years of operation as a monastery drew to a close, and a sixteen-year career as a prison camp began.

In its very first decades, at the beginning of the fifteenth century, the monastery was a simple place, an Arctic refuge for hermit monks, far from the worldliness of a Russia which then was a fragmented set of territories, threatened on all sides. From the foundation of the monastery by Savvaty and Herman in 1429, to the building programmes of the mid-sixteenth century, there was not a single stone building on the island.

But change and expansion, when it came, was fast. Abbot Phillip, son of a prominent Moscow family who either fled the political dangers of the capital or, in the less secular-minded version of the story, experienced a vision of Christ warning him against serving two masters, arrived on Solovki in 1537. Ten years later he had embarked on the transformation of the monastery into an economic and spiritual centre, building canals, mills, a harbour and the monastery's first cathedral, which he aptly dedicated to the transfiguration. The prestige of the monastery rose still further in 1566 when Phillip accepted the role of Patriarch, a post he held for barely two years before his criticism of Ivan the Terrible led to a show trial, exile to Tver and, finally, his murder in 1569.

The monastery became a fortress against Livonian, Swedish and German encroachment on Russian territory and then, in the mid-seventeenth century, a centre for Old Believers, those opposed to reform of the Orthodox church. Between 1668 and 1676 the monastery was under siege by Russian forces. But by the time that Peter the Great visited for the first time, in 1694, the monastery had become a mixed institution: serving both as monastery and as prison. To judge from the graves of the prisoners from this time, they were an eclectic group: aristocrats, priests, and the last ataman of the Zaporozhian Cossacks, who died in 1803 at the reported age of 113. They were also relatively few: some 400 prisoners in total from the late seventeenth century, to the last Tsarist-era prisoner in 1903.

The fortress itself was decommissioned in 1814, considered redundant after Russia's military victories against the Swedes in 1809 (as a result of which Finland became part of the Russian empire) and Napoleon in 1812. Apart from a brief episode in the Crimean War, when two British ships – the *HMS Miranda* and the *HMS Brisk* – demanded and were refused the surrender of Solovetsky, the islands played no further role in Russia's military history. What had once been an outpost of Russian Orthodoxy was now firmly within Russian territory.

At the same time, the era of serious competition between church and state was definitively over. Solovetsky could no longer serve as a renegade monastery in opposition to the wishes of central authority, as it had in the seventeenth century. The church itself was increasingly subsumed into the apparatus of the state, reconfirming the spiritual authority of the Tsar in the process and, in return, assuring itself of its property and of its protected role as the principal guarantor of Russian values and morality. When Tsar Alexander II visited Solovetsky in 1858, a small church was raised in his honour by the harbour side. Photographs of the monastery at the turn of the twentieth century by the famous Russian photographer Sergei Prokudin-Gorskii show gleaming white buildings, leafy trees and shining crosses: photographs of a world about to be shattered beyond recognition.

Now, at the beginning of the twenty-first century, after seventy years of Communism and twenty years of post-Communism, there are many endeavouring to return the Orthodox church to that privileged position in Russian society, and the Solovetsky monastery to its former glory.

They have already achieved significant successes. The monastery has been reborn. The Soviet star was removed from the top of the belfry in the mid-1980s and in 1988 a Moldovan monk, Herman, came to Solovetsky to re-establish the monastery. Officially reopened in 1990, the monastery negotiated a series of agreements with the state, achieving the return of parts of the monastery complex and the islands to the church, for a limited term. The bones of the fifteenth-century founders of the monastery were returned in 1992, having been stored for the duration of the Soviet regime in the anti-religious museum of church history and atheism which then occupied St Petersburg's deconsecrated Kazan cathedral.

In the eyes of the Orthodox church, the final religious purification of Solovetsky took place in 2000, when the sixty monks who died as camp inmates were beatified. But the monastery's political consecration took place a year later when then President Vladimir Putin, in a self-conscious retracing of the visits of the Russian Tsars, travelled to Solovetsky.

But as is the case with Russia's Arctic itself, the symbolic meaning of Solovetsky remains contested. For some, it is a symbol of the triumph of the Orthodox faith after decades of accommodation and oppression. For them, the story of the monastery's rebirth far outshines Solovetsky's sixteen-year interlude as an Arctic prison camp. As Father Simeon, one of Solovetsky's monks, crossly upbraids me, 'History is not so important – the most important thing is man's relationship to God.' For them, the revival of the Solovetsky monastery represents a first step towards the spiritual rebirth of the Russian nation after seven decades of Marxist materialism, and a decade of godless capitalism under President Boris Yeltsin. Anatoly, one of the monastery's staff, leaning conspiratorially over a table in the management office of the monastery, tells me of Russia's 'unseen war – on a mystical level' with Western civilisation. Having spent seven years in the United States, including two years in an Orthodox seminary, he sees the Arctic monastery of Solovetsky as a bastion of the Russian spirit.

But for others, Solovetsky's role – as the beginning of the Arctic archipelago of the Soviet labour camps – can never be finessed into a narrative of Russia's religious or political rebirth. They fear that, in such a dramatic national-religious narrative, the unfashionable history of the gulag may be lost. When Yuri Brodsky – Solovetsky's local historian – speaks about the gulag, in beautiful sentences half a stanza long, he seems haunted by the fear that the history of what happened here will be forgotten.[1] In the 1970s and 1980s he attracted the attention of the KGB for his attempts to put together a photographic and written record of what remained of the evidence of the gulag on the

Solovetsky Islands. But in the new Russia, the threat to remembrance comes not so much from a policy of deliberate obliteration of the past, as from a culture of wanting to forget its least pleasant aspects.

Solovetsky was not, of course, unique. It formed only part – the first part and in some ways the most important part, but only one part – of a system of prison camps which extended across the length and breadth of the Soviet Union's 8.5 million square miles. Alexander Solzhenitsyn's description of the gulag system as an archipelago referred to the specific geography of the Solovetsky, but it was also a schematic description of the overall gulag system.[2] Many of the camps were, like Solovetsky, located in the Arctic region – where labour was needed and escape was impossible – but not all. The summit of the gulag, in purely numerical terms, occurred near its end, in 1951, when Solovetsky had already been closed for several years. The NKVD statistics for the beginning of that year showed 2,561,351 registered inmates.[3]

But this sobering statistic in a sense conceals the conflicted purposes of the gulag, in the Arctic and elsewhere. For the true horror of the gulag went beyond the arbitrariness of who was sent there, who would survive and who would die; beyond its massive scale, comprising 476 camp complexes and many thousands of individual camps; beyond the millions of unspeakable acts of calculated dehumanisation, of pointless cruelty, or of out- and-out barbarism which marked its history.

A less codifiable horror of the gulag was that it was, in intention, a rational enterprise. Perhaps not at first, but certainly by the late 1920s, it was supposed to serve a dual purpose – not just the incarceration of elements considered to be politically or criminally dangerous, but a contribution to the industrialisation of the Soviet Union. In the mind of Josef Stalin, the gulag became part of a wider struggle: the conquest of Russia's geography and, in particular, of the Arctic.

Stalin was not unique in thinking that the Arctic – in particular the Russian Arctic – was undeveloped in its economic and strategic potential. Nansen, travelling down the Yenisei River in 1913, described 'new towns [which] have grown up with American rapidity' while expressing the possibility that Siberia would become a European resource base for the future.[4] Vilhjalmur Stefansson had pointed out multiple ways in which the Arctic, often overlooked by strategic or economic thinkers, would play a much more important role in the future of global history.

And Stalin himself, having lived close to the Arctic Circle from 1914 to 1917, knew something of the region. He had seen, at first hand, the difficulties of living in the Arctic and the relatively inefficient way in which the Tsarist government administered the region. Under his direction, Stalin believed, the Soviet Union could do much better.

An indigenous Russian belief that national salvation would come from Siberia provided an additional foundation to Soviet policy, as did the classical analysis of Russia's strategic vulnerabilities: an economic heartland too concentrated in the West and a land-bound empire unable to take full advantage of transport by sea. These were traditional and long-recognised features of Russia's strategic geography, though ideology and the eastward shift of Russia's European borders made their vulnerability seem more acute in the 1920s and 1930s. Similarly, some of the measures the Soviet Union undertook were projects with a long history before 1917. The economic development of Russia's eastern and northern land mass had been a dream for centuries. The establishment of shipping routes across the Arctic north had been suggested in the seventeenth century. Indeed, to a large degree, given that Siberia's rivers flow north into the Arctic rather than south, opening coastal shipping and the Russian interior were part of the same overall strategic design. Plans to build a canal linking the White Sea, in which the Solovetsky Islands lie, to the Baltic, had been on the drawing board since the eighteenth century.

But if the outline of Stalin's Arctic designs could be found in earlier Russian history, there was, nonetheless, something distinctly Soviet about the scale of ambition, ideological content and coercive nature of the Stalin-era conquest of the Arctic. It was inevitable that the Soviet Union's development of the Arctic territories, beginning as it did with a relatively blank slate in the Arctic, would be a reflection of the regime itself.

Contemporary observers commented that the Soviet transformation of the Arctic in the 1920s and 1930s would have been impossible under a different system of government. Often, this was intended as a compliment to Communist forward thinking. But it was also a reminder of the lack of external mechanisms in the Soviet system which might have checked its more wild designs or corrected its more extreme features. Without a market mechanism to signal economic mistakes the 'plan' decided in Moscow would be implemented regardless of real economic cost. Without a democratic mechanism to limit the will of the state, legitimate environmental and other concerns would be trampled on. Certainly, much could be achieved in this way in the short term. But would it be sustainable in the long term?

After the collapse of the Soviet Union, it would become obvious just how much the Soviet Arctic's economic survival depended on the skewed structure of the Soviet planned economy. And just how ill-equipped the Soviet-era Arctic was for the demands of a market economy. In the contemporary Russian Arctic, the Soviet Union is more than a memory – it is a reinforced-concrete-and-steel legacy. The very shape of the Russian Arctic in the twenty-first century is nothing but the faded lines of Soviet-era plans. The stranded, economically unviable cities of the contemporary Russian Arctic are pure Soviet legacy.[5]

Perhaps not surprisingly, Stalin himself was publicly closely associated with the signature projects of Soviet Arctic development of the 1930s: the *Belomorkanal* (the White Sea Canal), the development of Arctic aviation and the Northern Sea Route. It was not uncommon for particular achievements to be honoured with Stalin's name. For

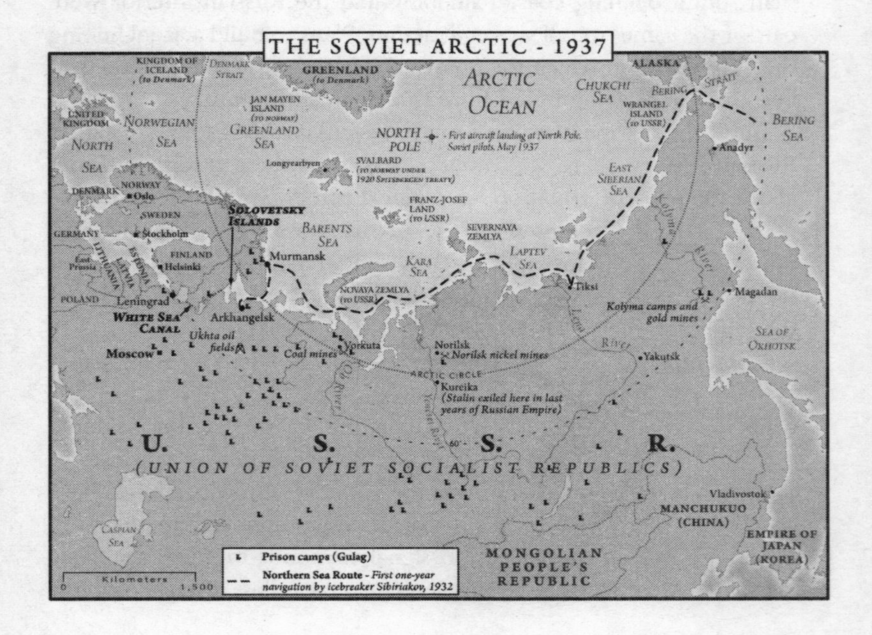

example, the route flown by Soviet pilots to the first ever landing at the North Pole, in 1937, was named the *Stalinskii Marshrut* (Stalin's Route). When the White Sea Canal was completed in 1933, the event was celebrated by the publication of a book entitled *Kanal imeni Stalina* (The Canal Named for Stalin). The Arctic gulag, meanwhile, was the acme of Stalin's world view, combining massive projects of economic development with a total lack of concern for human life. For him, the ends would always justify the means. The project of transforming the Soviet Arctic would outlive Stalin himself, but it would be impossible to forget its Stalin-era origins or even its Stalinist feel. If you visit Norilsk or Magadan or Murmansk you cannot escape it.

Stalin learned his lessons on the strategic vulnerabilities of Russia early in the 1920s, in the Soviet–Polish war and then as one of the leading members of the Politburo, the central policy-making organ of the Soviet Union. Emerging from civil war in the early 1920s geographically reduced, without allies and without official diplomatic recognition, the Soviet Union was fundamentally weak. Russia's strategic vulnerabilities – concentration of economic resources in the west and lack of sea transport – were made more acute by the loss of Russia's borderlands in the west (Poland, Finland and the Baltic region) and the rise of Japan in the east. The stamp of Lenin's domestic and international policy in these years was necessarily one of expediency in the service of a single objective: holding the Soviet Union together. But when his time came, Stalin sought a permanent escape from the instability and weakness which marked the Soviet Union's first decade by strengthening the regime – and his leadership of it – at almost any price. Under him, transformation of the Arctic – as one of the prime sources of the raw materials which would feed the Soviet Union's breakneck industrialisation – became a central front in a wider economic war.

Certainly, historical achievements of the pre-revolutionary past could, on occasion, be incorporated into a Russian–Soviet narrative of national occupation of the Arctic. The Soviet Union would always claim to be the moral and legal heir of the explorers sent by Russia's Tsars to map their northern and eastern territories. However, the Soviet Union would also seek to correct what it viewed as the errors and lack of consistent direction of imperial policy. Where the pre-Soviet history of the Arctic was viewed as a succession of missed opportunities and noble but scattered endeavours, the construction of the Soviet Arctic would be comprehensive and, above all, it would be planned:

> There is a revival of the heroic times of the great explorers, but it is no longer caravels that carry the Soviet scientists to the unknown . . . Soviet expeditions are not enterprises undertaken by determined individuals. They are complex organizations which methodically realize a single plan of investigation of the country.[6]

Before the revolution, Arctic affairs had rarely been a state priority. Russia's progress in establishing itself as an Arctic power had been haphazard, driven as much by strong personalities and by the particular interests of individual Tsars, as by any particular concept of Russia's strategic interests in the North. Perhaps it was inevitable under a system of government which produced many visionaries but few competent managers, that a comprehensive, long-term policy of state directed to the development of the Russian Arctic never really emerged.

There had been, of course, marked successes. In 1648 Semyon Dezhnev became the first European into the Bering straits (despite the claims of the Spaniard, Laurent Ferrer Maldonado). A century later, Peter the Great had established the Great Northern Expedition – a massive undertaking over sixteen years with a mandate to map the entire northern coastline of Russia, establish diplomatic relations with Japan and claim land in North America. Later, Soviet historians and lawyers adopted these expeditions as evidence of Russia's historical interests in the Arctic, and used them as justification for the Soviet Union's sovereignty over any land north of its continental coastline.

Peter the Great's intention – and that of some of the more visionary of his successors – was that the mapping of the Arctic would provide the basis for large-scale commercial exploitation. To that end, commercial companies such as the Russian–American company – which operated a monopoly on trade from Russian North America (Alaska) from 1799 to 1867 – were set up. Expectations had been high. One firm advocate of Arctic development, the great Russian polymath Mikhail Lomonosov, reportedly declared that 'it is in Siberia and the waters of the Arctic that Russia's might will begin to grow'.[7]

But those high expectations were largely unmet. In contrast to the role railways played in opening up the vast interior of the United

States, Canada or Australia, infrastructure across Russia remained inadequate.[8] Sources of capital for private citizens to exploit or develop Russian lands in the Arctic were largely non-existent. Even in its most easily accessible parts the Russian Arctic was poor and undeveloped, compared to the Arctic of its immediate neighbours. In the Barents Sea, for example, it was estimated in the mid-nineteenth century that there were four thousand Russian fishermen, but as many as ten times more Norwegians, with larger ships and far greater access to capital.

There were, of course, corners of Arctic Russia where the prevailing story was one of prosperity. Whether gold was struck in the Yukon or the Yenisei, the result was generally the same. As one contemporary account put it:

> The small town of Krasnojarsk, romantically situated on the Jenisei, is the chief seat of the rich miners. Here may be seen the choicest toilettes, the most showy equipages and champagne (which in Siberia costs at least 1 *l.* a bottle) is the daily beverage of the gold aristocracy. Unfortunately, Krasnojarsk had, until very recently, not a single bookseller's shop to boast of; and while thousands were lavished on vanity and sensual enjoyments, not a rouble was devoted to improvement of the mind.[9]

But overall, the Arctic was an economic backwater in the nineteenth century. The vast river basin of the Ob, Yenisei and Lena Rivers – draining an area several times larger than the mighty American Mississippi – was largely undeveloped. This was something that the Soviet planners of the twentieth century intended to reverse. Instead of being the country's neglected economic hinterland, they insisted, the Arctic must become an integrated part of the country's economic heartland. To them, the economic backwardness of the Arctic in imperial Russia was not just an inexcusable failure of government, but a demonstration of the inherent tendency of capitalism towards colonialism, even within one country.

The Soviet interpretation of the imperial economy ran as follows: the periphery was exploited for raw materials; those raw materials were sent thousands of miles to be processed into manufactured products; those manufactured products were then transported thousands of miles back to be sold. In the language of Soviet economic geography, this was to create 'a pool of historical stagnation' in Russia.[10] Such a system was an affront to logic, Soviet planners argued, as well as being patently inefficient. It was thoroughly out of keeping with the 1930s craze – in some parts of the West as well as in the Soviet Union – for a centrally managed economic strategy.

Instead, the Soviet Union would turn the imperial system on its head. Rather than allowing extraction, processing and manufacturing to take place in separate parts of the country, Moscow would direct the construction of industrial 'combines' – places where all activities would take place on a single site or in a single region. Instead of 'exploitation', the dominant model of the Soviet economy would be 'development'. And instead of the recurrent economic crises which Marxist economists considered an inevitable feature of high capitalism, twentieth-century socialist planning would result in constant and unbroken economic expansion. In the late 1930s, a Soviet geographer would write that, in the Soviet Union, 'the possibility of crises has been eliminated'.[11]

The Soviet image of a sclerotic imperial government, entirely uninterested in concepts of economic planning and, particularly, in the economic potential of the Arctic, was not entirely accurate. Towards the end of the nineteenth century, the imperial government did begin to pursue Arctic policy more actively. But the measures taken were largely defensive, in response to growing international interest in the Arctic. Oftentimes, they were reactions to passing threats rather than consistent strategies for Arctic development. One particular challenge stemmed from Norwegian activity in the Barents and Kara seas. From the 1860s onwards, Russia took steps to close the northern coastline to international trade.[12]

More positive action was taken in 1897, with the beginning of construction of the port of Alexandrovsk, on the Kola Peninsula, then almost completely undeveloped.[13] The following year, Finance Minister Sergei Witte agreed to back the construction, in British shipyards, of Russia's first ice-breaker. In keeping with the common analogy of Russia's northern frontier with its internal, Siberian, frontier, the 8,000-ton ship was named after the Cossack adventurer who had first conquered Siberia for the Russian state, *Yermak*.

Russia was still conspicuously absent from the race for the North Pole or the race to discover the remaining lands of the Arctic. The islands due north of the Siberian mainland, Franz-Josef Land, had been discovered by an Austro-Hungarian expedition in the 1870s. Whereas Nansen's *Fram* might drift across the north coast of Russia, there was no Russian equivalent drifting across the north coast of Alaska or Canada. The *Yermak*, meanwhile, was an almost incidental expense compared to the great Russian infrastructure project of the age: the construction of the trans-Siberian railway, linking European Russia to Vladivostok.

To Soviet eyes, the lack of any consistent imperial policy towards the Arctic had had severe geopolitical as well as economic consequences. Later, under Stalin, it was military and geostrategic considerations, as much as economic ones, which would shape how Soviet policy towards the Arctic developed.

In 1904–05, the cost of the imperial government's failure to pursue a long-term Arctic policy had been cruelly revealed. Russia's defeat in the Russo–Japanese war underscored the country's fundamental strategic weakness. It might dominate the Eurasian land mass, but it was unable to shift forces rapidly from one side of this immense empire to the other. Following the defeat of the Russian Pacific squadron by Japan, the lack of a usable Arctic route had forced the Russian commanders to sail their Baltic fleet halfway around the world to relieve the garrison at Port Arthur (now China's Lushun naval base, just north of Dalian).[14]

But when Admiral Rozhestvensky finally arrived in the Far East, the Russian navy was defeated a second time, and on a grand scale. Defeat was a source of national shame, accentuated by racial prejudices – similar to the Italian defeat at Adowa in 1896, or the French defeat at Dien Bien Phu in 1954. Tsushima brought to a head a simmering crisis of confidence in Russia's leadership. The 1905 revolution followed a month later.

Neither the timing nor the location of the final battle had been of Rozhestvensky's choice. Had the fleet been able to sail along a shorter Arctic coastline, and refuel fully at Russian coaling stations, the outcome might well have been different. Tsushima reminded imperial Russia's leaders of the military advantages of an Arctic sea route. Construction of the *Taimyr* and *Vaigach* ice-breakers followed soon after. Once finished, they were rushed to Vladivostok through the Suez Canal, partly to demonstrate that Russia would not give up possessions in the eastern Arctic, and partly to monitor a growing and unlicensed trade between natives along the Russian Chukchi coast and

American whalers. Later, in 1915, the *Taimyr* and *Vaigach* were successfully piloted through the northern sea route from Vladivostok to Arkhangelsk.

The post-Tsushima preoccupations of the imperial high command would persist in the outlook of the military planners of the Soviet Kremlin. As the Soviet Union looked at an increasingly imperialist Japan in the 1920s and 1930s, and at the military potential of the Arctic to counter it – both in terms of the Northern Sea Route and in terms of the use of air transport in the Arctic – the lessons of Tsushima retained all their salience. Under the Russian empire, money for a Northern Sea Route had become available after the defeat of 1904–05. It was a similar reminder of the rising military power of Japan – the occupation of Manchuria in 1931 – which prompted the creation of Glavsevmorput, a Soviet body charged with developing the Arctic as a commercial and military asset. The Assistant Chief of the new administration was appointed by the People's Commissar for National Defence. A Pacific fleet was established at the same time. In March 1939, defending his role and his record, the head of Glavsevmorput would tell his audience at the Eighteenth Party Congress that full exploitation of the Northern Sea Route would mean 'no more Tsushimas'.

In the west, the Soviet Union was no better off than in the east. Here too, the Arctic could play a role in mitigating the Soviet Union's strategic weaknesses. The problem was an old one: the concentration of economic resources and population close to the western border. But the problem had been exacerbated by the new borders which the Soviet Union had been forced to accept in the early 1920s.

The peace deal which Lenin had struck with Germany in March 1918 had been disastrous enough – with the loss of most of Poland, the Ukraine, Finland and the Baltic region. But Germany's defeat by the Allies invalidated it. In 1919, it was reported that, in order to secure a peace with those Allies who were supporting the 'White' resistance to the newly established Soviet regime 'the Soviet government was prepared to give up Western Belorussia, half of the Ukraine, all of the Caucasus, the Crimea, all of the Urals, Siberia and Murmansk'.[15] This was probably no more than a temporary Soviet ruse in preparation for a new campaign and in the hope that revolutions would soon take hold in central and eastern Europe.[16] In Hungary and Germany in particular the prospects looked good. But even as a ruse Lenin's offer reflected the calculations of a regime poised between the promise of world revolution and the fear of imminent collapse.

Everywhere, from east to west and from south to north, the Soviet government's control seemed under threat from both internal and external enemies. In 1918–19, British troops were in Murmansk, French and American troops were in Arkhangelsk – the gateway to Russia's Arctic – while Japanese, British and Canadian forces were present in Vladivostok. A British detachment took up position around the oilfields of Baku, Azerbaijan. Czechoslovak forces, trapped in Russia at the end of the First World War, took up arms against Soviet authority. Britain gave consideration to either indefinitely holding onto a large part of Siberia, or, alternatively, recognising Siberia as an independent state under the White government of General Kolchak. The head of the British military mission suggested that British forces there were 'as much a factor of trade and empire as Clive's men [in India]'.[17]

In 1919, the administrator of the British-run Siberian Supply Company, Leslie Urquhart, confidently reported to British shareholders that 'the question of ore reserves need never trouble us and can only be limited by our ability to handle and smelt it . . . This great mining and metallurgical enterprise is now established and ready to earn large profits'.[18] But, as it turned out, from the Soviet perspective, 1919 was the worst of it. In Siberia, Britain's interest in holding on quickly declined after Kolchak's defeat. By 1921, Britain had signed a trade agreement with the Soviet Union, though it stopped short of full diplomatic recognition.

The Soviet Union's final western borders were determined at this time too, as the product of a series of cessions, wars and treaties culminating in the Treaty of Riga, signed with Poland. (The Red Army had finally repulsed Polish forces after two wars which saw large tracts of Russia proper occupied by the Polish army.[19])

No one was happy with the result, least of all in the Soviet Union, which saw its loss of Poland and also Finland and the Baltic states crystallised in agreements with those newly independent countries. Like the losses of the Soviet satellite states in the 1990s, the losses of the satellite states of the Russian Empire were widely seen as an imposition from outside, born of temporary weakness. For some, they suffered from the stigma of geopolitical illegitimacy. Twenty years later – in the war with Finland, the Molotov–Ribbentrop Pact and the annexation of the Baltic states – the Soviet Union demonstrated its keenness to reverse the losses of 1918–21.

The Soviet Union's geographic vulnerability was acute. The Russian borderlands – the country's traditional defence, along with the harsh Russian winter, against invaders from Napoleon to Kaiser Wilhelm – had been removed. The physical border had been pushed east, and the mechanisation of transport threatened to further erode the defensive advantages of space. St Petersburg – renamed Petrograd in 1914 and Leningrad in 1924 – was within a few hours' drive of Helsinki, capital of newly independent Finland. (When Lenin arrived at Petrograd's Finland station in April 1917 – completing his return from exile by way of Zurich, Stockholm and Helsinki – he was only twenty miles within the borders of Russia proper.[20])

The decision to move the seat of Soviet government from Petrograd to Moscow reflected both the Soviet Union's new borders and the nature of the regime. Moscow was well inside the country's new frontiers, and the city was much less strongly associated with the aristocratic and European-minded *ancien régime* than was Petrograd. But for Stalin, moving the capital was only the first step. One of the key aims in both the First and Second Five Year Plans – spanning the late 1920s to the late 1930s – was to shift the Soviet Union's centre of economic gravity to the north and to the east. Siberia and the Arctic were destined to become more important to the Soviet economy than they had been to its imperial predecessor.

Grand strategy predominated in Stalin's mind. The potential advantages of economic concentration in the west – access to European markets, an existing transport network, high population density, a warmer climate – were outweighed by the need to mitigate strategic vulnerabilities of the Communist state. More than this, the autarkic objectives of Soviet economic planning – preferring economic self-containment to global market integration – made some of those potential advantages irrelevant.

It was no surprise that over half the capital investments of the Second Five Year Plan (1932–37) were in the eastern part of the Soviet Union or that the labour camps of the gulag, linked to Stalin's economic objectives, were concentrated in the north and east of the country. The major infrastructure projects of the 1930s were intended to fundamentally reshape the economic geography of Russia and provide it with additional strategic depth. The most famous infrastructure project of them all, the forced labour construction of the *Belomorkanal* (White Sea Canal), connected the old industries of the Baltic with what Stalin envisaged as a new zone of industrial development, out of danger from the Soviet Union's enemies, along the Arctic shores of the White Sea and beyond.

As a result of Stalin's plans, the official population of the Soviet Arctic region increased from 656,000 in 1926 to 1,176,000 in 1935. This included a doubling of the population of the Soviet Far East in just nine years: from 104,600 to 200,100. Meanwhile, the population of Siberia increased so much in the 1920s that the region was administratively split into two in 1930.

There was a final element to Stalin's Arctic policy which had nothing to do with strategy, but which was, perhaps, most distinctly Communist, as opposed to Russian or authoritarian. That element was the Soviet attitude to nature. Overcoming the obstacles of economic exploitation in the Arctic – the distance, the cold, the ice – appealed to a very Communist grand narrative: the victory of human science over the natural world, and over the superstitions which surrounded it. In the Arctic, the Soviet Union found a metaphysical adversary, to be tamed, defeated, and ultimately controlled.

Soviet writers tended to reject the traditional description of the Arctic as a lifeless emptiness. Like Vilhjalmur Stefansson, they presented the Arctic as full of opportunity and potential, waiting to be harnessed. In their view, to understand the Arctic as a 'white desert' was a colossal failure of imagination. It meant accepting nature as destiny, rather than challenging it. For a regime devoted to the idea of all-conquering scientific Communism, nature – or rather, Nature – was an affront.

Conquering Nature – rather than adapting to it – was at the root of even the most banal accounts of Arctic activity:

> The stones have been cleared away from the experimental fields . . . Acid soil has been neutralized with alkalis and fertilized . . . This is not adaptation to circumstances: it is an alteration of Nature.[21]

Perversely, where Nature was at its most challenging was precisely where the ideological achievement of economic development would be greatest. To ignore the costs, environmental or otherwise, of opening up the Arctic was a proof of ideological conviction.

Human needs were considered more important than those of nature. A speech made by V. Zazubrin to the First Congress of Siberian Writers in 1926 was not uncommon in its appeal to the destructive and aggressive instincts of economic exploitation in the service of human prosperity:

Let the fragile green beast of Siberia be dressed in the cement armor of cities, armed with the stone muzzles of factory chimneys, and girded with iron belts of railroads. Let the taiga be burned and felled, let the steppes be trampled. Let this be, and so it will be inevitably. Only in cement and iron can the fraternal union of all peoples, the iron brotherhood of all mankind, be forged.[22]

It followed from this that the protection of nature – Arctic or otherwise – would be an inexcusable, and ideologically obsolete, luxury.

In the USSR it is considered that human work is a powerful organic and inalienable factor in the evolutionary process of nature, and therefore the idea of non-interference with natural conditions in the sanctuaries has been rejected. The fetish of 'inviolable Nature' has been discarded.[23]

In the 1920s and 1930s, the notion of releasing human energies from the constraints of the natural world and from the fatalism of geography was as appealing in the United States, Germany and Australia, as it was in Russia. New technologies of the late nineteenth and early twentieth century – the aeroplane, the telephone, the radio – had already exploded notions of distance.

The distinction of the Soviet attitude to nature – compared to its European or American counterparts – was that rather than populist fantasy, it was an element of state ideology. As such, it was very difficult to dislodge. The consequences for Russia's environment and the health of its people were correspondingly much more long term.

As late as the 1970s, environmental degradation within the Soviet Union was considered an ideological tautology:

The Soviet citizen must simply believe the statement that socialism itself, by its very essence, guarantees harmony between man and nature, that 'universal ownership of the means of production and of all natural resources foreordains the successful resolution of ecological problems in the USSR'.[24]

Lurking in the background was the persistent idea that, given the extent of Soviet territory, the country could afford relatively greater pollution in the pursuit of material production.

As long as we do not invest very much [on ecology] we can spend more on boosting the economy and on other targets. The more we are forced to invest, the less is left for economic development. In terms of land mass and many resources, our situation is better; hence we should use this strategic advantage over the United States . . . hence we can wait and accumulate funds while capitalism suffocates in its smoke . . . Right?[25]

The first and most important tool available to Stalin in the Soviet project for the transformation of the Arctic was the coercive power of the regime. Violence had always been a central element of Lenin's theory of the revolution. But under Stalin, coercion became an organising principle of the Soviet state, with the gulag as its most feared expression. In the Arctic, Solovetsky was followed by the *Belomorkanal*, the nickel mines of Norilsk, the coal mines of Vorkuta and, most infamously of all, the gold fields of Kolyma.

The second tool available to Stalin was the administrative body formed in 1932 to manage the development of the Soviet Union's Arctic domains: Glavsevmorput (Main Administration of the Northern Sea Route). Glavsevmorput had not been created in a vacuum. It incorporated a set of Arctic institutions which had operated through the 1920s: the Committee of the North, the Arctic Institute and Komseveroput (the Committee of the Northern Sea Route). What made Glavsevmorput different was the administrative latitude which it acquired – allowing it to cut through the bureaucratic maze of Soviet economic planning.

Together, the gulag and Glavsevmorput constituted the pillars on which the Soviet Arctic would be built in the 1930s. The relationship between them evolved, with the gulag steadily growing in importance. In 1938, Glavsevmorput was effectively demoted as the primary agency for Arctic development in the Soviet Union, and many of its assets and functions passed to the Dalstroy corporation, a branch of the Soviet secret police, the NKVD. As a result, the balance between coercive and civilian development of the Arctic shifted dramatically towards the former over time.

The expansion of the Arctic prison camps began in the 1920s. A little more than eighteen months after it was opened, the Solovetsky prison camp alone held some 6,000 inmates from across the Soviet Union. But more important than the increase in numbers was the change in its internal operation. The boundaries between different types of prisoners were steadily eroded. Initially, 'political' prisoners had been relatively well treated, with access to books and newspapers. But over time, this changed. Conceived as a corrective institution with limited economic functions, Solovetsky soon became the testing ground for another model of mass incarceration: the self-sufficient labour camp.

How and when this shift occurred is not clear. In itself, the use of prison labour was not new, either in Russia or elsewhere in the world. Nonetheless, the Soviet gulag was quantitatively and qualitatively different from any predecessor. First, there was an ideological difference. When Maxim Gorky described his visit to Solovetsky in 1929 he wrote that camps in Soviet Russia were, by the very fact of their high-minded Soviet nature, different from what existed abroad – in intention and in structure.[26] Over time, there was also a series of practical changes – some originating in local innovation, some resulting from centrally established targets – which coalesced into a model for the specifically Soviet gulag, with its focus on self-sufficiency and exploitation of economic resources.

Partly the shift occurred as a matter of necessity. As the prison population grew through the 1920s and 1930s, particularly after the acceleration of agricultural collectivisation in 1929, the collective financial losses of the prison system were bound to increase. It became imperative to reduce or even eliminate the net costs of incarceration, as was already required in some prison camps.

Partly it happened as a matter of bureaucratic jostling in Moscow between the OGPU security agency, which had run the Solovetsky prison camp from the start, and the Commissariat of the Interior, which managed the ordinary prison system of the Soviet Union. The OGPU was keen to prove its usefulness by demonstrating its effective and economically productive management of prisoners.

But there were other factors, too. In Solovetsky, the dark figure of

Naftaly Frenkel played a significant and perhaps decisive role. A typical example of prisoner turned prison guard, Frenkel was said by Solzhenitsyn to have developed the concept of rewarding prisoners with different amounts of food according to their work.[27] This was a sharp deviation from Marx's aim of 'from each according to his abilities, to each according to his needs', but it was certainly in keeping with the lower phase of Communist society which the Soviet Union inhabited.[28]

Whether or not Frenkel was the inventor of the gulag system of food rationing, he was certainly a key figure in the development of the Solovetsky prison camp as a factor in the local labour market, hiring out prisoners for commercial activities in the process. As early as 1925, SLON (*Savernye lagerya osobogo naznacheniya*) was a shareholder in the local Karelian bank and had outbid a civilian forestry business in a contract to cut 130,000 cubic metres of wood.[29] When it came to the construction of the White Sea Canal a few years later, SLON was on hand to provide a part of the labour required. Just as the Solovetsky monastery had been a key part of the local economy in the White Sea, so SLON took on and expanded that role.

But the main reason for the transformation was the requirements of the Five Year Plan. Inevitably, additional labour requirements were greatest in those parts of the Soviet Union which were least populated, least attractive for free or semi-free citizens and most physically demanding to develop. That the gulag should develop most in the Arctic North and in the Far East was natural. Though lip service was still paid to the idea of prisoner reform in the 1930s – certainly in public documents, and certainly in publications intended for a Western audience – the organising principle of the Soviet prison system had moved towards economic development.

The most significant projects of the gulag in the Arctic regions of the Soviet Union were the *Belomorkanal* (the White Sea Canal, begun in 1931) and the mines at Kolyma (from 1931), Vorkuta (from 1931) and Norilsk (from 1935). Three of those projects were either entirely, or partly – in the case of the Kolyma complex – above the Arctic Circle. The White Sea Canal, while below the Arctic Circle, was nonetheless very much an Arctic project – aimed at opening up shipping between the Arctic Ocean and the Leningrad region, bypassing the Barents and Baltic Seas.

At their peak, each of these Arctic or sub-Arctic camps contained tens of thousands of prisoners. Norilsk, Vorkuta and Magadan – the main administrative centre of the Kolyma camps – all turned into substantial Soviet cities in their own right, which survived long after the camps themselves had closed. Both directly and indirectly the gulag became a vector of Soviet Arctic development.

The White Sea Canal, started in September 1931 and completed just eighteen months later, was the realisation of a historical project, expected to transform Russia's Arctic North-west by connecting it to the industrial and trading heartland of Leningrad. It is estimated that 25,000 prisoners died – from accidents, exhaustion or the cold – at the construction site itself. In a sort of parable of the entire history of the Five Year Plan, Maxim Gorky led the writing of an account of the construction of the White Sea Canal after it had been finished. Translated into English and released in Britain in 1935, the book – written in the form of short dramatic vignettes each focusing on a single event or individual – followed a familiar sub-Hollywood plot: an impossible task, an unlikely team, an unexpected problem and finally, an extraordinary achievement.[30]

There was success of a kind, despite the poor equipment, lack of planning and many interruptions in construction, reflexively ascribed to fifth-columnists and wreckers.[31] After all, the White Sea Canal was built. But the human costs were not calculated by the Soviet Union. And the original promise of the canal, to become the major artery of Arctic industry, was not realised.[32] Within a few years, the shallow canal had begun to silt up.

The White Sea Canal project was certainly the most publicised of the Arctic endeavours of the gulag. The most important – in terms of scale, suffering and economic impact – was Kolyma. A network of camps, settlements and mines in Russia's harsh Far East, the country's coldest region, Kolyma became a byword for death. Though a few of the Kolyma labour camps were far above the Arctic Circle, including the fisheries camp at Kresty, most were just south of it, clustered along the Kolyma River and along the road north of the town of Magadan. Climatically, however, the distinction made little difference. Conditions were summarised in an ironic camp rhyme:

Kolyma, Kolyma,
Wonderful Planet:
Twelve months winter,
The rest summer[33]

Stalin's interest in Kolyma was long standing and personal. He had apparently sent an engineer to the United States in 1926 to study mining in the hope of developing the Kolyma fields. The mineral resources of the region – principally gold – were well known. Gold had been discovered there in 1901 and a mine had operated with free labour since 1927. But its production was small. In the absence of manufactured exports with which to raise foreign exchange, an increase in the production of gold – from Kolyma and elsewhere – was seen in Moscow as the only way of raising the hard currency to purchase much-needed foreign capital equipment. One historian reports that, between 20 August 1931 and 16 March 1932, the geology of Kolyma was discussed eleven times in the Politburo.[34]

At least on the face of it, Kolyma operated as a fairly standard commercial enterprise. The Dalstroy (Far Northern Construction Corporation) had been set up in November 1931 by the Council of Labour and Defence, and by the mid-1930s most of the ships which carried workers across the Sea of Okhotsk to Magadan were decked out with the letters DS, to signify the ownership of the Dalstroy corporation. But most workers were only leased out to Dalstroy – in 1932 the ratio of prisoners to free workers was approximately three to one. And the power behind Dalstroy was the Soviet secret police, the OGPU/NKVD.[35]

By the mid-1930s Kolyma was producing 20 tons of gold a year. And from 1931 until 1937 – under the leadership of Eduard Berzin – Dalstroy was even considered one of the more liberal parts of the gulag. From 1937 on, however, things changed. Berzin was arrested on his way to Moscow in December 1937, a few months after his boss Yagoda. Accused of plotting a Japanese takeover of the Soviet Far East, he was shot a year later in the basement of the NKVD headquarters on Lubyanka Square, Moscow.

With Berzin gone, and the population of the gulag rising sharply as a result of the Terror, the conditions at Kolyma worsened. But the production of gold increased: it was estimated at 74.5 tons by the end of the decade, one quarter of the Soviet Union's total production. And the population of the Soviet Union's Arctic north-east grew with it. Magadan, gateway to Kolyma, grew from 15,000 people in 1936 to some 70,000 by 1939. In the 1940s the population of the now 100 Kolyma camps would soar again.

The rise of gulag – and Dalstroy in particular – helped to eclipse the primary civilian agency charged with developing the Soviet Arctic: Glavsevmorput, or the Main Administration of the Northern Sea Route. In the 1930s, as Glavsevmorput accumulated functions and control over large areas of the Soviet Arctic, the organisation began to suffer from bureaucratic overstretch. At its height Glavsevmorput had responsibility for everything from scientific exploration to geological investigation to the production of coal.

As so often in Stalin's Soviet Union, responsibility cut both ways. Burdened with the responsibility for grandiose targets under the Five Year Plan the agency was bound to underperform. The inflated ambitions of Stalin's project to transform the Soviet Arctic always ran ahead of the means available to achieve them.

An extract from the report submitted by the head of Glavsevmorput, Otto Schmidt, to Stalin in November 1936, gives a feel for the demands placed on Glavsevmorput, and for the central significance of quotas in economic planning:

> Coal extraction was 363,000 tons, which constitutes 107% of the total output planned for nine months, with a reduction of 1.4% in cost and an increase in productivity of 9%; spar extraction was 8,217 tons, which is 91.9% of the year's plan; the catch of fish and the canning industry during nine months amounted to 92% of the year's plan; the fur industry yielded 11,937,000 rubles instead of 8,535,000 rubles, which was the estimate for the nine months; reindeer farms up to October 1st had an increase of 25,637 head as compared with 5,412 head during the same period in 1935.[36]

In fact, Glavsevmorput came remarkably close to fulfilling the second Five Year Plan. (Quotas for the third Five Year Plan (1938–42) were yet more unattainable.)[37] But in 1937, when twenty-six of the agency's ships became trapped in Arctic pack ice – the last of which was only freed in 1940 – the agency's fate was sealed. Glavsevmorput's predecessor, the much weaker Komsevmorput, had been dissolved in 1932 because of failure to meet its objectives in the first Five Year Plan. Now it was Glavsevmorput's turn.

Despite the exploration successes of 1937, including the first aircraft landing at the North Pole, the agency was stripped of most of its powers, surviving until 1970 in a much weakened form and with a much more limited mandate. As Stalin had envisaged, Glavsevmorput played a key role in keeping the Arctic route open for some forty-one convoys along the Northern Sea Route during the Second World War, with losses of less than one in ten. But, by then, the bulk of the agency's responsibilities had been usurped by Dalstroy and the NKVD.

To many Western observers, the Soviet Union's Arctic policies looked to be an extraordinary success. Vilhjalmur Stefansson wrote approvingly of Soviet progress in the Arctic. But it was not always easy to separate truth from propaganda, and hyperbole about Soviet achievements from reality. Vladimir Zenzikov, who had himself been sent into political exile in the Arctic in 1911 under the government of Tsar Nicholas II, explained the need to interpret Soviet sources on the Arctic with an understanding both of the likelihood of exaggeration and the certainty of deceit:

> When speaking of the Soviet Arctic it is necessary to keep in mind that during the last few years many achievements have been concealed for military reasons . . . There are reasons to suppose that during the last ten-fifteen years enormous progress has been made in this field [gold mining], especially in the far North East of Siberia (primarily at Aldan and Kolyma). But there is no information about gold mining available to the general public.

But at the same time:

> The government never makes a sharp distinction between information and propaganda . . . The whole atmosphere of this work is so unusual, so picturesque and exciting, that it undoubtedly stimulates imagination at the expense of respect for the facts. But even if one has to make allowances for exaggerations, it still must be admitted that the actual achievements have been very great indeed.[38]

Halford Mackinder, the celebrated British geopolitical writer, made a similar observation in his introduction to the English translation of the standard Soviet work on the Soviet Union's transformation of its economic geography, though he added a perceptive insight into the Soviet mind:

Altogether it affords a noteworthy insight into the dynamic mentality of those who, with command of a working population of 160 millions, claim to be re-making the geography, physical as well as human, of one sixth of the land on this globe. That our author does not discriminate nicely between fact and prophecy was to be expected; to a revolutionary, plans may appear more important than achievement. Geography is a study of the present, but the present is the past flowing into the future. There is no such thing as a static present; your description of the present depends on whether your face is turned to the future or the past. Only the future will tell whether these 'engineers' both of society and environment have underrated the momentum of human values from the past.[39]

The role of the gulag in Arctic development was largely absent from rather starry-eyed Western accounts of the Soviet Arctic in the 1930s, despite the fact that the existence of prison camps was by then well known.

In a July 1937 speech given to the Royal Institute of International Affairs – the main foreign-policy think tank of the United Kingdom at the time – journalist H. P. Smolka recounted a trip through the Soviet Arctic. He claimed to have barely experienced the gulag at all, and this at a time when already well over 1.5 million Soviet citizens were incarcerated there, with many more having already passed through.

In Igarka, Smolka came across some 4,000 'kulaks' – previously landowning peasants, considered as a class in the Soviet Union – who had resisted collectivisation of their lands. But he reported that:

They receive normal wages for their work, are free to move about the town, and their children study side by side with the children of free workers . . . The only point where they differed from free workers was that they had no access to political meetings and were not allowed to take part in voting and discussion . . . After a few years, when they have proved themselves to be what the Russians called 'de-kulakised', they are restored to civil rights and are allowed to return to their homes.[40]

On another occasion Smolka met a convict who claimed to be a 'kulak', but accepted the subsequent explanation of an officer of state security that he was, in fact, a murderer and that he called himself a 'kulak' only to sound better. In a longer description of his Arctic experiences – published in 1937 as *Forty Thousand Against the Arctic: Russia's Polar Empire* – Smolka described the situation which arose when he left one prison camp. The tone is more one of heavy farce than anything else:

'What, are we going to sleep in jail?'

'No!' She roared with laughter, and Sadkov grinned like a Chinese god of mirth.

'We have just come out of the camp. Now we are in the town of Dudinka.'

'So we have been in the camp all the time?'

'Of course, what did you think?'

'But are the prisoners on holiday during the summer, or what?'

'Why should they be on holiday? Did not you see them? There were plenty about on the way till we got to the barbed wire.'

She could hardly control her amusement at my ignorance, and Sadkov began to tease me: 'Did not you see the heavy chains they carried, and the guards with the knouts, beating the slow ones on the railway track?'[41]

It is impossible to know whether Smolka's impressions were simply naive or whether they were, in effect, Soviet propaganda. Certainly, Smolka had Communist sympathies – in the 1930s he was a close friend and business partner of British double agent Kim Philby. Later, during the Second World War, Philby claimed that Smolka was well aware that he was spying for the Soviet Union and, despite Smolka's position as Head of the Soviet Relations Division of the British Ministry of Information, chose not to denounce him.[42]

It is also true that Soviet authorities were entirely capable of showing foreign visitors only a very partial view of the Soviet Union's Arctic reality. More often than not, in continuance of the Russian tradition that Potemkin had made famous, visitors would receive a distinctly abridged version. One Finnish inmate of the gulag described the lengths to which Soviet authorities went before one visit of a foreign delegation:

A secret code telegram was received from the head office in Moscow, instructing us to liquidate our camp completely in three days, and to do so in such a manner that not a trace should remain . . . Telegrams were sent to all work posts to stop operations within twenty-four hours, to gather the inmates at evacuation centres, to efface marks of the penal camps, such as barbed wire enclosures, watch turrets and signboards; for all officials to dress in civilian clothes, to disarm guards, and to wait for further instructions.[43]

In any case, whether the result of deliberate obfuscation or not, Smolka's work was not unusually pro-Soviet compared with most Western writing about the Soviet Arctic. Outside the Soviet Union, as inside the Soviet Union, development of the Arctic was presented largely as a positive and even heroic accomplishment, rather than as a shameful and ultimately destructive episode in Arctic history. Some external observers wondered aloud why their countries had not been so successful in Arctic development. After the Second World War, Vilhjalmur Stefansson speculated that, had US President Franklin Roosevelt or Canadian Prime Minister Mackenzie King been exiled to the Arctic, as Stalin had been, perhaps the United States and Canada too would have undergone what he described as the 'northward surge of development' which occurred in the Soviet Union in the 1930s.[44] It was something of which a Soviet citizen could be proud.

ST PETERSBURG, RUSSIA, 59° North: On the corner of Ulitsa Marata in St Petersburg a short flight of steps leads to a white-fronted church in the neo-classical style which dominates the architecture of so many Russian cities. Like most Orthodox churches – and unlike most churches in the West – the church is the same on all four sides. A set of five columns supports a plain beige pediment. Above that, between the top of the pediment and the base of the dome, there is another level of windows. There are seven on each side of the church, facing north, south, east and west.

These days, most of the church is hidden behind scaffolding and plastic sheeting. To the casual observer, it looks as if the church is either closed for renovation or being prepared for demolition. But should you venture as far as the thick double doors of the church and push, quite hard, the doors will give way, and a blast of overheated air will rush out to greet you, bearing the sickly smell of cheap wood polish.

Inside, there are no icons on the wall and no altar. A ticket counter stands where the baptismal font should be and, directly under the central dome, there is a massive globe – or, rather, there is that part of the globe which lies north of the Arctic circle. Summarily converted into the Russian Museum of the Antarctic and Arctic in the 1930s, the church has not functioned as a church for over seventy years.

After the collapse of the Soviet Union, a religious sect attempted to reclaim the church for its original use.[45] The 'vedinovertsy' or Followers of One Faith – a halfway house between the mainstream Orthodox church and the schismatic Old Believers who once occupied the Solovetsky monastery – launched a legal and administrative battle against the museum. In 2000, the parish priest told a local newspaper reporter, 'All those white bears and penguins – they're nice, but they stand on the blood of martyrs we all need to atone for.'[46] But even as the Russian economy was collapsing and as St Petersburg faced financial ruin, the museum survived.

Though the church is no longer a church, it is a temple of a kind. It is a pantheon of the citizens of the Soviet Union who conquered the Russian Arctic, laying claim to the North in a way that no country had ever done before. This museum celebrates the heroism of the Soviet Union's Arctic adventurers – the flip side of its Arctic gulag.

When I call at the museum, on a weekday afternoon, there are only a few visitors – mostly the old and the very young – methodically working their way around the exhibits, whispering to one another as if they were, in fact, in a place of worship. The only sound above their hushed voices is the squeak of the guards' shoes as they circle round, waiting to pounce on anyone standing too close to a model of a Soviet ice-breaker or attempting flash photography of a facsimile of a Soviet hero's diary.

On one side, a young child is being led round by her grandmother, who leans heavily on a walking stick and lists from side to side as she walks. On the other, a father explains each exhibit to his son, occasionally stooping to put his arm around the boy's shoulders. I can't help wondering whether Vladimir Putin and Dmitri Medvedev, both of whom grew up in Leningrad, were brought here as children.

Perhaps it's just that the middle generation is at work now. Perhaps the museum is full to bursting with young Russian couples at the weekend. But I have the sense that there is more to the demography of the museum's visitors than that. This is a place for the older generation to show to their children and grandchildren.

For any Russian under the age of thirty, the Soviet Union is barely a memory. For them, the Soviet Union lives on only in a legacy of bad plumbing in communal apartment blocks. But for Russians over thirty there is a sense of continuing loyalty to the Soviet Union – a country that, even at its worst, was theirs. Their stake in a positive memory of it increases with the proportion of their life that they spent obeying its diktats and believing the half-truths about its power and prosperity. In celebrating the feats of Soviet Arctic explorers there is the possibility of salvaging something from the wreckage of the Soviet Union, and of salvaging something from the years spent under its tutelage.

As Leon Trotsky wrote, 'The Arctic flights are known to everybody – these high points speak for a whole mountain chain of Soviet achievements.'[47] The Museum of the Antarctic and Arctic is at least as Soviet as it is Russian. The popularisation of the Soviet Union's Arctic exploits goes some way to explaining why, for many Russians, the Arctic is by rights a Russian preserve. In terms of lives, money and endeavours, Russia's exploits far outstrip those of any other Arctic nation.

And at one time, the names of the Soviet Union's Arctic explorers were as familiar as the names of the members of the politburo. Names such as Otto Schmidt, the latter-day Lomonosov who built his fame around the first one-year navigation of the Northern Sea Route in 1932, at the helm of the ice-breaker *Sibiriakov*. Or Valerii Chkalov, the pilot who proved the Soviet Union's technological prowess and military reach by landing a converted Tupolev bomber at the North Pole in May 1937. Or Evgeny Fedorov, one of the four men who occupied *Severny Polus I* (*North Pole I*, abbreviated as *SP-1*) – the first drifting Arctic research station – over the winter of 1937–38. Not to mention the names of the seven heroic pilots of the *Cheliuskin* saga who, after that ship sank in February 1934, managed to rescue all 104 of the survivors, stranded for two months at 'Camp Schmidt'.[48]

These men were, for the most part, members of the 'narod', the Soviet people.[49] Their avowed aims were not personal glory, but the realisation of Soviet policy in the Arctic. As Otto Schmidt wrote, 'We in the Arctic do not chase after records (though we break not a few upon the way). We do not look for adventures (although we experience them with every step). Our goal is to study the North in order to settle it economically . . . for the good of the entire USSR.'[50]

And if their names were unfamiliar outside Russia and the world of Arctic exploration, they were hugely celebrated at home. Years before the Great Patriotic War (1941–45), these were the first Heroes of the Soviet Union.[51] They were phenomena of Soviet mass culture – the subject of songs, poems, murals, films, plays and posters.

Some – including three of the four inhabitants of *SP-1* – wrote books recounting their adventures, eagerly published by the state publishing house.[52] One, Mikhail Vodopianov, wrote a play entitled *Mechta pilota (A Pilot's Dream)*. A fictional account of a flight to the North Pole, the play was staged at Moscow's Variety Theatre in 1937 just as Vodopianov was undertaking a real mission to the North Pole as the chief pilot of the *SP-1*.

As part of a propaganda campaign in which these men were willing participants, their actions were represented in popular culture as proof of the Soviet Union's technical capacities, evidence of Stalin's grand vision for his country's future and a demonstration of Soviet values. Here, in Arctic exploration, was unassailable proof of what the Soviet 'narod' could achieve when they worked together. The inevitable personal failures and petty rivalries were carefully hidden from public view.[53]

Otto Schmidt – also known as the Commissar of Ice – was one of only two people ever to be pictured embracing Stalin on the cover of *Pravda*. The other was the polar pilot Valerii Chkalov, so devoted to Stalin that he wrote an essay about him entitled '*Nasha Otets*' ('*Our Father*'). When Chkalov died in December 1938 Stalin was a pallbearer at his funeral, though there were suggestions that Stalin himself may have ordered Chkalov's death.[54] An aircraft company was formed in Chkalov's honour, operating continuously until it became part of Russia's United Aircraft Corporation in 2006.

Even as the organisation for which most of these men worked – the Glavsevmorput – was acting as 'the taxi service of the gulag'[55], these men represented the gulag's flip side: as publicly praised as the gulag was privately feared. Indeed, in at least one case, that of Otto Schmidt, Arctic celebrity may have actually saved him from the Arctic gulag.[56]

The truly heroic period of the Soviet Union's Arctic exploits ended in the late 1930s. After a disastrous year in 1937–38, the freeing of the ice-breaker *Sedov* from the polar ice pack in 1940 did not fire the same enthusiastic response as the rescue of the so-called 'Cheliuskinites' six years previously. Two great successes of 1940–41 received less attention as the world collapsed into conflict.[57] The first, the guiding of the German raider *Komet* through the Northern Sea Route in a record twenty-one days in 1940, was not publicised for what it was, effective Soviet military assistance to its then ally, Nazi Germany. The result of the transit was that the *Komet* became a threat to British shipping in the Pacific. The second success, almost entirely eclipsed by the war, was Cherevichny and Akkuratov's aircraft landing on ice at the pole of relative inaccessibility – the farthest point from land in the Arctic – in May 1941.

From the point of view of political propaganda, the heroes of Arctic exploration became considerably less important after the Great Patriotic War, which furnished the Soviet Union with a whole generation of heroes to exalt.[58] The war was formative of the Soviet Union in a way that Arctic exploration could never be.[59] Later still, the USSR could rely on the successes of its space programme as a propaganda replacement for the Arctic heroism of the 1930s. Yuri Gagarin replaced Valerii Chkalov. And like Chkalov, Gagarin died in a plane accident.

But even if the propaganda function of the Soviet Union's Arctic exploration receded, its Arctic achievements retained their hold on the public imagination. Today, as Russia seeks to recreate a heroic self-image for itself in the twenty-first century, it is natural for it to return to the glorious successes of the Soviet Union, just as the Soviet Union appropriated the exploration successes of Imperial Russia.

The continuity of present-day Russian Arctic ambitions with the historical glories of Soviet Arctic development is more than just a matter of propaganda. In the Museum of the Arctic and the Antarctic, a wall is covered with a chart listing the technical details of the Soviet Union's most famous ice-breakers – presented as a series of bar charts showing the inexorable upward trend of Soviet achievement. First, for comparison, the *Josef Stalin*, a steam ice-breaker built in 1938, with 10,300 horsepower, famous for releasing the *Sedov* from the ice pack in 1940. Then the first nuclear ice-breaker, the *Lenin*, from 1959, with 44,000 horsepower. And, finally, the *Arkitka*, which came down the slipway on the Leningrad shipyards in 1975, and which, with some 75,000 horsepower, remains the pride of the Russian fleet. In 1977 the *Arktika* became the first surface ship to reach the North Pole. Even today, after the prolonged economic crisis and underinvestment of the 1990s, Russia's ice-breaker fleet, a Soviet inheritance, remains the world's largest.

On another wall a display tracks the passages of all the Soviet scientific floating stations which followed the *SP-1*. Their trajectories, sometimes smooth, sometimes jagged, contain within them a thousand individual stories of Soviet endeavour. For all the propaganda impact of *SP-1*, the use of floating stations by the Soviet Union was not renewed until after the war, and then with the greatly enhanced purpose of measuring meteorological conditions and ice thickness relevant for the Soviet Union's military. From the 1950s until the mid-1980s there were always at least one or two floating scientific stations in or around the Soviet Arctic. In July 1991, as the Soviet Union collapsed, the last Soviet floating scientific station, *SP-31*, was closed. And for the rest of the 1990s, at a time of economic crisis and shrinking budgets, the Russian Federation's Arctic programmes remained on hold. But then, in 2003, under the presidency of Vladimir Putin, Russia returned to the Arctic. Consciously taking up where the Soviet Union had left off, the Russian Federation launched its first floating scientific station and, in the Soviet tradition, named it *SP-32*.

Upstairs on the Ulitsa Marata, under the curve of one of the side arches of the church, is the office of Victor Boyarsky, Russia's most famous Arctic and Antarctic explorer, and the director of the Museum of the Arctic and the Antarctic. Boyarsky – along with Arthur Chilingarov – embodies the Russian continuation of the Soviet Union's heroic Arctic. Like Otto Schmidt before him, he wears a thick beard. His hands are coarse and his manner is direct. If anyone can explain to me what the Soviet Arctic means in contemporary Russia, it is Boyarsky.

His office is an extension of the museum and a store of personal polar memorabilia. In one corner there's a stuffed penguin. In another there's a stuffed polar bear, its paws lifted in warning of attack. A German code-machine stands in a glass cabinet to one side. Mugs, medals, photos and crossed flags from numerous national and international Arctic and Antarctic missions adorn the room.

Like so many who have been drawn to the Arctic, his path there began as a child. He read the books of Jack London, hugely popular in the Soviet Union. And he immersed himself in the tradition of the Arctic explorers. 'I wanted to be a sailor like my father,' he explains, 'or an explorer like Nansen.' In the end it was an expertise in radio physics and a doctorate in mathematics which provided Boyarsky with his ticket into the world of Arctic exploration.

Later, faced with the abandonment of the Soviet Union's infrastructure of Arctic and Antarctic exploration in the 1990s, Boyarsky turned to private enterprise. After all, even if the state was unable to fund Arctic exploration, the equipment and the skills were still there. In 1993 Boyarsky became involved in a company to take tourists to the North Pole, using Soviet-era equipment – Antonov-74 short landing aircraft, Mi-8 helicopters – and Soviet-era experience. Boyarksy is proud of what they were able to achieve: 'No one else could have done this.'

Even now, Russia's Arctic community – degraded by years of under-investment in the 1990s, but beginning to experience an increase in funding now – remains, in Boyarsky's view, the equal of any other country. Partly it's a matter of culture: 'For Russians, the North is not a place for vacationing, but we are more adapted to the cold,' he tells me. But it's also a matter of capability. Three years previously, Boyarsky reports, he was telephoned by a newspaper and asked to comment on a report that Denmark was about to claim the North Pole: 'My response was "Can they even get there?".'

Isn't this part of the Soviet Union's Arctic legacy, a confidence that, in some sense, the Arctic is a Russian preserve, built by Russians and inhabited by Russians? After years of scientific stations, Soviet-built ice-breakers, a literature of Soviet heroism in the Arctic and the countless explorations of the ice pack by Soviet scientists, the Arctic is, to most Russians, their national back yard. This is the imprint of the Soviet Arctic on the Russian mind. 'The memory still keeps them going,' Boyarsky tells me. 'It's passed from generation to generation.'

But there's more to the legacy of the Soviet Arctic than a sense of ownership over the North. In its mines, its cities and its pollution the Soviet Arctic has left its indelible mark on the Russian landscape, too. At the time of the 1917 revolution, Norilsk did not exist and Murmansk – today the biggest city north of the Arctic Circle with a population of over 330,000 – was a small port. Millions of Russian citizens live and die in the Arctic. For better or worse, the Soviet Union achieved a large part of what it set out to do in the Arctic: it permanently transformed the country's economic geography. Now, the Russian Federation must live with the environmental, social and political consequences of large cities built to serve an economic system which no longer exists.

Finally, with the gulag, the Soviet Arctic has left its stain on the Russian soul. However much the narrative of Russia's economic, political and spiritual renewal may sideline the history of the gulag, the words of the survivors remain. My mind wanders back to Solovetsky, to the low sunlight skimming towards me across the White Sea, blinding me with its Arctic brightness.

Boyarsky sighs. 'People are more pragmatic these days,' he says. 'It's not like it was in the 1940s and 1950s – people were more idealistically thinking, then; this generation is motivated in a different way.' Many of the illusions which guided the development of the Soviet Arctic have died – not least the capacity of a new political system, Communism, to conquer nature.

But the memory of Soviet success in the Arctic – even the perverse successes of the Arctic gulag – remains. As Russia re-emerges from decades of economic stagnation and considers its national future, the Arctic is, once again, a central focus. The Russian vision of the Arctic as a source of material strength and national power – rather than simply a wilderness of ice – remains very much alive.

PART TWO
Power

'Possession is nine points of the law'
Legal maxim

NORTH AMERICA - 1867
CHUKCHI SEA
BERING STRAIT
Extent of the Arctic islands unknown as of 1867
BEAUFORT SEA
GREENLAND (to Denmark)
BAFFIN BAY
RUSSIAN AMERICA
NORTH-WESTERN TERRITORY
DAVIS STRAIT
BAFFIN ISLAND
GULF OF ALASKA
BRITISH NORTH AMERICA
ATLANTIC OCEAN
Sitka
BRITISH COLUMBIA
Mackenzie River
HUDSON BAY
RUPERT'S LAND
PROVINCE OF CANADA
NEWFOUNDLAND
PACIFIC OCEAN
Vancouver
PRINCE EDWARD ISLAND
Montreal
Ottawa
NOVIA SCOTIA
NEW BRUNSWICK
UNITED STATES
Kilometers
0
1,000

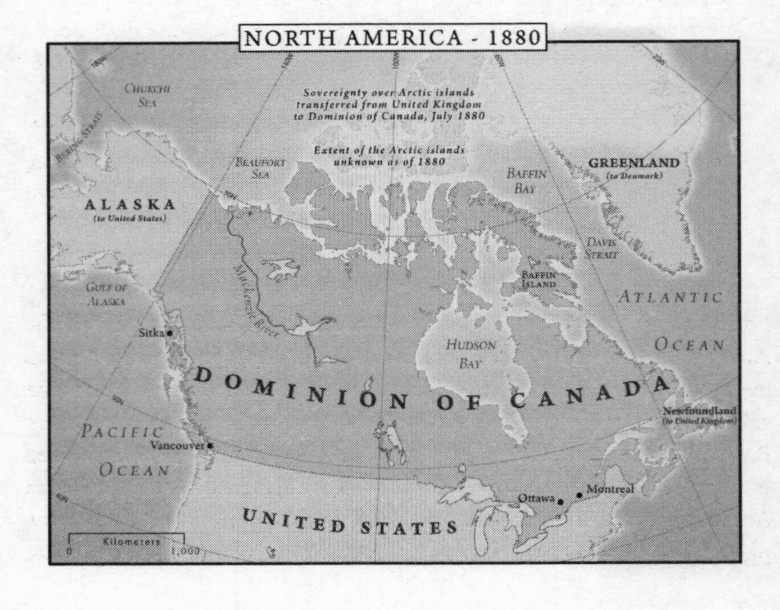
NORTH AMERICA - 1880
CHUKCHI SEA
BERING STRAIT
Sovereignty over Arctic islands transferred from United Kingdom to Dominion of Canada, July 1880
Extent of the Arctic islands unknown as of 1880
BEAUFORT SEA
BAFFIN BAY
GREENLAND (to Denmark)
ALASKA (to United States)
DAVIS STRAIT
BAFFIN ISLAND
GULF OF ALASKA
ATLANTIC OCEAN
Sitka
Mackenzie River
HUDSON BAY
DOMINION OF CANADA
Newfoundland (to United Kingdom)
PACIFIC OCEAN
Vancouver
Ottawa
Montreal
UNITED STATES
Kilometers
0
1,000

3

Northern Designs: The Making of the American Arctic

'O Canada!
Our home and native land.
True patriot love in all thy sons command.
With glowing hearts we see thee rise,
The True North strong and free!'

Words of the Canadian national anthem

'The rising States and nations on this continent, the European nations, and even those of Eastern Asia, have exhausted, or are exhausting, their own forests and mines, and are soon to become largely dependent on the Pacific. The entire region . . . seems thus destined to become a ship-yard for the supply of all nations'

William H. Seward, former US Secretary of State, Sitka, Alaska, 12 August 1869

From 1867 to 1880 the mantle of responsibility for a large part of the Arctic passed from the Old World to the New. The Russian flag was brought down over Sitka, Russian America, in 1867, an event that subsequent Soviet historians would view as a geopolitical disaster. British-held territories in the Arctic and sub-Arctic were transferred to Canada over the course of the 1870s. It was then that the American Arctic was created, the result of a legal, diplomatic and political game played out in London, Moscow, Ottawa and Washington. At the beginning of 1867 none of the Arctic was in American hands; by 1880, nearly half of the Arctic was American.

The emergence of Canada and the United States as Arctic powers was far from a foregone conclusion. The United States might easily have failed to buy Alaska, and Canada might easily have failed to obtain its Arctic territories from Great Britain.

For both countries, northern expansion was a seminal moment in their national histories – confirming the continental ambitions of both. Some American expansionists cursed the failure of the United States to acquire yet more northern territory; Canadians viewed the establishment of their own trans-oceanic nation as precisely a defence against any such American design.

An Arctic without a strong North American component – both from Canada and the United States – is hard to imagine now. In Canada, the North has come to occupy a central place in Canadian national identity – particularly Anglophone national identity – as 'The True North strong and free!'.[1] Defining his country's global stature in terms of its commitment to its Arctic possessions, Prime Minister Stephen Harper has placed himself in a succession of Conservative leaders striving to exploit Canada's huge geography.[2] In the United States, meanwhile, the once widely ridiculed purchase of Alaska in 1867 has come to be seen as a geopolitical masterstroke, providing the United States with a forward position in the Pacific, a vital source of domestic oil and gas, and a point of access to the Arctic Ocean.[3] When, in 2008, an Alaskan was picked as candidate for US Vice-President, the Republican party portrayed the state as more American than America, a bulwark of US national security and US energy strategy. In 2012, the United States may well find itself with a presidential candidate from the state.

In the United States and Canada, the Arctic has come to be an accepted part of national identity and an integral part of national territory. But the making of the American Arctic at the end of the nineteenth century was fraught. William H. Seward and John A. Macdonald, two men with radically different views of the world, dominated the drama.

Seward, born to a wealthy family in New York State in 1801, was the older of the two. He was also the greater visionary. Though he became renowned in Washington for opulent entertaining, photographs of Seward show a serious man, his evident high-mindedness matched by the upturned collar of the day. The scale of his facial features alone would have made him a candidate for Mount Rushmore. But he never got quite as far as the presidency. Governor of New York in the late 1830s, Seward served as a senator from 1849 and, in 1861, was appointed President Lincoln's Secretary of State. Over the course of his political career Seward became one of the first true globalists, envisaging a world bound together by global communications networks, free trade and financial integration. But he was also a nationalist and an ardent exponent of American power. In his model of the world, the United States assumed the central role.

Like Seward, Macdonald was a lawyer and an expansionist. Born in Glasgow in 1815, his loyalties lay with the British Empire and, throughout his life, he maintained a certain idea of Canada's place within it. As Canada's Prime Minister for most of the country's first quarter-century, Macdonald defined himself and his country against the United States, advocating a protectionist 'National Policy' and seeking to prevent the union of British territories in North America with their republican southern neighbour. From his beginnings as a Conservative politician in Canada West to his end as Prime Minister, accusations of 'treason' and warnings of threatened 'annexation' were never far from his lips. Seeking the vote of his fellow Canadians in 1891, the last political campaign of his career, he famously proclaimed, 'A British subject I was born, and a British subject I shall die.' (Macdonald's political influence survived his death, drafted in to support the Arctic policy of his twentieth- and twenty-first-century Conservative successors, John Diefenbaker and Stephen Harper.)

In the 1850s and 1860s, few were far-sighted enough to see the potential of the Arctic region, despite the fortunes being made by American whalers in the Bering Strait.[4] Despite Seward's own best efforts at government-funded propaganda, aimed at easing the passage of the Alaska purchase through an initially sceptical Senate, questions were asked as to why the United States had paid $7.2 million for an 'icebox' or a 'polar bear garden'.[5] Meanwhile, in Canada, the transfer of large tracts of Arctic and sub-Arctic territory from Great Britain and the Hudson's Bay Company was seen as something of a sideshow compared to the western expansion which those transfers secured. Much more significant than Canada's Arctic acquisitions was the fact that Canada had broken out of its Laurentian enclave and now spanned a continent.

In the United States and Canada, the course of empire continued to be to the west rather than to the north, running in the same direction as the railway tracks of North America which would be built to secure it. The Pacific Railroad across the United States would be completed in 1869; its Canadian counterpart was completed, at great expense, in 1885, 'a British girdle around the world'. It was the settlement and expansion of the western frontier which defined both the United States and Canada in the mid-nineteenth century.

But in establishing the United States and Canada in their Arctic territories, Seward and Macdonald laid the foundations for their countries to become Arctic powers.

William H. Seward's interest in expansion of the United States to the north was long-standing. Alaska was far from being the only prize in his sights. It was just a part of a wider plan of territorial aggrandisement by which he hoped to create the basis for an American commercial empire, turned as much towards Asia as towards Europe.

Seward's creed was simple: expansion, commerce, the Republic. As early as 1844, twenty-three years before the Alaska purchase, he laid out his vision of American territorial expansion in a speech to the Phi Beta Kappa society at Union College, his alma mater in upstate New York. Although he did not actually use the phrase 'manifest destiny', his argument amounted to the expression of its secular counterpart.[6]

Looking at the United States in 1844, Seward saw a country 'unsurpassed in general salubrity, in varied fertility, in forest and mineral resources, in natural channels for inland navigation, and in advantages for external commerce'. He concluded that 'the home allotted to us has capacity for four times the present number of organised States, and for more than two hundred millions of People'.[7] As he saw it, such an extensive and potentially populous land was far from complete.

The US had, after all, already demonstrated the appetite and capacity for enlargement, through the purchase of Louisiana in 1803 and the acquisition of Florida in 1819. Throughout the presidential campaign of 1844, debate raged on the issue of whether or not to annex the Republic of Texas. (Seward, like many in the North, opposed the annexation of Texas, but only because it would cause the expansion of slavery within the United States.) The Monroe Doctrine was already being interpreted by many as an expression of the United States' continent-wide ambitions. Had he chosen to, Seward might have referred back to one of the founders of the Republic itself, John Adams, who had written that the United States was 'destined to spread over the northern part of that whole portion of the globe'.[8]

Nonetheless, the title of Seward's 1844 lecture, 'The Elements of Empire in America', was something of a provocation. The United States had only recently freed itself from the British Empire. Were the principles of the American republic compatible with such an old-fashioned notion as empire? As Seward understood it, they were, for America would be a solidly republican empire, driven by Jeffersonian principles of government rather than by tyrannical principles of empire.[9] Whereas traditional empires depended on coercion for control of new territories, an American empire would accumulate new territories either by virtue of the attractions of its system of government, or by buying them. Prefiguring the purchase of Alaska twenty-three years later, Seward argued that 'peace is more propitious to the ruling passion of empire, than WAR, and . . . provinces are more cheaply *bought* than *conquered*'.[10] An American empire would be built on solid economic grounds – and united by free trade and geographic contiguity. (This principle, contiguity, would be broken in 1867 with the purchase of Alaska, separated from the Union by a stretch of British territory along the Pacific coast.)

Seward saw American expansion as inevitable:

> I am considering not what ought, but what probably will be, the future policy of our country; nor do I calculate how fast or how far this national passion will unfold. I am endeavoring to look into an uncertain FUTURE that lies far away beyond the reach of our footsteps, and practically is INFINITE.

Perhaps naturally for a man who lived in upstate New York, a few leagues from Canada, Seward's gaze turned north and west as much as south and east. He reminded his audience at Union College of a standing invitation to the British colonies of North America in the Articles of Confederation of the United States, pointing out that 'we have strenuously maintained our Northern frontier at the highest attainable latitude'.[11]

Elected to the United States Senate in 1849, Seward co-sponsored, with Senator William Gwin of California, legislation to map the Northern Pacific 'for naval and commercial purposes', resulting in the passage of the first American government ship north of the Bering Strait.[12] In 1857, Seward travelled to British North America, which trip undermined his confidence in the inevitability of the British colonies joining the United States – a view he now dismissed as a 'national conceit' – but nevertheless reaffirmed his belief in the territory's prospects:

> A region grand enough for the seat of a grand empire. In its wheat-fields in the West, its broad ranges of the chase at the North, its inexhaustible lumber lands, the most extensive now remaining on the globe, its invaluable fisheries, and its yet undisturbed mines, I see the element of wealth.[13]

Seward's conclusion was unequivocal: the United States, instead of developing itself towards the south – what Seward called the 'decaying Spanish provinces' – must develop good relations to the north.[14]

Losing to Abraham Lincoln in the race for the Republican nomination for the presidential elections of 1860, Seward nonetheless accepted the position of Secretary of State in Lincoln's administration. Any immediate hopes of territorial expansion were, however, necessarily put on hold by the Civil War. From the moment of his appointment, his time and energy would be involved primarily in preventing foreign intervention and parrying the diplomatic manoeuvrings of the European powers and the Confederate states.

A statement of British neutrality in North America was widely interpreted as providing the Confederacy with international legitimacy. France offered herself as a mediator, but was rejected by the Union. Only Russia, still smarting from its own defeat at the hands of Britain and France in the Crimea, unequivocally supported the Union. Eduard de Stoeckl, Russia's long-serving chargé d'affaires, himself married to American Elisa Howard, would find himself one of Secretary of State Seward's few diplomatic friends in Washington. As the relationship between Russia and the United States improved through the 1860s, so did the diplomatic basis for a sale of Alaska to the United States.

Once the last shot of the Civil War had been fired in mid-1865 – off the coast of Russian America – Seward quickly returned to his grand designs of American empire. Free at last to pursue his vision of American expansion, Seward engaged in a flurry of diplomatic activity. Navy Secretary Gideon Welles claimed that Seward 'has become almost a monomaniac on the subject of territorial acquisition'.[15]

Though in retrospect the Alaska purchase would seem a one-off, it was in fact part of a wider set of targets for Seward's expansion. And despite the general tendency of Seward towards the north and west, he also had aims to expand into Nicaragua – where he hoped to build a canal – and the Caribbean, where he failed in an attempt to purchase the Danish Virgin Islands.[16] (Fifty years later, the Panama canal was built and a subsequent American government purchased the Danish Virgin Islands for $25 million.)

The most obvious targets for acquisition in the north were the British possessions in North America. Some Canadians, looking at a United States now armed as never before, feared annexation. It was partly to counteract this perceived threat that political leaders in British North America moved towards confederation in 1864, which ultimately led to the establishment of the Dominion of Canada three years later.

The proposed confederation of Canada – comprising the British colonies around the St Lawrence River, including most of present-day Ontario and Quebec – did not include Britain's Pacific colonies, hundreds of miles to the west: Vancouver Island and the vast territory of British Columbia, stretching far into the north.[17] Tied economically to the United States, and entirely cut off from the rest of British North America by the territories of the Hudson's Bay Company, this was the land which most attracted Seward's attention.

As Seward saw it, the acquisition of British Columbia would dramatically extend the United States' Pacific coastline and its potential for trade with Asia. Britain, he thought, could be prevailed upon to give up British Columbia in return for settlement of outstanding legal claims arising from the Civil War.[18]

Throughout 1866 things seemed to be going Seward's way, with intelligence from British Columbia suggesting that the local population would be willing to join the United States. In mid-September that year, Seward spoke again of his ambitions for the establishment of American empire, telling an audience in St Louis, Missouri, 'I do not exclude the region that lies between us and the North Pole.'[19] A few weeks later, the American consul in Victoria, Allen Francis, reported a public meeting where a majority present voted that Britain be requested to permit American annexation.[20]

But it was in Russian North America that Seward finally achieved his vision of American expansion. Acquisition of Alaska had seemed a possibility for some years, but, in conversation with Eduard de Stoeckl, by then the doyen of Washington's diplomatic corps, it became clear that Russia was ready to strike a deal. Sensing his opportunity, Seward acted quickly and – partly to forestall the possibility of Britain entering the bidding – out of public view.

Initially, Seward offered de Stoeckl $5 million for Alaska. When de Stoeckl held out, Seward raised his offer to $7 million. In return for winding up various positions of the Russian–American Company for an additional $200,000, a deal was reached. A treaty was signed at the State Department building in the early morning of 30 March 1867. Characteristically, the native population of Alaska had not been consulted.

A later painting of the event showed a grand event on an imagined backdrop of the American flag and the double-headed eagle of Russia. De Stoeckl was shown standing by a globe of the earth, his hand hovering above an exaggerated Alaska. American diplomats – including Seward's son, Fred Seward – were arranged behind him. Charles Sumner, Senator from Massachusetts and chair of the powerful Senate Foreign Relations Committee, was seated at the back. (The reality of the event was more prosaic. Sumner, invited at the last minute, missed out on it entirely, going to Seward's private residence rather than the State Department.)

Whether or not the moment of signature was quite as grand as the artist's depiction, the occasion was a crucial event in the history of the North: the United States had joined the Arctic.

The geographic boundaries of the territory now belonging to the United States were defined by an 1825 treaty between Russia and Britain. In a northerly direction, Alaska's eastern border was considered to lie along the 141st line of longitude and to extend 'in its prolongation as far as the Frozen ocean'. Subsequently there would be debate as to what exactly was meant by this clause – whether or not, as some jurists suggested, it comprised an extension of the border as far as the North Pole.

To the south, the border with British territory was murkier. The 1825 treaty had stipulated that the frontier lay along an inland mountain range. But the area had not yet been surveyed well, and the true border was not settled until 1902. Alaska's western border with Russia – one which would be of tremendous significance in the mid-twentieth century but which, in 1867, was considered a border between allies – was set by a line running roughly down the middle of the Bering Strait.

As part of its purchase of Alaska, the treaty stipulated that the United States had purchased all public property on its territory. All of Alaska's Russian inhabitants were to be given the option of returning to Russia within three years or else becoming full citizens of the United States of America with the rights that this entailed. 'Uncivilized native tribes', meanwhile, were denied American citizenship and made subject to 'such laws and regulations as the United States may, from time to time, adopt in regard to aboriginal tribes of that country'. It would take more than a century before the native population of Alaska would be able to settle land claims in the territory.

Soviet historians would later view the sale of Alaska as 'shameful', a mark of 'the depravity of the rotten tsarist regime in Russia'.[21] Others took it as a craven American exploitation of Russian weakness. Even at the time, some Russians asked whether the sale of Alaska was simply the first part of a fire sale of more traditionally Russian territories: would Crimea, Transcaucasia and the Baltic provinces be next?[22]

But, to many, the sale made sense. Russian America was not well developed economically, and the territory's trade monopoly – the Russian–American company – was losing money. In 1863, a particularly low year, there were just 672 Russian settlers in Alaska. The focus of the Russian–American Company was on furs, as it always had been.[23]

Russian estimates of the economic value of their possessions in America were not far off the final sale price agreed with the United States. In 1857, acting on instructions from Grand Duke Constantine, Baron Ferdinand Wrangel – explorer and former manager of the Russian–American company – had valued Russian America at 7,442,800 rubles (approximately $5.6 million). The future value of the territory might be three times higher, Wrangel suggested, but this would depend on investment which was unlikely to be forthcoming in the wake of the Crimean War. It was, perhaps, only out of loyalty to his former company that Wrangel advised against a sale.

A more important factor to Russia in determining the sale of Russian America may have been the simple fact that it was hard to defend. During the Crimean War, in order to prevent a British annexation, Stoeckl had been involved with a plan to transfer the rights of the Russian–American Company to a neutral entity – the confusingly named American–Russian Company, founded with the support of San Francisco mayor Charles J. Brenham.

By 1867 the strategic situation in the North Pacific was, in any case, different from that which had prevailed ten years previously. The 1860 Treaty of Peking confirmed Russian control over the Amur Valley, as well as the port of Vladivostok. In selling Alaska, Russia provided itself with the financial means to strengthen new Pacific territories which were, unlike Russian America, contiguous with the rest of the empire.

In the United States, the economic value of America's newly purchased territory was much debated. Some, such as Bret Harte, the poet who had made a name for himself writing about the settlement of gold-rush California, were wildly enthusiastic. In *An Arctic Vision*, Harte drew a picture of a land (to which he had never been) bursting with wealth and possibility:

> There's a right smart chance for fur-chase
> All along this recent purchase,
> And, unless the stories fail,
> Every fish from cod to whale.
> Rocks too – mebbe quartz – let's see,
> 'Twould be strange if there should be,
> Seems I've heard such stories told,
> Eh! – why, bless us – yes, it's gold![24]

Some of that enthusiasm rubbed off in Washington, where Senator Charles Sumner argued that Alaska's 'hyperborean Eldorado' would be a source of future American wealth and strength.[25] But many saw the purchase of Alaska as an ill-thought-out extravagance. Sumner's view that Alaska would extend 'a friendly hand to Asia', stimulating trans-oceanic trade, was highly speculative. Accurate information about the territory was sketchy. Thirty years later, Jack London described

Alaska at the time of purchase as 'a vast wilderness . . . as dark and chartless as Darkest Africa'.[26]

It was partly to chart the uncharted that the Western Union Telegraph Company expedition was sent to Russian America, a year before the Alaska purchase was confirmed. The primary aim of the expedition, strongly and personally supported by Seward, was to put the final link in an overland telegraph cable from St Louis, Missouri, to St Petersburg, with a short undersea cable across the Bering Strait.[27] But the information from Western Union's survey served a secondary purpose as well – as a broader source of information on Alaska's economic potential. In 1867, Seward made public a letter from the leader of that mission, Perry McDonough Collins, in which it was argued that 'in the hands of Americans [Alaska] would soon grow to wonderful proportions'.[28]

A more formal source of information on Alaska was the first official American scientific expedition to the territory, led by George Davidson. Between August and November 1867 the expedition followed the coastline of Alaska, travelling on board the United States Revenue steamer *Lincoln*. Davidson's report was principally an account of the topology of the southern Alaskan coastline, as an aid to navigation.[29] Published in book form in 1869, the report provided a wealth of additional information from the vocabulary of various languages spoken in Alaska to accounts of the agricultural conditions. The supply of fish was compared to that of the Atlantic, while the trade in furs with the native population was viewed as an easy source of income. Davidson wrote that 'the almost absurdly small amount of trading articles paid to the Indians for their most valuable skins was ascertained to be marvellously low compared to their prices in our markets'.[30] When Seward visited his acquisition in 1869, the former Secretary of State ventured up the Chilkat River to visit Davidson personally.

In 1870, William Dall, the scientific director of the Western Union expedition, published a still more thorough report on Alaska. Whereas Davidson's expedition had been sea based, Dall's was conducted inland. Dall pointed to the existence of coal, amber, sulphur and gold in Alaska, and noted that Alaska had already developed a trade in ice with the sweltering west coast. (Up until the 1850s ice for California had had to be shipped from the east coast, round Cape Horn and up the entire western coast of South and North America.) In 1868, ice worth some $28,000 was shipped from Alaska.

With some considerable prescience, Dall pointed to the natural resource with which Alaska is now most commonly associated:

> Among the other mineral products of Alaska, probably of this age, is petroleum. This is found floating on the surface of a lake near the bay at Katmai, Aliaska peninsula. It is of the specific gravity of 25° (Beaume), quite odorless, and in its crude state an excellent lubricator for machinery of any kind.[31]

Overall, while Dall accepted that 'our knowledge of the geology, minerals, and rocks of Alaska is extremely meagre' he argued that 'it only remains for the irresistible energy of American citizens to hasten their development. Time alone can prove their ultimate value.'[32] While Seward's prediction that Alaska would be soon the world's 'great fishery, forestry and mineral storehouse' was overblown, the deal he'd struck in the Arctic looked to be a sound investment.

In purchasing Alaska for the United States, Seward had realised at least some part of his dream of American empire. But elsewhere he faced a formidable obstacle to the full realisation of his vision of northern expansion, and the still-greater prize of the entire Pacific seaboard. Seward's ploy – to settle Civil War claims with Britain in return for the cession of British Columbia – might have come off, were it not for the persistence and the foresight of one of Canada's founding fathers, John A. Macdonald.

On 29 March 1867, Queen Victoria gave royal assent to the British North America Act. John A. Macdonald, Prime Minister of the much-expanded Dominion of Canada, had come to London to witness the creation of a country, sitting anxiously on the sidelines as the British North America Act passed the British House of Commons and House of Lords.

The culmination of several years of discussion, the British North America Act provided a constitutional basis for the new Dominion of Canada, encompassing the colonies of Canada (essentially central Ontario and Quebec), Nova Scotia and New Brunswick. Like the Constitution of the United States, the act was a brief and elegant document outlining the role and structure of government. Unlike the

American constitution, where powers not specifically given to the federal government were given to the states, the British North America Act gave specific powers to the provinces, and reserved the rest for the Dominion. Written at a time when the United States was emerging from a civil war over the meaning of 'states' rights', Canada's constitution was designed to learn from the errors of its southern neighbour.[33] John A. Macdonald was one of the key figures in the writing of the act.

The final two parts of the act hinted at the future of Canada envisaged by Macdonald and others. The tenth part of the act – 'Intercolonial Railway' – which outlined a plan for a railway line to run from St Lawrence to Halifax, Nova Scotia, had persuaded the Maritime Provinces to join the new Canada.[34] The eleventh and final part of the act – 'Admission of Other Colonies' – demonstrated the expectation that other British colonies in North America would join the confederation in the future. Canada was, after all, still a work-in-progress. The colonies and territories of Newfoundland, Prince Edward Island, British Columbia, Rupert's Land and the North-Western Territory – over half the territory of modern-day Canada – were not yet part of the country. All, with the exception of Newfoundland, would join under John A. Macdonald's premiership.[35]

Only a few days after royal assent had been given to the British North America Act, rumours surfaced of Seward and Stoeckl's deal for the sale of Russian America (Alaska) to the United States. At first, British Foreign Secretary Lord Stanley was forced to admit that he did not know whether the rumours were true, informing Parliament that:

> When I left the Foreign Office a quarter of an hour ago no dispatch had been received from our Minister at Washington either confirming or contradicting the rumour. I telegraphed this morning to St. Petersburg for information upon the subject, but sufficient time has not elapsed for me to receive an answer.[36]

As details of the deal emerged, Macdonald's suspicions of American expansionism were confirmed. It was clear that the long-term intention of the deal was to force Britain's hand in British Columbia – to cede territory to the United States. No loyalist reading a report of Charles Sumner's speech to the United States Senate on the acquisition of Alaska could have been left in any doubt but that Queen Victoria's realm in North America was the next target:

> By it [the purchase of Alaska] we dismiss one more monarch from this continent. One by one they have retired – first France, then Spain, then France again, now Russia – all giving way to that absorbing Unity which is declared in the national motto, 'E Pluribus Unum'.

A century earlier, in 1763, towards the end of the Seven Years' War, many Britons had called for Canada to be partly abandoned to France in favour of the then much richer Caribbean colony of Guadeloupe. (Such a course had been averted, ironically enough, by the contributions of pre-revolutionary American writers such as Benjamin Franklin.[37]) In 1867, would Britain sacrifice Canada's future for the sake of broader imperial policy?

Macdonald fought to avert that outcome. The passage of the British North America Act strengthened Macdonald's hand, not least by the financial authority it provided. Over the next four years Macdonald would engage in a complex dance of diplomacy, negotiations and political deals to secure British Columbia, the North-Western Territory and the territories of the Hudson's Bay Company for Canada, eventually extending the Dominion's borders west to the Pacific and north into the Arctic. Seward's vision of a United States extending across the top of the American continent would be defeated as a result.

The first obstacles to Macdonald were the areas controlled by the Hudson's Bay Company: Rupert's Land and the North-Western Territory. These lay to the west and north of the new dominion of Canada, occupying the largely empty spaces between the 1867 Dominion of Canada and the crown colony of British Columbia. If Rupert's Land and the North-Western Territory could be made part of the Confederation, then the task of securing British Columbia would be made far simpler.

Rupert's Land had been defined by the 1670 royal charter which gave the commercial Hudson's Bay Company its rights. The North-Western Territory was a more recent creation. The area had been exploited from the late eighteenth century on, but its borders remained undefined until 1859.[38] Spanning what was left of continental British North America after Rupert's Land and British Columbia, the Territory extended up to the Arctic coastline of North America.

In principle, the British government agreed that both Rupert's Land and the North-Western Territory should ultimately be ceded to Canada. From London's perspective, the fact that Rupert's Land was not adequately controlled by the Hudson's Bay Company was a headache in Anglo-American relations.[39] But the land could not simply be given away. Money would have to be paid to the Hudson's Bay Company in compensation for their Charter rights. (At one point, a United States Treasury Department Official, James W. Taylor, cheekily suggested the US intervene with an offer of $10 million for the whole lot.[40])

Macdonald believed that if anyone should pay anything it should be Britain rather than Canada. Under his direction, the Canadian Senate and House of Commons addressed Queen Victoria directly in December 1867 arguing that the admission of Rupert's Land and the North-Western Territory would 'promote the prosperity of the Canadian people, and conduce to the advantage of the whole Empire' and that it could not wait.[41] Ever wary of American intentions, Macdonald also noted 'the avowed policy of the Washington government'. A response was requested by cable.

The deadlock on the Hudson's Bay Company was broken in 1868 by an Act of the British Parliament which authorised British acquisition of Rupert's Land, and the subsequent assignment of both it and the North-Western Territory to Canada. But compensation would have to be agreed. At first, the Hudson's Bay Company suggested a figure of £1 million, an exemption from certain taxes and one tenth of any subsequent grants of land. But under pressure from the British government, the Company buckled.[42] By March 1869 they had accepted considerably reduced terms: a £300,000 payment and one twentieth of the territory. On 23 June 1870, Queen Victoria issued an Order-in-Council granting admission of Rupert's Land and the North-Western Territory into Canada.[43]

British Columbia was not far behind. The immediate threat of the colony being ceded by Britain to the United States as compensation for the 'Alabama claims' had been removed in 1869, when William H. Seward struck a deal with Britain whereby America's Civil War claims – known as the 'Alabama claims' – would be settled by arbitration. This convention was criticised by some of the more ardent American expansionists, including Charles Sumner, as letting Britain off the hook and weakening American leverage for territorial expansion northwards. But Seward still believed that, over time, commerce would drag British Columbia and other British territories into the United States, with or without the leverage of the 'Alabama' claims.

Once out of office, Seward took matters into his own hands. In summer 1869, while professing admiration for their loyalty to Britain, the former Secretary of State told an audience in Victoria, British Columbia that:

I have never heard any person, on either side of the United States border, assert that British Columbia is not a part of the American continent, or that its people have or can have any interest, material, moral, or social, different from the common interests of all American nations . . .'[44]

Back in Oregon a few days later, Seward was still more candid: 'A permanent political separation of British Columbia from Alaska and Washington Territory [is] impossible.'

Macdonald would no doubt have been appalled by Seward's confidence in the inevitability of British Columbia becoming part of the United States. But Seward had a point. The agreement to settle the 'Alabama claims' by arbitration might have killed the immediate risk of British Columbia simply being given to the United States over Canadian objections. At the same time, the accession of Prince Rupert's Land and the North-Western Territory undercut the argument that British Columbia would be perpetually isolated from other British possessions in North America. But what could Canada do to actively secure the colony for itself?

As late as February 1869 the Legislative Council in British Columbia had voted against Confederation with Canada. But the majority of British Columbia's inhabitants were still positively predisposed towards a Canadian future for the colony. John A. Macdonald in Ottawa and supporters of Confederation in Victoria, including the new Governor, Anthony Musgrave, gave new direction to the process of British Columbia's accession in 1869.

Accession hinged on finding a solution to British Columbia's financial situation, and to its relative isolation. Public expenditures linked to the Cariboo gold rush of the early 1860s had left British Columbia fiscally weak. The Confederation would pay off the resulting debt. To improve British Columbia's accessibility to Canadian markets –

Governor Musgrave claimed that, as it stood, 'free commercial intercourse would be easier with Australia than with Canada' – a promise was made that construction on a Canadian Pacific Railway would begin within two years.[45] In July 1871, British Columbia joined Canada.

By 1871, Canada had become a trans-continental country, bounded by the Atlantic, Pacific, and Arctic Oceans. But the exact extent of Canada's northern borders was unclear. Were the islands in the Arctic Ocean north of the continent Canadian or not? What if new Arctic islands were discovered?

In early 1874 two applications were made for grants of land on Cumberland Sound, part of Baffin Island. One was made by a British citizen, A. W. Harvey, the other by an American, Lieutenant William A. Mintzer of the US Navy Corps of Engineers. Both applications were sent to the British colonial minister in London, Lord Carnarvon. The problem he faced in considering whether to accept or reject the applications was that sovereign title over the area around Cumberland Sound was not entirely clear. Did Britain own the land? Canada? Was the land, indeed, owned by anyone?[46]

After enquiries to the Admiralty, the British government concluded that while it believed that it did own Cumberland Sound, the area was not covered by the transfer of Rupert's Land to Canada made in 1870. In other words, Cumberland Sound was probably British, but certainly not Canadian.

While Britain was debating Cumberland Sound, Lieutenant Mintzer was taking the initiative. On 31 May 1876 a news dispatch from Norwich, Connecticut, reported that the 160-ton schooner *Era* was about to sail for Cumberland Inlet. This would be Mintzer's third trip north, but this time his ship was outfitted with mining equipment and explosives, and he was reported to be operating 'under [US] Government auspices'. The goal was to mine for graphite and mica whereby 'a new branch of commerce between that section of the world and the more civilized quarters' could be created.[47]

In October, the *Era* headed home. It appeared that Mintzer's 'visions of wealth, as presented to Aladdin in his cave' had become reality. Mintzer's earlier public claims, that he had travelled north to acquire samples for the Smithsonian Institute, were, it was conjectured, 'a veil to cover the principal aim of the expedition'. Intelligence reported that his ship returned 'laden with minerals of rare value, worth an immense fortune to those who sent her on this expedition'.[48]

Reports of Mintzer's voyage were forwarded to London by the Governor General of Canada, Lord Dufferin. In Ottawa, the issue of ownership in the Arctic lands north of mainland Canada became increasingly urgent. In early 1878 a resolution was raised by Canadian Interior Minister David Mills in which he asked the Queen for an official transfer of the islands from Britain to Canada.

Sir John A. Macdonald, now leader of the opposition, strongly supported Mills's resolution, extolling the mineral resources of the islands and warning that were Canada 'so faint-hearted as not to take possession of it, the Americans would be only too glad of the opportunity, and would hoist the American flag'.[49] (In 2007, Stephen Harper quoted Macdonald word for word in support of his own Arctic policy.[50])

In London, uncertainty over reasonable northern boundaries of a British claim persisted. (One government hydrographer suggested the British claim should stop at 78°30′ north.[51]) In the end, perhaps to forestall the publicity and possible American opposition which might have resulted from an Act of Parliament on the matter, the government acted by an Order-in-Council issued by the Queen. But after six years of differing legal opinions and communications between Ottawa and London, the exact delineation of the Arctic territories covered by the Order was left vague, perhaps so as to not foreclose the possibility of ultimately claiming part of Greenland as well:

> ... all British territories and possessions in North America, not already included in the Dominion of Canada, and all islands adjacent to any such territories or possessions [are now the possession of Canada].[52]

This was diplomatic obfuscation of the highest order.

Macdonald's project of the western development of Canada by railway line, meanwhile, had been suspended. Despite the promises of 1870, no line had been built.[53] Macdonald blamed the Liberals. Prefiguring the rhetoric of Diefenbaker in the twentieth century and Harper in the twenty-first, Macdonald argued that expansion into Canada's open spaces could only be guaranteed by a Conservative government:

The pledge made to British Columbia . . . will be carried out under the auspices of a Conservative Government, and with the support of a Conservative Party. That road will be constructed, and, notwithstanding all the wiles of the Opposition . . . the road is going to be built and proceeded with vigorously, continuously, systematically and successfully to a completion, and the fate of Canada will then, as a Dominion, be sealed. Then will the fate of Canada, as one great body, be fixed beyond the possibility of the hon. Gentlemen to unsettle.[54]

By the time of his death in 1891, the railway line had been built and Sir John A. Macdonald was a Canadian icon. One of the Conservative campaign posters of that year showed Macdonald, looking much younger than his seventy-five years, borne aloft on the shoulders of a Canadian farmer on one side and an idealised Canadian worker on the other. Dressed more as a man of the mid-nineteenth century than the late nineteenth he sat at the apex of the composition, wrapped in the old flag of Canada, looking into the future, with pale, determined eyes. In the background western prairies and the smokestacks of Canadian industrialisation were visible.

The foreground was all about Macdonald himself. There was no need to outline his political positions or political party. The campaign slogan of 1891 was simple: 'The Old Flag, The Old Policy, The Old Man'. To many of his compatriots, Macdonald was the embodiment of the Canadian past and future, the man who had defined Canada's western and Arctic frontiers, the man who had built a country.[55]

In the flood of diplomatic and political activity between 1867 and 1880 which ushered the United States and Canada into the Arctic, two European colonies – Greenland and Iceland – escaped the tide of American expansion.[56]

In 1868, encouraged by former US Treasury Secretary Robert John Walker, Seward requested a State Department report on the economic resources of the territories in order to weigh their potential value for the United States. Compiled by Benjamin Peirce, a young mining engineer from an illustrious and well-connected political and scientific family, the report lamented the meagreness of available hard facts.[57] 'Facts and statistics', Peirce complained, 'come in as incidental observations generally, and need to be disentangled from a mass of useless matter.' But Peirce's trawl through the material resulted in a surprisingly comprehensive picture of Greenland and Iceland, supporting the case that the territories were worthy of consideration as American investments.

At first sight, Peirce described Iceland as 'totally devoid of interest from a material point of view'. Grain needed to be imported from abroad. Iceland was sometimes surrounded by sea ice in winter. The few trees on the island were stunted. One source described Iceland as 'nothing but bogs, rocks, precipices; precipices, rocks, bogs; ice, snow, lava; lava, snow, ice; rivers and torrents; torrents and rivers'.

But this impoverished impression was, Peirce argued, ultimately deceptive. Some of the island's natural phenomena – volcanic activity and glaciers, for example – could be considerable natural economic resources in the future. Peirce recognised the island's geysers as potential sources of medicinal salts and valuable sulphur, and noted the wealth of hydraulic power available.[58] Today, virtually all of Iceland's electricity is derived from the sources (geothermal and hydroelectric) that Peirce had outlined in the nineteenth century.

From agriculture and fisheries to horse farming, Peirce saw an enormous potential for development.[59] The Icelanders themselves were presented as poor but well educated (through home education), and, through their love of 'constitutional liberty' as almost ideal American citizens (a theme that would recur in the later writing of Vilhjalmur Stefansson).[60] That Iceland was economically underdeveloped was, in Peirce's view, squarely the fault of the Danish government. America could do better:

> The future of Iceland is closely connected with the future of her mother government . . . [If] a more liberal system of government should be adopted, and the natural energy of the people be encouraged, the world would be surprised at the rapid advance of this little northern island.

In an argument that would be reprised for the Internet age in the twenty-first century, Peirce pointed to Iceland's potential, by virtue of its location, to be a strategic communications hub. He envisaged that a telegraph cable might be laid from Britain to Iceland, from Iceland to Greenland, and from Greenland to America. 'The advantages of such a route,' he noted, 'are too evident to be insisted on.'

Greenland – though even less well known – was counted a similarly attractive opportunity. Besides seals, whales and the products of the sea, Peirce referred to Greenland's geology as offering 'much promise'. In 1868, the only substance mined in any exportable quantity was cryolite – a mineral used in making soap and, more importantly, in making aluminium. Peirce was insistent: 'We have within our nearest reach a metal whose properties are most precious to us.' The interior of Greenland, while 'perfectly unknown' might yet offer further minerals of value to American industry.

Robert John Walker expanded on Peirce's conclusions in his own introduction to the report, arguing that the purchase of Greenland and particularly of Iceland would encourage the cultivation of the sea and the growth of an experienced American merchant marine. This, in turn, could furnish sailors for the navy and help maintain the security of America's sea-borne commerce. The purchase of Greenland, meanwhile, could extend Seward's grander northern designs:

> . . . the Dominion of Canada was gotten up in England in a spirit of bitter hostility to the United States . . . [and is] intended to embrace all British America from the Atlantic to the Pacific . . . By this great purchase [of Alaska], we have flanked British America on the Arctic and Pacific, cutting her off entirely from the latter ocean from north latitude 54°40′ to 72°, leaving the new dominion but 5°40′ on the Pacific, pressed between Alaska on the north and California, Oregon, and Washington Territory on the south, with even British Columbia now being rapidly Americanized. Now, the acquisition of Greenland will flank British America for thousands of miles on the north and west, and greatly increase her inducements, peacefully and cheerfully, to become a part of the American Union.

Purchasing Greenland would, in effect, force Canada to join the United States.

But Peirce's report came too late. Seward left his position as Secretary of State within a year of the report's publication. The chances of a US purchase of Greenland and Iceland were low. America's relationship with Denmark, which owned both Greenland and Iceland, had been harmed by Seward's failed attempt to purchase the Danish Virgin Islands. After the bruising Congressional battle to purchase Alaska, buying Greenland and Iceland was simply a step too far for a country still recovering from the Civil War. The issue would arise again in 1916 and at the end of the Second World War, but for the time being, the case for Greenland and Iceland was closed.[61] America's northward march would halt at Alaska.

As a private citizen, sixty-eight years old, Seward finally visited Alaska in 1869. What first struck him was the clearness of the skies: 'Of all the moonlights in the world, commend me to those which light up the archipelago of the North Pacific ocean.'[62] Alaska, so different from the post-colonial East Coast in which Seward had been raised, struck him as the place to build an American future.

He was amazed by the territory's natural wealth. Looking at the sea's observable bounty from his boat, Seward confessed that he was almost converted to 'the theory . . . that the waters of the globe are filled with stores of sustenance of animal life surpassing the available production of the land'. But ever the politician, Seward saved his greatest praise for the men and women who would build Alaska into an outpost of American empire.

Seward spoke to them as 'the pioneers, the advanced guard, of the future population of Alaska'. Within a few years, he said, many would follow in their wake. Their future was assured by the same technologies – the steamboat and the railroad – which had, in Seward's own lifetime, transformed the United States from a vulnerable republic to a nation on the verge of global power. These, he told the citizens of Sitka, 'are the guaranties, not only that Alaska has a future, but that that future has already begun'.

4
Scramble: Dividing the Arctic

'Have honor place North Pole at your disposal'

Commander Robert Peary to US President Taft,
8 September 1909

'I don't give a damn what all these foreign politicians there are saying about this . . . Russia must win. Russia has what it takes to win. The Arctic has always been Russian'

Arthur Chilingarov, Russian explorer and parliamentary deputy,
7 August 2007

Use it or lose it is the first principle of Arctic sovereignty

Canadian Prime Minister Stephen Harper, 2007

On 2 August 2007, at 1.36 p.m. Moscow time, a Russian-led expedition placed a titanium canister containing the Russian flag on the seabed at the geographic point where all lines of longitude converge, the North Pole. Ninety-eight years previously, it had taken five months before news of Robert E. Peary's conquest of the North Pole reached New York. In 2007, news of the deposit of Russia's flag on the seabed at 90° North had spread from one end of the world to the other within minutes.

Peary's message, sent by telegraph from Indian Harbor, Labrador – then a British colony not yet part of Canada – on 6 September 1909, had been unequivocal in its expression of national pride. It read, 'Stars and Stripes nailed to the pole – Peary.'[1] The nationalist overtones of planting a Russian flag at the North Pole were similarly unmistakable, to be backed up in the coming weeks by the tub-thumping language of Arthur Chilingarov, Russia's foremost polar explorer, parliamentary deputy and one of the leaders of the 2007 expedition. A few days after completing his polar exploit, Chilingarov told the press, 'The Arctic has always been Russian.'[2]

The global reaction to the flag-planting of 2007 was quick and largely dismissive. Canadian Foreign Minister, Peter MacKay, retorted, 'You can't go around the world these days dropping a flag somewhere. This isn't the 14th or 15th century.' US State Department spokesman Tom Casey told newspapers that he was 'not sure whether they've put a metal flag, a rubber flag or a bedsheet on the ocean floor', but that, 'either way, it doesn't have any legal standing or effect on this claim'.[3] Other Arctic states agreed that while the descent of the Mir submersible 4,200 metres below the sea was a great technological achievement, it would have no impact on the validity of any legal claim to the Arctic.

Just as Peary's offer to put the North Pole at the disposal of the United States in 1909 had elicited a rather uncertain reply from President Taft – 'Thanks for your interesting and generous offer. I do not know exactly what I could do with it . . . William H. Taft' – so Chilingarov's bold assertions of Russian sovereignty were tempered by Russian diplomats.[4] A legal adviser to the foreign ministry confirmed that planting the flag was a 'symbolic gesture'.[5] Sergei Lavrov, Russia's foreign minister, clarified that the purpose of the *Arktika* expedition – taking the same name as the famous Russian ice breaker – was not to stake a claim by itself, but to collect evidence to back up its pre-existing claim to the seabed at or near the North Pole. The international reaction, he said, was overblown and overly suspicious of Russian motives. 'Whenever explorers reach some sort of point that no one else has explored, they plant a flag,' he argued, pointing to the American moon landing as such an instance.[6] A flag at the North Pole might have no legal force, Lavrov was saying, but neither did an American flag on the moon.

There was considerable justification to Lavrov's line. Flags had been planted at the North Pole many times before, with the United States and Russia vying for historic firsts. Peary reached the Pole first, in 1909.

In 1937, a Soviet team was the first to land an aeroplane at the North Pole. In 1959, the submarine *USS Skate* broke through the polar ice to surface at 90° North. In 1977 the Soviet ice-breaker, the *Arktika*, was the first surface vessel to reach the North Pole. In 2004, Nikolai Patrushev, head of the Russian security services – the FSB – raised a flag when he landed at the North Pole. While each of these events had demonstrated the capacity to reach the North Pole, none of them had conferred ownership over it.

In any case, the 2007 *Arktika* expedition was not the consequence of a decision taken in the Kremlin, though it was initially presented as such. The initial promoters of an attempt to reach the sea floor at the North Pole were a retired American submarine captain, Alfred McLaren, and an Australian entrepreneur, Mike McDowell. In a neat role inversion from an earlier episode in Arctic history, it was Swedish money, from the pharmaceuticals millionaire Fredrik Paulsen, which put the expedition on a sound financial footing.[7] Although it was clear from the start that Russian expertise and Russian equipment would be used, from ice-breakers to the submersibles themselves, the expedition initially looked like an international endeavour rather than a particularly Russian one.

Although the expedition had been conceived of internationally, the people who actually got to the seabed at the North Pole first were all Russian, in a Russian submarine, launched from a Russian ship.[8] In January 2008 President Putin awarded the title of Hero of the Russian Federation to Chilingarov, Anatoly Sagalevich and the pilot of the second submersible, *Mir-2*, Evgeny Chernyaev.

Despite the Russian crew and context, and the breathless news reports to the contrary, the *Arktika* expedition did not amount to a Russian claim on the North Pole. The scramble for the Arctic will not be settled by a single act. The law and process of claiming territory is, these days, far more complex and far more subtle. Facts on the ground certainly count, but so do rarefied legal and scientific arguments made in air-conditioned rooms in New York, London, Moscow and Washington D.C.

The *Arktika* expedition did, however, highlight Russia's pursuit of her legitimate ownership of a large part of the sea and sea floor, to which the country has a right under the internationally recognised law of the sea, the United Nations Convention on the Law of the Sea

(UNCLOS).[9] Russia, Canada, Norway, Denmark, Iceland and 152 other countries are all parties to UNCLOS.[10] (At the time of writing, the United States is not.) The process which will define the nature and extent of Russia's rights – and those of all coastal states – is long and complicated. It requires states to submit huge amounts of scientific data to an international commission. It involves intense diplomatic, legal and scientific argument, bounded by the many stipulations of UNCLOS, some of which are very specific (though never enough to prevent disagreement over interpretation).

If there is a scramble for the Arctic, it is a scramble in slow motion. The *Arktika* expedition was only one scene, in one act, in a long-running drama.

Deciding who has legal rights to what in the Arctic is complicated not because there is an absence of law, but because there is a surfeit of it. Different legal regimes apply to the land, the sea and the seabed. The result is a palimpsest, with each set of rules overlaying a previous set of rules, but not quite effacing them.[11] But none of the coastal states in the Arctic – Canada, Denmark, Norway, Russia, the United States – wants to open the Pandora's box which would result from trying to negotiate some new overarching deal for the Arctic.[12] To do so would invite non-Arctic states to muscle in.

In some cases – such as that of ownership of land in the Arctic – matters are broadly settled, bar a few still-disputed parcels of territory.[13] In other cases – such as that of ownership of the seabed – matters are more open, despite the existence of international law on the issue. As in the past, each advance of economic and political interest in the Arctic – on land, on sea, on the seabed – forces international law to catch up with emerging realities. The gaps are filled with politics.

Peary's 1909 conquest of the North Pole took place against a backdrop of European and American territorial expansionism driven by two forces: the reassignment of territories belonging to declining imperial powers (notably Spain and the Ottoman Empire) and a scramble for what remained of the globe's 'unclaimed' spaces.[14] The process climaxed in the so-called 'scramble for Africa' in the 1880s and 1890s.[15] At the end of the nineteenth century, empire building continued beyond Africa. Even the United States, a country built on its rejection of British colonialism at the end of the eighteenth century, had succumbed to the colonial temptation, acquiring Cuba, Guam, the Philippines and Puerto Rico at the end of the Spanish–American War in 1898.

The drive to conquer the North Pole was another expression of this global process. By 1909, the North Pole was one of the very last places on earth which had not been mapped, explored or claimed by a 'modern' state.

Most land in the Arctic was already ruled by those of European descent, normally with little regard to the notion of native rights or ownership. Russia, through a mixture of exploration, treaties and outright military conquests, had taken possession of all of Arctic Asia and a fair portion of Arctic Europe. What remained of the European Arctic was held by either the kingdoms of Sweden and Norway or the kingdom of Denmark, which controlled both Greenland and Iceland. (Finland was, until 1917, still part of the Russian Empire.) The American Arctic was divided between the United States and Canada.

Indigenous people were generally considered incapable of self-government and therefore incapable of exercising sovereign rights. The United States acquired Alaska from Russia without any consultation of the native population, and indeed failed to offer citizenship to them until 1924.[16] Canada failed to consult the native population of Prince Rupert's Land or the North-Western Territory when they were acquired in 1870. The principle of 'terra nullius' – or 'empty land' – was applied freely to territories that were home to native peoples. On occasion native populations were forcibly moved from one place to another so as to boost states' claims of effective occupation.[17] Only much later in the twentieth century would the claims of sovereign states in the Arctic and the rights of native populations be put into some kind of balance, and even then the process would prove difficult, contested and drawn-out.[18]

In 1909, the slim possibility remained that new discoveries of land might yet be made in the Arctic. (As late as 1926, when Roald Amundsen travelled over the North Pole in the airship *Norge* with a mandate to claim any new discoveries for Norway, it was still thought the map of the Arctic might be incomplete.) But for the most part the process of dividing the Arctic had moved from discovery of land, to asserting and 'perfecting' claims to it, through occupation of the land and notification of ownership to other states.

The principles which governed how a state could assert sovereignty in the Arctic derived, in part, from agreements made at the Berlin conference in 1884–85, held to regulate the division of spoils in Africa. Notification of ownership to other states was easy enough. But effective occupation was more difficult to prove. At what level was occupation considered sufficient? Did the inherent uninhabitability of much of the Arctic mean that the level of occupation required to 'perfect' a land claim in the Arctic could be less than it would be, say, in equatorial Africa? If the conditions for sovereignty were changing over time – from mere discovery to effective occupation – did the new conditions apply to long-governed areas? Would states be required effectively to occupy not just newly discovered territories, but land they considered they had owned for decades, if not centuries?

One potential solution, proposed by Canadian Senator Poirier in 1907 and adopted by the Soviet Union in the 1920s, was to divide the Arctic along parallel lines of longitude, converging on the North Pole. For Canada and the Soviet Union – both of which had potentially huge Arctic possessions and limited resources with which to assert their sovereignty – a purely geographic division of the Arctic was attractive.

According to Poirier's 'sector theory', states would have sovereign rights (or already had sovereign rights) 'to all the lands that are to be found in the waters between a line extending from its eastern extremity [of a state's territory] north, and another line extending from the western extremity north'.[19]

There were three main precedents for a sectoral division of territories along a line of longitude. The first was the 1494 Treaty of Tordesillas, by which Spain and Portugal had determined the extent of their empires. In essence, Spain and Portugal agreed to divide the world from pole to pole, along a line of longitude running through present-day Brazil.[20] The second was the treaty which defined Canada's own western border – an 1825 treaty between Britain and Russia. That treaty had defined a boundary between Russian and British possessions in North America simply by a line of longitude, 141° West, which would extend to the north 'as far as the Frozen Ocean'. The third and final precedent came from the

1867 Russo-American treaty which followed the American purchase of Alaska, much of the language of which was adopted from the 1825 Anglo-Russian treaty. According to the terms of the sale of Alaska, the border between Alaska and Asiatic Russia would extend along a line of longitude running through the Bering Strait, which would then simply proceed 'due north, without limitation, into the same Frozen Ocean'.

Twenty years after Poirier first suggested it, the 'sector theory' was enthusiastically adopted by the Soviet Union, in a 1926 declaration from the Central Executive Committee of the USSR:

> All lands and islands, both discovered and which may be discovered in the future, which do not comprise at the time of publication of the present decree the territory of any foreign state recognised by the Government of the USSR, located in the northern Arctic Ocean, north of the shores of the Union of Soviet Socialist republics up to the North Pole between the meridian 32°04′35′′ East from Greenwich . . . and the meridian 168°49′30′′ West from Greenwich . . . are proclaimed to be territory of the USSR.[21]

The chief Soviet proponent of the 'sector theory', W. Lakhtine, argued it was the only way to divide the Arctic peacefully and neatly.[22] Not incidentally, it was also very much in Soviet interests.[23] First, it would pre-empt the possibility that one of the airship expeditions of the late 1920s (none of which were Russian) might discover and lay claim to a new Arctic land.[24] Further, the 'sector theory' would settle any outstanding disagreements with other Arctic states, particularly with Norway over Franz-Josef Land, in the Soviet Union's favour. And if the Soviet Union could persuade other states that a territorial sector applied also to the sea and the air, then it would obtain sovereignty over the major part of any future air route across the top of the globe.[25] (The law of the air had been a non-issue up until the beginning of the twentieth century; now it looked as if it might be of crucial strategic importance.)

As for the North Pole itself, under the sector theory, Lakhtine argued that the point 90° North would be owned by no one. Instead, he suggested, 'it might be represented as an hexahedral frontier post on the sides of which might be painted the national colors of the State of the corresponding sector'. It was an elegant solution to the problems of Arctic sovereignty, but one which never took off in the North, even as it was being applied by all of the claimant states to part of the Antarctic continent, in the South.[26]

The comparison of 90° North and 90° South raised a more fundamental problem with Peary's apparent claim for the North Pole: unlike at the South Pole, there was no land to claim. Might there be a way around that? What of the polar ice? Could the 'freedom of the high seas' be applied to an area which could not be navigated?[27] Or should, instead, solid ice be assimilated to land, in which case it could surely be claimed on the same basis?

Writing about the problem in an American legal journal in 1910, the jurist Thomas Balch remembered an episode in the Russo-Japanese war of 1904–05 when the Russians had built 'a railroad on the ice over Lake Baikal and established a station mid-way across the frozen lake'.[28] The railway had served as the means to transport 'many thousands of men and great quantities of stores and implements of war'. Might not humankind 'permanently occupy the ice cover of the Polar Sea', Balch asked. And if not functionally different from land, then why should solid ice be considered any differently from the point of view of international law? Balch ultimately decided that it should – or at least the North Pole should – principally because, as Nansen had demonstrated in the 1890s with the voyage of the *Fram*, polar ice was constantly moving. As a result, 'such possible occupation would be too precarious and shifting to and fro to give any one a good title'. However, as states' claims moved into regions with more or less permanent ice cover the question of how to deal with it under international law would continue to test and expand legal theories of sovereignty.[29]

There were several corners of the Arctic where, ice or no ice, the question of who owned what was still unclear in 1909, or had only recently been resolved.

On the North American continent, a dispute over the exact course of the land border between Alaska and Canada, based on an 1825 treaty between Britain and Russia, had been resolved only recently, mostly to the advantage of the United States, in 1903.[30] But Ottawa feared that its northern borders might be under threat. Canada's Arctic claims were essentially based on two transfers from prior British ownership – Prince Rupert's Land and the North-Western Territory, in 1870, and the islands north of the North American continent in 1880. But, as a confidential government report explained in 1905, the northern extent of Canada's sovereignty might be vulnerable.[31] Some of the lands then known to exist had not been discovered by British explorers – the primary basis of Canadian sovereignty – let alone occupied by Canada. To allay doubts about the strength of Canada's claims, Ottawa undertook a series of initiatives to assert sovereignty in the last years of the nineteenth century and the first years of the twentieth.[32]

Just a few months after Peary had reached the North Pole, Captain Joseph Elzéar Bernier, a Quebecker, unveiled a plaque on Melville Island, far beyond the Arctic Circle, that read, 'The memorial is erected today to commemorate the taking possession for the Dominion of Canada of the whole Arctic Archipelago lying to the north of America from long. 60W to 141W up to the latitude of 90N.' Over the next few decades, Bernier's assertion of Canadian sovereignty served as an insurance policy against rival claims from Denmark, Norway or the United States.

To the east, meanwhile, the extent of Denmark's sovereignty over Greenland was put under the legal microscope. The Danish claim to Greenland derived essentially from the 1814 Treaty of Kiel, at the conclusion of the Napoleonic Wars. By that treaty, the defeated Danes had ceded to the victorious Swedes the entire territory of Norway, save her dependencies (Greenland, Iceland and the Faeroe Islands). While the Kingdom of Norway had been transferred to the Kingdom of Sweden, Norway's prior dominion over Greenland, Iceland and the Faeroe Islands had been passed to the Kingdom of Denmark. It was a sort of Scandinavian musical chairs.

The idea of native sovereignty over Greenland did not seriously occur to the lawyers in the nineteenth century. For them, the transfer of Greenland from one European people to another was entirely valid. But what did 'Greenland' mean, anyway? How far did it extend? Was it to be assumed that a single parenthesis in the Treaty of Kiel and the presence of a few small Danish settlements on the west coast, inhabited by no more than a couple of hundred Danes, conferred sovereignty over this huge land mass, the exact extent of which was not yet known?

By 1909 there were two serious possible challenges to the Danish position, from the United States and Norway. The American challenge concerned north-west Greenland, which had been unknown to the drafters of the Treaty of Kiel. It had only been discovered, and (briefly) occupied, by Charles Francis Hall, an American explorer, in the early 1870s.[33] And Denmark had only established a permanent settlement there in 1909, when a Greenlandic church mission was set up on the shores of North Star Bay, close to the site of the current United States air base at Thule. To counter the risk of an American claim for northern Greenland, Denmark needed an American declaration that it entertained no such claim. On 4 August 1916, it got what it needed. As one of the conditions for the sale of the Danish Virgin Islands to the United States, US Secretary of State Robert Lansing issued a declaration that 'the Government of the United States will not object to the Danish government extending their political and economic interests to the whole of Greenland'.[34] (A few years later Britain tried and – because of American opposition – failed to obtain a right of pre-emptive purchase of Greenland from Denmark, should Denmark ever consider a sale.)[35]

The Norwegian challenge did not fully materialise until the 1920s and 1930s. From a legal point of view, the fact that Greenland had been 'discovered' by Norse sailors (Norwegians rather than Danes), and that Hans Egede, the missionary who re-established a European colony on the island in 1721, was a Norwegian, was of limited relevance.[36] While it was true that the Treaty of Kiel had indeed specified Greenland as a dependency of Norway, the same treaty had transferred it to Denmark.[37] Whatever historical rights Norway might have had, they had passed, at that point, to Denmark.

However, Norwegians had long accessed the east coast of the island for commercial purposes, principally hunting. The Danish administration had not attempted to exercise jurisdiction over this area; indeed the so-called 'prohibited area' which Denmark actively controlled was quite limited.[38] Despite pressure from Copenhagen, Norway refused to issue a declaration similar to that which the US had issued in 1916. Instead, when Denmark unilaterally declared sovereignty over the whole of Greenland, in 1921, Norway strenuously objected, claiming East Greenland was a no man's land.[39]

In 1924, after several months of negotiations, Denmark and Norway reached an agreement on what activities were to be permitted to Danish and Norwegian citizens in East Greenland. Norway expressly stated that this understanding did not amount to a recognition of

Danish sovereignty; Denmark countered that such a recognition was implicit. The sore remained. When, in June 1931, five Norwegian trappers raised the country's flag on the east coast, the government in Oslo backed the actions of their citizens and proclaimed occupation of a part of East Greenland. The matter was only settled by the case being taken to the International Court of Justice in The Hague, with a decision in favour of Denmark in April 1933.

Beyond Greenland, there were four other land masses in the Arctic where sovereignty remained unclear in 1909: Jan Mayen, Franz-Josef Land, Wrangel Island and, finally, the Svalbard archipelago.

Jan Mayen, a tiny spot of land lying between Norway and Greenland, was too small to excite much international interest. Norway annexed it in 1930, recognising Canadian sovereignty over the 'Sverdrup Islands' in exchange for British recognition of its annexation. Franz-Josef Land was more complicated. The islands had been named after the Austro-Hungarian Emperor by two Austrians, Carl Weyprecht and Julius Payers, in the early 1870s. But the islands had never been formally claimed by Austria. (Indeed Italy was later to claim that the expedition should more properly be thought of as Italian.) Explorers and whalers from several other countries, particularly Norway, had visited Franz-Josef Land towards the end of the nineteenth century. The first recorded Russian visit was a landing party from the ice-breaker *Yermak*, in 1901. So, in 1909, who actually owned the islands was unclear. Norway and the Soviet Union claimed the islands in the 1920s. The Soviet claim won out, largely on the back of the sector theory.

The ownership of Wrangel Island, in the East Siberian Sea, was similarly complex. The United States and Britain had potential claims by right of discovery, both of which would subsequently be exploited by Canadian explorer Vilhjalmur Stefansson in his attempt at freelance empire-building on the island in the 1920s. In 1911 the Russians sent the ice-breaker *Vaigach* round the island, leaving a beacon signifying Russian occupation. In 1916 the Russian claim was backed up by a diplomatic note sent to the Allied powers. Again, with a combination of legal arguments and, in the 1920s, a demonstration of effective authority, the island would be claimed successfully by the Soviet Union.[40]

The final Arctic land mass over which sovereignty was unclear in 1909 was Svalbard, an archipelago of islands strategically located between the north coast of Norway and the North Pole.[41] The difficulties began with the question of who had discovered the archipelago. Some Norwegians claimed that a reference to 'Svalbard' (meaning 'cold coast') in a twelfth-century Icelandic annal, demonstrated that the land was known to the medieval Norse. The Dutch, meanwhile, claimed that Willem Barents had discovered the archipelago in 1596, and pointed to long-standing Dutch whaling settlements on the islands in the seventeenth century as proof of occupation. Russia tended to disagree with the Dutch assertion of first discovery, suggesting that hunters from the Russian 'Pomor' region had reached Svalbard as much as a century before Barents.

To the question of who had first discovered the archipelago could be added the question of who had first claimed it. Here, Britain had a case to make. Robert Fotherby had claimed Spitsbergen for King James I in 1614, erecting two crosses on the islands, one of which was subsequently removed by a Dutch explorer.[42] Finally, in 1909, while most of the inhabitants of the island were Norwegian, the main employer on the island was the Boston-based Arctic Coal Company, co-founded by American John Munroe Longyear, who gave his name to the main town on the archipelago, Longyearbyen. Norway, Russia, Britain, the Netherlands and the United States could all claim an interest in who was to be given sovereignty.

The resolution to the Svalbard question, when it came, was the Spitsbergen Treaty, the terms of which were finalised at the tail end of the Paris peace conference which followed the First World War.[43] The nine original signatories – more would join the treaty over time – recognised the 'full and absolute sovereignty' of Norway over the archipelago, and all the lands situated between 74° to 81° North and 10° to 35° East.[44] But the treaty also granted certain rights to the contracting parties, including access, settlement and various commercial rights, on a basis of 'absolute equality'. Were any minerals to be exported – with the expectation that coal mining would continue on the archipelago and might very well expand – they would be taxed at a very low level: no more than 1 per cent of the value of the exports, up to the first 100,000 tons, with total tax revenues in no case exceeding the direct costs of administering the archipelago.

Though it only signed the treaty in 1924, the Soviet Union was the only country other than Norway to fully exercise its rights in Spitsbergen.

Having lost the opportunity to claim the islands outright, and recognising their strategic importance, Moscow jealously guarded what little hold it had. A permanent Soviet settlement was established in the 1930s, a constant source of low-level tension with the archipelago's Norwegian administrators. In the 1970s and 1980s the Soviet population of Svalbard numbered some 2,000, including officers of state security. When Norway and the Soviet Union countries found themselves on either side of the Iron Curtain, Spitsbergen was frozen in a Cold War stand-off between West and East.

Things have since thawed – the Russian presence is a shadow of its Soviet past, and Norway's basic assertion of sovereignty is unquestioned. But the thaw has opened an uneasy question, too: how should a treaty made in the months after the end of the First World War be applied in the twenty-first century?

LONGYEARBYEN, SVALBARD, 78° North: Just twenty-eight people govern the Svalbard archipelago, a land mass covering 60,000 square kilometres, a fraction smaller than the Republic of Ireland, and about twice the size of the state of Massachusetts. Most of the people who live here carry Norwegian passports. The current Governor, Per Sefland, is Norway's former chief of police. His Deputy, Elisabeth Bjørge Løvold, is a civil servant from Lillehammer, a town in the uplands north of Oslo. There are Norwegian flags for sale in the tourist shops, along with polar bears, keyrings and Arctic-styled bottle-openers. In 1995 tourists spent 32,695 nights in Longyearbyen; by 2006 that number had reached 83,049.[45]

Despite the prominence of the colours of Norway on Svalbard – and the firm insistence from any government representative that Svalbard is an integral part of the kingdom of Norway – there are reminders that the archipelago is both something more and something less than that. There are Russians and Ukrainians who live here, some in Longyearbyen, though most are at the Russian settlement at Barentsburg. The girls at the supermarket check-out counter speak Thai. Somewhere in town is an Iranian who came here six years ago and, under the terms of the Spitsbergen Treaty, was able to settle here. If he were to return south to the Norwegian mainland he would almost definitely be forced to leave the country, his asylum claim having been refused. Import duties are non-existent on Svalbard: Cuban cigars cost less in Longyearbyen, at 78° North, than they do in Oslo, three hours' flight to the south.

Meeting in the pine-and-glass administrative headquarters in Longyearbyen, Deputy Governor Elisabeth Bjørge Løvold explains how it all works. The main function of the Governor's office is to uphold Norway's sovereignty, while ensuring compliance with the Spitsbergen Treaty. Many of the government regulations on the archipelago concern the environment, an area over which the treaty specifically accepts Norwegian rules. Those rules are strict – covering everything from pollution to taxidermy.[46] Some 65 per cent of the archipelago is a protected environmental area, where even stricter laws apply. What little crime there is on Svalbard tends to be violations of environmental regulations, such as 'driving a snowmobile in the wrong place'.

But not all infringements of Norwegian-imposed regulations are so minor, and on Svalbard local problems can quickly escalate into disputes over sovereign rights. In May 2008 Løvold flew to Barentsburg and ordered a Russian fishery patrol vessel to leave port because it lacked proper documentation to dock. Though the *Mikula* left, Russia protested that it had been well within its Spitsbergen Treaty rights, granting equal access to all signatories. As Russian interest in Spitsbergen returns such disputes are likely to become more frequent.[47] Norway's right to impose strong environmental regulations threatens to collide with signatory states' rights of equal access.

Norwegian diplomats argue that the potential for disputes onshore is limited by the fact that their country not only has sovereignty over the archipelago, but that the Norwegian state owns the land. However, the legal status of Svalbard's offshore areas, where any oil and gas deposits are likely to be, is far more uncertain, challenged by Russia as well as by a number of other signatories to the Spitsbergen Treaty.

Offshore, the Norwegian position is that the treaty only applies to a narrow strip of water around the archipelago. 'The treaty is very specific when it comes to the geographic limitation of its effects,' says a senior official in Oslo. 'It says exactly what area it [the treaty] covers – exactly.' Norway argues, accordingly, that the Spitsbergen Treaty – and its provision on equal access – does not apply to Svalbard's 200-mile exclusive economic zone or to its continental shelf, legal concepts which had not been codified in international law at the time of the

Spitsbergen Treaty. In those areas, there is no treaty limitation to Norwegian control.

This is, however, a source of disagreement. A Norwegian-maintained fisheries protection zone around the archipelago is disputed by Russia and has led to several run-ins between Russian and Norwegian ships. Britain has expressed its doubts over Norway's position over ownership of the continental shelf, arguing that it risks depriving the parties to the treaty of potentially hugely valuable rights to the sea floor. (Norwegian diplomats are still sore over a meeting of the treaty signatories called by Britain to discuss the issue – without informing Norway – warning that such actions could harm long-standing alliances.[48]) In July 2008 the Russian geological agency Rosnedra issued plans to map parts of the Spitsbergen continental shelf for oil and gas resources. The Russian navy resumed a military presence near Spitsbergen around the same time.

But one way or the other, Norwegians argue the world will have no choice but to play by their rules in Svalbard. It's Catch-22. As a Norwegian government official put it to me:

> Say that British Petroleum went up there and said, 'Hey, listen, we feel we are entitled to do what we want here under the Svalbard Treaty.' Well, I guess they would try to get the support of the UK government and get the UK to say, 'Yes, we support that.' Fine. But Norway could then say, 'Well, we disagree with you, but if that is the position that you take with regard to the continental shelf [that it is subject to the conditions of the treaty] then you must also accept that we can establish the rules under which any exploitation is undertaken. In any case, you will have to get a licence from Norway. So, you are entitled to equal access? Okay, fine, then we will ban all oil companies working up here, including Norwegian ones. That's equal!'

The conflict over the application of the Spitsbergen Treaty is symbolic of a wider shift in the contentious areas of international law. Most of the potential areas of dispute over sovereignty in the Arctic today are over the sea and the seabed, rather than the land.

The law of the sea used to be a matter of customary international law. State practice frequently differed. While one country might claim that it owned a territorial sea which extended three nautical miles beyond its coastline, another might claim a territorial sea twelve miles wide. Over time, as states unilaterally extended their claims, the idea of the seas as an international space – the *Mare Liberum* described by Hugo Grotius in 1609 – was eroded. What had been manageable under a small clique of like-minded European states became untenable when the number of coastal states exploded at the end of colonial rule in the 1940s and 1950s. In order to prevent increasingly extensive and contrary claims over the sea from destroying free maritime passage and causing open conflict, a framework of international maritime law was established.

The process took a long time.[49] It was not until 1982 that the United Nations Convention on the Law of the Sea (UNCLOS) – which pertained to both the sea and the seabed – was signed. And only in 1994 did it come into force. Today, UNCLOS is the single-most important legal reference for determining proprietorship of the sea and seabed. Even countries which have failed to ratify UNCLOS, including the United States, recognise its force as a matter of customary law.[50]

UNCLOS limits a coastal state's territorial sea to twelve nautical miles from the coastline.[51] It guarantees transit rights for all-comers through international straits, thus maintaining the principle of 'freedom of the high seas', particularly important for naval powers such as the United States.[52] It recognises the entitlement of coastal states to an Exclusive Economic Zone (EEZ) extending 200 nautical miles out from the coastline, over which they have economic rights (including to the seabed) but not unlimited sovereignty. It establishes that, in cases where a particular set of geographic and geological conditions are met, a state's economic sovereignty of the seabed may extend beyond 200 nautical miles (though with a mechanism for royalties on resource development to be distributed internationally).[53] UNCLOS further decrees that the seabed beyond any state's jurisdiction be managed by the International Seabed Authority (ISA), headquartered in Kingston, Jamaica.[54]

Still, UNCLOS does not solve every problem in maritime law. Maritime borders, for example, can still be drawn in different ways, and it is up to the adjacent states to determine which borderline they are willing to accept. That applies as much to the seabed as to the sea itself. Even if UNCLOS sets the outward limits of a state's claims over the sea and the seabed, these claims are likely to overlap.

At present, there are two major outstanding disputes over maritime borders in the Arctic – between Norway and Russia in the Barents Sea, and between Canada and the United States in the Beaufort Sea – and a third area where a border has been agreed but not fully ratified.[55] (The Barents and Beaufort Seas are particularly important because the seabeds are likely to contain oil or gas.) In the Beaufort, the disputed wedge of sea is quite small.[56] The contested part of the Barents Sea (some 176,000 square kilometres, or 68,000 square miles) is three times as large.[57] On-and-off negotiations between Russia and Norway have resulted in several interim agreements, but no final settlement.[58] 'If there is a willingness, we could find a solution quickly . . . or it could take another twenty-five to thirty years,' says one diplomat close to the negotiations.[59]

Recently, the status of the Northwest Passage has attracted more attention than any of these long-standing disputes.[60] Up until the 1960s the question of who owned the Northwest Passage was academic. Canada claimed a three-mile territorial sea around its territory, including its Arctic islands. Apart from one place where the passage between two Canadian islands was less than six miles across, the Passage could potentially be viewed as part of the 'high seas'. However, given that the Northwest Passage was unused by international shipping its exact status under international law was not considered important.

That changed in 1969, when an American oil tanker, the *SS Manhattan*, sailed from Prudhoe Bay, through the Northwest Passage, to the east coast of the United States. Although the *Manhattan* had been accompanied throughout the voyage by a Canadian ice-breaker, the *CCGS John A. Macdonald*, the question of who actually owned the Northwest Passage was suddenly a live issue. Ottawa responded to the uncertainty by announcing a unilateral extension of its territorial seas out to twelve miles and by passing the Arctic Waters Pollution Prevention Act (AWPPA), providing for Canadian environmental regulation of Arctic seas up to 100 miles from the country's coastline.[61] (In December 2008 the Canadian government extended the coverage of the AWPPA to 200 miles.)

The United States protested that these changes amounted to unacceptable 'unilateral extensions of jurisdiction on the high seas' and could provide an excuse, elsewhere in the world, for abusing territorial rights.[62] A State Department memorandum to US National Security Advisor Henry Kissinger stated that discussions between the

US ambassador in Ottawa and the Canadian Department of External Affairs had never amounted to 'bilateral consultations'. Instead, the United States was being presented with a 'fait accompli', the consequences of which 'are critical for national security interests and seriously degrade the entire United States law of the sea posture on which military mobility depends'.[63]

In 1985, Canada went one step further. In response to the voyage through the Northwest Passage of a United States ice-breaker, the *USCG Polar Sea*, Canada made an expansive claim to sovereignty over the area.[64] In effect, the country redrew the map. Instead of looking at sovereignty from the point of view of each island in the Canadian Arctic, Canada redesignated the area as an archipelago, with a single line drawn around it. As a result, the Northwest Passage could be considered part of Canada's 'internal waterways'.[65]

Officially, the United States does not accept this position. A 2009 Presidential directive on US Arctic policy was unequivocal: 'The Northwest Passage is a strait used for international navigation.'[66] But America's practical approach to the issue has been to 'agree to disagree'. With some reason, Canada's policy-makers argue that it is in their southern neighbour's best defensive interest that the Northwest Passage is controlled by them, rather than being open to all comers such as China and Russia.

When it boils down to it, ownership of the North Pole is really a question of who owns the sea floor at 90° North, and ownership of the sea floor – in the Arctic and elsewhere – largely depends on UNCLOS.

According to UNCLOS, a coastal state has rights over the sea (and the seabed) up to 200 miles out from its coastline. Beyond that point, however, states are only entitled to that part of the sea floor which can be demonstrated to be the natural prolongation of the continental shelf – the submerged continuation of the land.[67] To work out where, exactly, the continental shelf ends and the deep seabed begins is no easy matter. It's not just about the depth of the sea. The geology and shape of the sea floor must also be taken into account. Some sub-sea features count as part of the continental shelf, others don't, and distinguishing between them is not necessarily straightforward. Even once these geological and geo-morphological conditions are satisfied, there is a hard outer limit to the claims to the seabed that a coastal state can make. No coastal state can claim the seabed more than 350 miles out from the coastline, or more than 100 miles beyond the 2,500 metre isobath (a line under the sea joining points of equal depth), whichever is further. In most cases, a state can decide which of these two rules to apply – but not always.[68]

There is another constraint on the claims of coastal states: time. Once a state has ratified UNCLOS it has ten years in which to submit its claims to the seabed.[69] Norway submitted its claims in 2006 and the final recommendations of the Commission on the Limits of the Continental Shelf were issued in 2009. Canada has until 2013 before it is required to submit its claims; Denmark has until 2014. The US has not yet ratified UNCLOS, so there's no pending deadline – but preparations are being made for a claim nonetheless. One official close to the US programme describes it as a 'huge project operating on a shoestring'.

The areas of the seabed to which states have already submitted claims are eye-watering. In 2004 Australia submitted a claim to economic sovereignty over 14.2 million square kilometres of the world's sea floor (approximately 5.5 million square miles). That's as large as the entire Arctic Ocean basin.[70] No one really knows what future wealth the sea floor may provide. 'We used to think of the seabed as just a giant bathtub,' says Peter Rona, a distinguished American marine geophysicist.[71] No longer. Oil and gas provinces tend to be close inshore, in areas where, because of the 200-mile rule, coastal states are unquestioned in their sovereignty. But as offshore technologies progress, the ability to mine the sea floor will extend further and further out from the coastline.

The idea of the continental shelf – and its economic potential – is not new. Six weeks after the end of the Second World War, as thoughts turned to building the peace, President Harry Truman issued a proclamation including the following lines:

> Having concern for the urgency of conserving and prudently utilizing its natural resources, the Government of the United States regards the natural resources of the subsoil and sea bed of the continental shelf beneath the high seas but contiguous to the coasts of the United States as appertaining to the United States, subject to its jurisdiction and control.[72]

Norway and Britain have been drilling for oil on their continental shelf in the North Sea since the 1960s.

But if the existence and economic potential of the continental shelf has been known for some time, determining its geographic extent is tough. This is particularly the case in the Arctic, which, until relatively recently was much less well studied than most other coastal seas. Except, that is, by the military. In the 1990s, under the auspices of the SCICEX (Scientific Ice Expeditions) programme, the US navy allowed the civilian use of nuclear-powered attack submarines to gather data on the Arctic. Since that programme stopped, in the late 1990s, more standard techniques of data gathering have had to be used. But some of these, such as multibeam sonar, are far more difficult when ice conditions require the use of noisy ice-breaking vessels. 'You could break ice, back up and go through again to get a few pings', says Larry Mayer, the chief scientist for the US mapping project.[73] But it's time-consuming. The exact route any mapping project can take will depend on prevailing ice conditions. In summer 2007, the general reduction in ice cover of the Arctic Ocean was favourable to the US programme, allowing their project to get as far north as 83°. But, the same year, Denmark's planned cruise had to change course because what Arctic ice remained was pushed towards Greenland and northern Canada.

As the mapping process begins, oceanographers are often surprised by what they find – unexpected submarine ridges, unexplained rises in the sea floor – all of which may have a bearing on subsequent claims. Mayer shows an audience in Oslo an eerie multicoloured three-dimensional seascape, taken from research north of Alaska. In one case, Mayer and his team discovered what they called the Healy Seamount – named after the coastguard ice-breaker the *USCG Healy* – which rises abruptly 3,000 metres from the sea floor in the Chukchi Sea, and which was previously unknown. 'This wasn't on US submarine bathymetric charts,' says Mayer, 'because although two subs had been over, they didn't know exactly where they were because their inertial navigation system was out by several miles.' The results of this discovery and others like it is that the scientific assessment of what the US may be able to claim in the Arctic is being 'optimised'. 'In the Arctic we're really in the exploration mode,' says Mayer, 'and in this case it really paid off.'

Once the science has been done, at the cost of millions of dollars, a claim has to be submitted. That, too, is complex and uncertain. In March 2006 the United Nations tried to simplify the matter, by publishing a manual explaining the science, law and administrative process of making a claim.[74] It was several hundred pages long. Claims are assessed by the Commission on the Limits of the Continental Shelf (CLCS), a group of international experts elected by the signatory states. UNCLOS is their guide. But with room for debate over definitions, the origins of certain geological features and even the amount of data needed to make a convincing case, it is hard to know ahead of time which way the CLCS will rule. As one observer has put it, 'At the end of the game, a coastal state may discover not only did it misjudge the value of the cards that it was holding, but it played them all wrong.'[75]

That is more or less what happened to Russia in 2001, when it first submitted its claims to the commission, in New York, covering nearly all the 'sector' of Russian interests, up to and including the North Pole. Several states, including the United States withheld their support for Russia's claim or pointed out its deficiencies.[76] The commission agreed, telling Russia it would need more evidence to support its case. Some of the original data points – the basis for Russian claims over certain areas of undersea territory – were considered confidential, and were not released. Russia had wanted to get the matter closed quickly – perhaps, one diplomat suggested to me, the Russians had hoped that the fact that the chairman of the commission was a compatriot would help.[77] But it wasn't to be.

Partly to reduce the costs associated with making a claim, and partly to avoid the risk that the commission will tell them to go back and do it again, countries have, on occasion, pooled their survey resources in the Arctic. The United States and Canada did so in the summers of 2008 and 2009. Canada and Denmark did so in 2007. That expedition, known as the LOMROG (Lomonosov Ridge Off Greenland) expedition hired a Russian ice-breaker, *50 Let Pobedy* (*Fifty Years of Victory*) – because only Russian nuclear ice-breakers could give the expedition the sustained ice-breaking capacity needed. Over time, as the complexity of making the claim has become clearer and as the number of areas where additional mapping is required have increased, the Danish budget for mapping the continental shelf (principally, off Greenland) has more than doubled.

In Copenhagen, Christian Marcussen, principal scientific investigator on the Danish team, shows off a bathymetric map of the Arctic Ocean. 'It looks nice,' he grins, 'but it's only based on a few points.' That's one reason why LOMROG was important. As Denmark moves towards submitting its own claim, before 2014, Marcussen is working on the assumption that one of the major subsea features of the Arctic Ocean, the Lomonosov Ridge, is an extension of either Greenland or Canada's Ellesmere Island. If he can prove it, then both Canada and Denmark may be able to claim parts of the sea floor up to and beyond the North Pole. If that happens, their claims will overlap with those of Russia. It will be up to the diplomats to open negotiations.

Over time, it is inevitable that climate change too will affect legal regimes in the Arctic.

It's not really that climate change will change who owns what in the Arctic. Although coastal erosion, the disappearance of some land masses and the appearance of others from under the ice could, in theory, have some marginal impacts on the extent of ownership, the impact of climate change on maritime borders will be far greater in other areas of the world, such as the Indian Ocean or the Pacific, where low-lying land masses may be doomed to disappear under the waves.[78] Coastal states in the Arctic are unlikely to accept a redrawing of their maritime borders, even if some areas of land, say in northern Siberia or Alaska, become submerged as a result of global warming. The changes would be too small to be worth the diplomatic and political cost of reopening discussions. Maritime borders in the Arctic will probably remain as they are.

But climate change is highly likely to provoke legal questions over rules and regulations applicable to Arctic shipping.[79] One article of UNCLOS – Article 234 – allows coastal states to enforce non-discriminatory environmental rules on shipping in ice-covered areas.[80] But what happens when areas cease to be ice-covered? There are some rules for shipping in the Arctic, generated by organisations such as the International Maritime Organisation, but these are much less stringent than, for example, the rules that Canada would wish to enforce in the Northwest Passage. More generally, of course, if climate change does result in an increase in Arctic shipping – something which, in time, seems almost inevitable – it is likely that the laws governing Arctic shipping will be more frequently challenged, simply by a failure on the part of users to adequately observe them.

And there's the rub. Whatever laws are applied in the Arctic – in terms of claiming sovereignty or in terms of managing international shipping – any increase in activity will require additional surveillance and government presence. Who is going to provide it?

5
Parade Ground: War and Peace in the North

'If there is a Third World War the strategic center of it will be the North Pole'

General Hap Arnold, US Air Force, 1950

'One can feel here the freezing breath of the "Arctic Strategy" of the Pentagon. An immense potential of nuclear destruction concentrated aboard submarines and surface ships affects the political climate of the entire world and can be detonated by an accidental political–military conflict in any other region of the world'

Mikhail Gorbachev, Murmansk, 1 October 1987

'The Arctic is upon us, now'

Rear Admiral Gene Brooks, US Coast Guard, 2008

KIRKENES, NORWAY, 69° North: The border with Russia starts a few miles due east of Kirkenes, where the European land mass and the Barents Sea meet. From there, the border runs south, twisting and turning along the course of the Pasvik River, through a landscape of woods and rocky outcrops. It is difficult ground, good for defensive actions and hard for would-be attackers. Further south the border kinks briefly around the Boris Gleb chapel – a Russian Orthodox chapel on the west side of the river – before breaking off its southern course and turning west, then north-west, in a direct line towards Kirkenes itself. Just before it gets there, the border turns south again, inland towards a three-way meeting point with Finland, once a part of the Russian Empire and only now breaking out of Soviet-enforced neutrality.

The border, from the Barents Sea in the north to Finland in the south, is just 120 miles long, a Norwegian conscript soldier tells me. But the strategic importance of the border is out of all proportion to its length. It is, after all, the only land border between Russia and NATO above the Arctic Circle. In 1968, when military exercises brought Soviet tanks up to the border, they could hear the revving of tank engines in Kirkenes. If the Soviet Union had ever invaded Europe, Kirkenes would have been one of the first towns to be wiped off the face of the map. For twenty years, up until 1983, the border was closed entirely.

'I realise now it was a kamikaze mission,' says Leif-Arne Ljøkjell, one of the soldiers whose job it was to defend the border from Soviet attack. In the 1970s, he led a contingent of twenty-five soldiers, armed with three heavy machine guns, two anti-tank guns and rifles; his orders were to delay any Soviet advance, before escaping by boat to rejoin the Norwegian defensive lines further east. Ljøkjell and his men, one of many Norwegian patrols along the Soviet border, could never have hoped to repel a concerted attack. Their role, he now concedes, was to act as a trip-wire for wider NATO intervention. Norwegian troops, unsupported in the first instance by either American or British soldiers – the idea was to re-enforce Norway rapidly in time of war, without retaining a permanent foreign presence on Norwegian soil in time of peace – were NATO's Arctic front line.

Tensions on the border are much lower now. The current task of Ljøkjell – a senior officer in the Norwegian armed forces who led his country's contingent in Afghanistan in 2002 – is mostly about co-operation rather than military defence. His job as Norwegian Border Commissioner is to oversee the border with his opposite number from the Russian border service, part of the FSB apparatus which includes Russia's intelligence service. The arrangement dates back to 1949, at the very beginning of the Cold War. These days, there's a ritual when the Norwegian and Russian Border Commissioners meet, Ljøkjell tells me: 'salute and hug'. After their meetings – some twenty of them every year, supplemented by daily phone calls – 'there's normally a meal or something; it's very informal'. In 1990 approximately 3,000 people crossed the border: in 2008 that number was around 100,000. Over the last few years, observers from Armenia, Azerbaijan, Georgia and Turkey have all come to Kirkenes to see this 'model' border co-operation in action.

Nonetheless, the border remains heavily guarded on both sides, involving a significant surveillance commitment. A few miles outside Kirkenes, the garrison base at Sør-Varanger is still a major military installation. An electronic listening post, of great importance in the Cold War, remains in the mountain above Kirkenes. For all the good-will on the ground between Ljøkjell and his Russian counterpart, the strategic relationship between NATO and Russia has been unsteady in recent years. Russia has resumed bomber patrols in the Arctic, and a key confidence-building treaty between Russia and the West has been suspended.[1]

Publicly, the Norwegian government characterises the increase in Russian military activity in the Arctic as a 'return to normal', intended to affirm the strategic importance it places on the Arctic rather than to threaten its neighbours. Nonetheless, the increase in Russian activity has been sharp and it has consequences. In January 2009, the Norwegian Foreign Minister, Jonas Gahr Støre, briefed the North Atlantic Council – the governing body of NATO – on security in what Norway calls the 'High North'. He told them that Russian fighter activity near Norwegian air space had increased by a half in 2008, though it was still well below the level of the 1980s. In December 2007, as a result of naval exercises involving the Russian aircraft carrier the *Admiral Kuznetsov*, flights to Norwegian oil and gas installations had to be cancelled for several hours.

A significant upgrade of Norway's military capabilities is under way – with a major refocus on Arctic security. Two F-16 fighters are on permanent fifteen-minute standby at the Bødo air base to intercept any Russian probing of the country's air defences. Norway's naval capabilities have been dramatically enhanced by the purchase of five brand new *Fridtjof Nansen* class frigates, the last one of which is due to enter service in 2010. On 1 August 2009 Norway moved its national command centre from Stavanger to Reitan, well above the Arctic Circle, the first country to do so. Meanwhile, Norwegian security officials have waged an intensive diplomatic campaign urging the country's NATO allies to involve themselves far more in Arctic security. Permanent foreign bases are not allowed in Norway, but NATO forces are encouraged to train there.

At the same time, non-NATO members Finland and Sweden have increased their cooperation with NATO, leading to the possibility that they might even join, a prospect unthinkable as recently as fifteen years ago, when NATO was almost a 'swearword' in Helsinki.[2] Attitudes towards defence cooperation with other countries are changing. One reason is that the Cold War is over, and Finland is no longer bound by the Treaty of Friendship, Cooperation and Assistance with the Soviet Union. But Russia is still Finland's neighbour.[3] In September 2007, the Finnish Defence Minister Jyri Häkämies told an audience in Washington D.C. that while it was 'crazy' to think that Russia would directly threaten Finland's security, the three main policy challenges his country faced, nonetheless, were 'Russia, Russia, Russia'.[4] In February 2009 proposals for beefed-up defence cooperation between Nordic countries were presented to a Nordic ministers' meeting in Oslo.[5]

On the ground near Kirkenes, the impacts of any strategic tensions with Russia are indirect. Despite some suggestions that Norway should renegotiate the 1949 border arrangements, as Finland renegotiated its agreements in the 1990s, the complexity of Norway's overall relationship with Moscow has meant that that has not yet happened. Ideas for the establishment of a far more open border zone – a so-called 'Pomor' area, in reference to the early Russian inhabitants of the area – have been held up, partly because the Russians are uneasy about providing access to a militarily sensitive area housing a number of installations and training grounds.

The main concern for Ljøkjell, apart from minor and accidental border violations, is illegal immigration. In the Cold War, the standing policy of the Norwegian government was that anyone who made it across the border would not be repatriated to the Soviet Union. On average, only one or two people were successful each year. These days, the numbers are much greater and the motivations tend to be economic rather than political. Norway's fear of mass emigration from Russia in the early 1990s has been replaced by apprehension over the possibility of organised criminals trying to ferry illegal immigrants into Norway, part of Europe's open Schengen area.[6] 'They [the Russians] are telling us that some people are paying considerable sums of money to get across,' Ljøkjell says. In June, a person was stopped on the Russian side with a secret compartment in their car, large enough for a human being. The Russians, for their part, worry about infiltration of 'terrorists'.

On the Norwegian side, the border is observed from six main stations, most of which have a helicopter pad attached and all of which have a number of smaller installations as well, dotted along the border from OP-247 – the northernmost installation – to the southern meeting point with Finland. Outside one of the stations, at Elvenes, just a mile or so from the border, three dozen white skis hang next to a tracked vehicle designed to carry troops across the snow. One soldier shows off his AG-3 assault rifle, made in Norway under licence from Heckler & Koch. A Quick Reaction Force of eight soldiers is ready to leave on fifteen minutes' notice to deal with any border incidents. The soldiers here are conscripts, mostly in their late teens or early twenties. They are proud of their role. 'We're Norway's first line of defence,' one tells me.

At the beginning of the twentieth century, the Arctic's strategic military importance was largely discounted. To the leading geopolitical thinkers of the age, the Arctic – largely inaccessible to land or sea forces – was irrelevant.

When, in 1904, Sir Halford Mackinder, one of the fathers of geopolitics, wrote the paper which made his name, 'The Geographical Pivot of History', the North barely warranted a mention.[7] Mackinder's central thesis, that sea power was declining relative to land power, led him to conclude that the Euro-Asian 'Heartland' – essentially Russia – was destined to geopolitical pre-eminence. In Mackinder's view of the world, at the end of what he called the 'Columbian epoch' of major discovery, power would stem from the relative efficiency of the use of geographic attributes. For him, the key to unlocking the power of the 'Heartland' lay in developing its internal transport routes and in exploiting its natural wealth:

The spaces within the Russian empire and Mongolia are so vast, and their potentialities in population, wheat, cotton, fuel, and metals so incalculably great, that it is inevitable that a vast economic world, more or less apart, will there develop inaccessible to oceanic commerce.

In Mackinder's view, the Arctic was an impenetrable fortification, further strengthening the natural defences of central Eurasia.[8]

By the 1920s and 1930s some analysts questioned Mackinder's logic. The militarisation of the air, they believed, would fundamentally redraw the strategist's map. Whereas Mackinder had viewed the Arctic as a strategic barrier, enthusiasts of air power conceptualised the Arctic as a strategic corridor. Their key insight, as valid to the Arctic now as it was then, was that while geography itself was more or less immutable, the military significance of geographical features was a function of the technologies available to master them.

In 1916, America's greatest living polar explorer, Robert Peary, protested at plans to relinquish America's claims over Greenland as part of the purchase of the Danish Virgin Islands, arguing that in the future Greenland 'might furnish an important North Atlantic naval and aeronautical base'.[9] His views found vocal support in 1920 when General William E. 'Billy' Mitchell, America's leading air strategist, testified to the Senate on the strategic importance of the territory.

Mitchell's crusade to develop American air power – and to gain recognition of its implications for the North – continued through the 1920s, through his advocacy of the use of the *USS Shenandoah* airship to explore the Arctic, and his consistent support for the establishment of military bases on both Greenland and Iceland.[10] Ever outspoken, Mitchell criticised the shortsightedness of American military strategists in their attitudes to air power and the Arctic. In the air age, he argued, Alaska was even more important to American defence than the Panama Canal, completed just a few years previously.[11]

It was only in the Soviet Union that the Arctic was substantially integrated into national strategy, with air power at its core. The Soviet aircraft industry briefly led the world in the 1920s, building more than any other country. Maps of the Soviet Arctic showed a region criss-crossed with airways, some real, some projected.[12] In 1937, Soviet pilots were the first to land at the North Pole, using converted military planes. Arctic airfields, and the experience of Arctic flight, served the more or less explicit military purpose of ensuring the security of the Northern Sea Route and thus the link between the Soviet Union's Northern and Pacific fleets.

Despite Mitchell's warnings, the Arctic remained relatively low on America's military priorities. In the summer of 1940, one of Mitchell's protégés, Major General Henry H. 'Hap' Arnold, then the Chief of the Air Corps of the United States Army, toured Alaska by air, partly to inspect the impact of some $25 million appropriated for enforcing the defences in the territory, and partly to promote the need for much more investment to secure what he called 'a vital part in the scheme of national defense'.[13] Up to then, it was often corporations' interest in a Trans-Polar air route, rather than military interests, which drove discussions of how flight was changing the northern map.

In parts of the Arctic the Second World War was principally about supply and logistics. The oilfields of the North American Arctic were put into the service of the war effort, leading to a vast northward extension of the transport and supply infrastructure on the American continent. Greenland and Iceland were both occupied by American forces, helping to ensure the security of trans-Atlantic convoys. Arctic sea and air routes allowed for Allied resupply of the Soviet Union.[14] In the east, Russian pilots flew American aircraft from Ladd Field, in Alaska, to the Soviet Union. In the west, in the Barents Sea, British supply ships carried American jeeps and American tanks across the top of Norway to the Soviet port of Murmansk.

Elsewhere, war in the Arctic was about the weather. German and Allied forces established weather stations in Greenland, Iceland and Svalbard, with the aim of helping to better predict the following day's weather in north-western Europe. The importance of these stations to the course of the war was out of all proportion to their size. Most consisted of no more than a few men equipped to live without outside support for twelve months or longer. A low-intensity war was fought across the Arctic to identify and destroy them.[15]

But around Kirkenes, in northern Norway, a more traditional form of high intensity conflict prevailed. The area was turned from Arctic borderland into Arctic battlefield, fought over by Soviet, German and Finnish troops.

In 1939, the land which lay directly east of Kirkenes was not yet the Soviet Union, but Finland.[16] Under the terms of a 1920 treaty between the Soviet Union and Finland, concluded at the height of the Russian Civil War, the Soviet Union had recognised Finnish independence – won with the backing of German troops – and confirmed Finnish access to the Barents Sea by way of a narrow slice of mineral-rich territory known as the Petsamo Corridor.

To Soviet strategists, the 1920 peace was both shameful and dangerous. Forced to give up control of Finland in 1920, the Soviet Union had lost a buffer zone around Leningrad, making its second city vulnerable to attack. In 1939, Stalin saw an opening to redraw the map and, having issued a set of territorial demands to which Helsinki could not reasonably be expected to accede, Soviet forces duly swarmed over the border into Finland. In the north, the Petsamo Corridor was occupied, but Stalin's primary war aim was to the south, around Lake Ladoga.[17] Despite strong resistance led by Marshal Mannerheim, himself a former general in the Russian Imperial Army, in March 1940 the Finns had little choice but to accept a major loss of territory in south and central Finland – thus enhancing the security of Leningrad – and a tiny area of Arctic territory on the Barents Sea. The Petsamo Corridor was returned to Finland.

But peace in the north did not last long. In April 1940, Germany launched a blitzkrieg operation first against Denmark and then Norway. The German strategy had several elements. One aim was to provide a staging ground for air attacks on Britain and, by the same token, to deny Norway to British forces. Another aim was to secure the port at Narvik, through which Swedish iron ore, vital to the German war machine, was exported in winter. Within a matter of weeks, and despite the landing of Anglo-French forces in Norway, Germany had achieved its goals. For the first time, German troops occupied Kirkenes.

With most of Europe under Berlin's control, German military planning began to focus on an invasion of Germany's nominal ally, the Soviet Union. The main thrust of attack would inevitably take place through Soviet-occupied Poland and European Russia, but a secondary northern front would divert Soviet resources from the main attack and might deny the Soviet Union access to the deep-water harbours of the Arctic Barents Sea. To attack in the north would require co-opting the Finns, whose territory had already been used in 1940 in the invasion of Norway. Reluctantly, the Finns signed up to Germany's plans, put into action in June 1941. The Petsamo nickel mines came under full German control, while Finnish forces were used to push into Karelia and the south, and to cut off the railway line to Murmansk.

With the outbreak of war between Germany and the Soviet Union, control of Arctic shipping became an essential military objective for Germany, and blocking assistance to the Soviet Union through convoys into the port of Murmansk became a strategic necessity. Partly as a result, German garrisons in Norway, including the facilities of Kirkenes, were strengthened. There were more occupying German troops in Norway per head of local population than in any other part of occupied Europe. In 1944, one journalist wrote of Hitler's desire to create a 'northern Gibraltar' in Kirkenes, an impregnable fortress and a guarantee of German control of the Barents Sea.[18]

In 1944, everything changed again. Under the pressure of a Soviet assault, the Finns switched sides, ceding Petsamo in the process. From being allies of Nazi Germany, supposedly dedicated to helping Germany defeat Russia, the Finns became allies of the Soviet Union, required to help push German troops out of Finland. For Kirkenes, the final phase of the Second World War was the worst of all. Having learned the tactical lessons of the 1939–40 war against the Finns, the Soviet assault was merciless. The German retreat was worse. The German general, Rendulic, gave orders that no building of any worth should be left to the invading Soviet forces. Kirkenes, after four years of German occupation, was destroyed by German sappers.

In 1945, the constellation of global power was transformed. Germany, the pre-eminent European power, was crushed. France was chastened by occupation and tainted with collaboration. The leading global power of the pre-war period, the United Kingdom, was financially ruined. Despite – or perhaps because of – its millions of war dead, the Soviet Union emerged from the Great Patriotic War with its prestige enhanced and its geopolitical legitimacy beyond question. Meanwhile, the United States, which before the war had oscillated between isolationism and engagement, stood at the summit of global power. Not only was the United States the only country to have the atomic bomb, but its economy produced approximately half the world's goods.

This reconfiguration of world power vastly increased the importance of the Arctic; the technological advances which had proved the possibilities of naval and aerial operations in the Arctic during the Second World War further reinforced its strategic significance after it. The two great post-war strategic winners – the Soviet Union and the United States – faced each other across the Arctic Ocean. Within a few years of the victory of 1945, their wartime partnership had turned into the superpower rivalry of the Cold War.[19]

The Soviet Union sought to capitalise on its new-found military prestige to push forward its presence in the north. As early as November 1944, months before the guns had fallen silent in Europe, and with Soviet troops occupying the north of Norway, Foreign Minister Molotov suggested joint sovereignty over Spitsbergen to his Norwegian counterpart, Trygve Lie. The Soviet border in the Arctic moved westwards with the annexation of previously Finnish Petsamo. In 1948, the Soviet Union concluded a treaty with Finland whereby that country agreed to resist any invasion by 'Germany and her allies', and to call on Soviet forces to defend Finnish sovereignty if necessary.

The United States was slower off the mark, but the mechanisms set up to advance its presence in the Arctic proved to be at least as successful. The US strategic relationship with Canada was more or less assured by geography. The Second World War had already seen an increase in military cooperation between the two countries, including the construction of military infrastructure in the Arctic. This continued, albeit with a greater emphasis on Canadian ownership of defence installations.

Some Canadians expressed concern. In 1946, Canada's future foreign minister and Prime Minister, Lester B. Pearson, wrote that there is 'an unhealthy preoccupation with the strategic aspects of the North; the staking of claims, the establishment of bases, the calculation of risks'.[20] Canada, he wrote, 'does not relish the necessity of digging, or having dug for her, any Maginot Line in her Arctic ice'.[21] But, uncomfortable as it might be to be stuck between what Pearson called 'the two greatest agglomerations of power in our world, the USSR and the USA', with the inevitable pressures this would place on Canada's independence, the transition from unconditional wartime alliance with the United States to conditional peacetime alliance was almost inevitable.

Elsewhere in the Arctic, the United States proceeded by a variety of bilateral and multilateral treaties, some of which were new and some of which confirmed or rebalanced wartime arrangements. Having cemented its position in Greenland and Iceland immediately after the war, the United States established the North Atlantic Treaty Organisation in 1949, with a guarantee of mutual defence at its core. Canada, Denmark, Iceland and Norway were all members from the start, sometimes in the face of intense domestic opposition and sometimes only after alternative defensive options had failed to materialise.[22] Sweden and Finland stayed out. Iceland, the largest recipient of Marshall Plan aid per head, signed a more far-reaching US–Icelandic Defense Agreement a few years later, the subject of political controversy in Iceland ever since.[23]

By 1950, the entire coastline of the Arctic Ocean was either controlled by the Soviet Union or by the United States and its allies.[24] That same year, a report from President Truman's National Security Council warned, 'the issues that face us are momentous, involving the fulfillment or destruction not only of this Republic but of civilization itself'.[25] Just five years after the end of the Second World War, the Cold War had already reached fever pitch, with the Arctic as one of its principal strategic arenas.

Before 1941, US forces had only rarely fought outside North America – in the Philippines at the end of the nineteenth century, and in Western Europe at the end of the First World War. Between 1941 and 1945, however, American troops fought in the deserts of North Africa and in the jungles of the Pacific. When Japanese forces occupied part of the Aleutian chain, an arc of islands stretching south-west from Alaska, the US military was forced to consider the possibility of warfare in or near the Arctic.

In 1942, the Arctic, Desert and Tropic Information Center (ADTIC) was set up, with a mission to research how these new environments might impact the equipment, tactics and strategy of US armed forces.[26] By far the most important area of study was the Arctic, involving everyone from Vilhjalmur Stefansson to the French explorer Paul-Emile Victor. After the Second World War, as the United States took on yet more global security responsibilities, the need for information on war-fighting in non-temperate environments was confirmed. Again, study of the Arctic was a priority.

The result was a proliferation of reports, papers and articles discussing military operations in the North. In 1945, Stefansson published *Arctic Manual*, originally written as a two-volume manual for the air corps of the United States Army in 1940, which contained information on everything from how to build an igloo to the calcium content of reindeer meat.[27] A shortened version was reproduced as a War Department manual, TM-1-240. In 1947, the Office of Naval Research (ONR) set up an Arctic Research Laboratory at Point Barrow and Moises C. Shelesnyak, head of the ONR's Environmental Physiology Branch, published a primer on the Arctic.[28] The Arctic Institute of North America was established in 1945 at McGill university to bring together Canadian and American scientists. A permafrost field division focusing on military engineering was created at the Northways Airfield in Alaska in February 1945. Eventually that initiative would evolve, with several others, into the Cold Regions Research and Engineering Laboratory of the US Army.[29]

Together with scientific research and training materials on the Arctic there was an increase in American military exercises in the Arctic, mostly in cooperation with the Canadian armed forces, then amongst the largest in the world. The land-based *Operation Musk-Ox*, otherwise known as the Canadian Army Winter Arctic Expedition, covered 3,100 miles in 1945, experiencing temperatures as low as -47° Celsius (-52° Fahrenheit) on the way. Setting off from the shores of Hudson Bay, the 48-man expedition headed north to the Arctic islands before returning south via the Great Slave Lake, Port Radium and the border of British Columbia and Alberta.[30] The Royal Canadian Air Force (RCAF) flew 792,000 miles in support of the expedition, intended to ascertain the difficulties of Arctic land operations and the best means of supporting them. In 1950 the United States Air Force and the RCAF cooperated on *Operation Sweetbriar* to test North America's 'back door'.

It was not that the Arctic was thought of as a likely route for a Soviet land invasion. The difficulties of *Operation Musk-Ox* had demonstrated how difficult (and expensive) land operations in the Arctic could be. The key to Arctic warfare was more likely to be operations by 'relatively small, highly trained and completely equipped forces' used to secure specific sites of strategic importance, such as airfields and weather stations.[31] Whatever land operations there were would be in support of air power, rather than as a projection of land forces across a broad front. The key weapon in Arctic warfare would not be the tank or rifle, but aircraft and the atomic bomb.

The growing importance of nuclear weapons in military strategy in the late 1940s and early 1950s was driven principally by the relative strengths of NATO and Soviet forces. While Soviet forces in Europe had a massive conventional superiority, the United States remained (until 1949) the world's only nuclear power – and retained a numerical advantage over the USSR for several decades to come. If war were to break out in Europe, it seemed unlikely that Soviet forces would march to London or that NATO forces would fight their way to Moscow. It was assumed that nuclear weapons would short-circuit any major land war. The ability to conquer and hold territory would therefore be less important than delivering nuclear devices to target.

The same geography which had made the Arctic the natural route for the delivery of aircraft from the United States to the Soviet Union when the two were allies in the Second World War, now dictated the direction of the risk of nuclear annihilation. In the absence of effective missile technology, nuclear weapons would have to be delivered by strategic bombers. The shortest route from the United States to the Soviet Union – or vice-versa – would be across the Arctic. Controlling that airspace would, therefore, be vital to the outcome of any war in which the use of nuclear weapons was considered.

In 1954, partly in response to the Soviet Union's testing of a hydrogen bomb the previous year, President Eisenhower gave the go-ahead for the construction of a Distant Early Warning (DEW) Line of radar stations, far to the north of the existing radar stations on the so-called Pinetree Line. Completed in 1957, in the same year that the Canada–US North American Air Defence Command (NORAD) agreement was established, the line ran from Alaska to Baffin Island, in Canada. The radars, it was thought, would provide the United States and Canada with several hours' notice of an imminent air attack from the Soviet Union across the Arctic. With fighters stationed across the Arctic – from Thule Air Force Base in Greenland to bases in northern Canada and Alaska – a significant number of those bombers might be destroyed before they reached their target, negating their effectiveness as a deterrent.

The primacy of Arctic bombers in nuclear strategy was brief. In 1957 and 1958 the Soviet Union and the United States demonstrated their mastery of inter-continental ballistics and nuclear submarines, two more advanced technologies which would dominate the strategic balance for the rest of the Cold War.

The launch of *Sputnik I*, on 4 October 1957, was a powerful symbol of the Soviet Union's technological leap forward since the 1940s. And it was worrying evidence of the West's diminishing technological lead. There had been a long five years between the first US nuclear weapon and the first Soviet weapon. The gap between the development of the American hydrogen bomb and its Soviet counterpart had been significantly shorter, just over three years.[32] Now the Soviet Union had stolen a march on the United States in one of the key areas of advanced military technology.

By launching *Sputnik* the Soviet Union was serving notice that it had the capability of developing, in short order, Inter-Continental Ballistic Missiles (ICBMs). Once that happened, avoiding nuclear annihilation through the application of air defences would be near impossible. While an appreciable percentage of strategic bombers might have been destroyed in-flight, the interception of even a single ICBM would be remarkable. An Arctic early-warning system might provide warning of attack, but it would offer no defence against annihilation.

In 1958, however, the United States demonstrated that while it might have temporarily given up its lead in missile ballistics, it had extended its lead in another area: the development of nuclear submarines. The *USS Nautilus* had been launched by the President's wife, Mamie Eisenhower, in 1954. Four years later, the Nautilus completed its transit of the Arctic Ocean, sailing directly underneath the North Pole. It was an American triumph. On his return to Washington, the man who had captained the *USS Nautilus*, Commander William R. Anderson, was publicly feted by President Eisenhower.[33] Just as *Sputnik* had alarmed Washington, Moscow was alarmed by the strategic implications of the voyage of the *USS Nautilus* and its sister ship the *USS Skate* a year later.

The *USS Nautilus* had demonstrated the navigability of the Arctic Ocean to submarines. In 1959, the *USS Skate* went one step further, fulfilling the vision of Australian explorer Sir George Hubert Wilkins and surfacing at the North Pole. (When it did so, Sir George's ashes were scattered to the winds and the Australian flag, along with the British naval ensign and the Stars and Stripes, was hoisted in his honour.[34])

This success hinted at the future possibility that submarines might launch nuclear-armed missiles from the Arctic Ocean, thus providing flexibility and surprise.

The launch sites for ICBMs were fixed. They offered an obvious target to the enemy. Even if heavily defended, they could, in principle, be destroyed. Submarines, on the other hand, would be easy to conceal and harder to destroy in the troughs of the Arctic Ocean and under the ice. Possession of a fleet of nuclear-armed submarines could offer a further guarantee of destruction for any side that chose to begin a nuclear exchange. Worse, a country relying on a land-based deterrent faced the risk that a surprise attack from submarine-based weapons might destroy those missiles before they had had time to launch.

The range of the first American submarine-based missiles in the 1960s was short. (The Polaris A-1 had a range of around 1,000 miles.) American submarines would have to sail quite close to the Soviet Union in order to launch their missiles. As a result, it became vital to know the bathymetry of the Arctic Ocean and the thickness of Arctic ice, through which a submarine might have to surface to launch its missiles. A number of dangerous and top-secret missions were devoted to the task of finding the answers to those questions.[35] But whatever the environmental and geographical challenges, at least access to these firing positions – whether by way of the Canadian Arctic or through the North Atlantic – was in friendly hands. NATO continued to dominate the seas.

Initially, the Soviet situation was considerably worse. Because of the short range of Soviet missiles, Soviet submarines would only be able to hit targets in the centre of the United States by sailing out of the Barents Sea and into the Atlantic. In doing so, they would expose themselves to NATO's Anti-Submarine Warfare (ASW) and naval defences in the waters between Greenland, Iceland, the United Kingdom and Norway, the so-called GIUK gap. By the 1970s, however, an increase in missile range meant that Soviet submarines could launch their weapons from anywhere in the Arctic. The GIUK gap could be avoided.

Instead, the Arctic Ocean itself became a Soviet nuclear bastion, with around forty-six strategic nuclear submarines (SSBNs) attached to the Soviet Northern Fleet in 1980, compared to around fourteen in 1968.[36] The Arctic was increasingly thought of as a 'Mare Sovieticum',

Above: Our mental picture of the Arctic is defined, above all, by notions of emptiness, space, isolation and by the majesty of apparently unchanging nature. Greenland, July 2008.

ons cela un ice-field, c'est-à-dire une surface continue de glaces dont
ı'aperçoit pas les limites.

—Et de ce côté, ce champ brisé, ces longues pièces plus ou moins ré
ies par leurs bords?

teras, produisirent un effet indescriptible. L'équipage, surexcité par l'
tion en présence de ces terres funestes, s'écria tout d'une voix :

« Au nord! au nord!
—Eh bien! au nord! le salut et la gloire sont là! au nord! le ci

Above: Jules Verne's *Les Anglais au Pôle Nord* (The English at the North Pole), 1866, describes the voyage of Captain Hatteras and the *Forward* to an imaginary land at the North Pole. Merging contemporary science with heroic fiction, hugely popular in Europe and America, the book depicts the Arctic as exotic and tragic. On the left, Édouard Riou imagines icebergs as mosques and minarets, borrowing from the contemporary European fad for the Orient. On the right, he represents the Arctic in terms more familiar to a public exposed to the noble failure of the British Franklin expedition of the late 1840s: desolate, cruel, unforgiving.

Left: Fridtjof Nansen (1861–1930) the Norwegian explorer, scientist, writer and national hero. The 1893–1896 voyage of the *Fram* – 'Forward' in Norwegian, following *Les Anglais au Pôle Nord* – proved his theory of polar drift and disproved the ancient idea of land near the North Pole. Combining the status of heroic explorer and crusader for science, Nansen stood midway between the romance of the Arctic in the nineteenth century and the idea of its conquest as a modernising impulse, pushing humanity towards the future.

Right: Vilhjalmur Stefansson (1879–1962) had no time for romantic notions of the Arctic hero-explorer struggling against the elements, arguing that this reinforced false preconceptions about the difficulties of the Arctic. In the future, he argued, the North would be transformed by aircraft, submarines and the world's hunger for natural resources. Following the failure of his attempt to set up a viable Canadian and then American colony on Wrangel Island – due north of the Soviet Union – in the early 1920s, he left Canada, criticising Ottawa for failing to develop the country's Arctic hinterland. His ideas were taken more seriously in the USSR. In the 1940s and 1950s, he became an adviser to the US military on Arctic affairs.

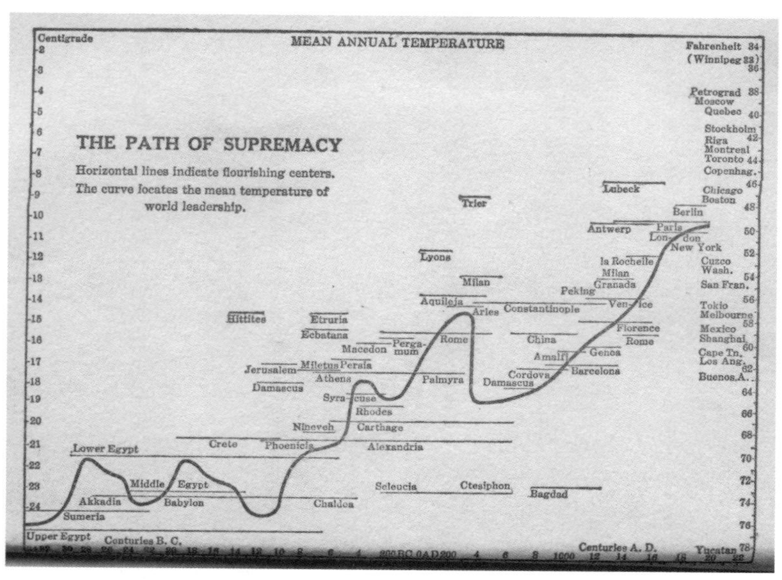

Above: From *The Northward Course of Empire* (1922). Stefansson sought to show that the course of human history ran north, replacing the ancient notion of the uninhabitable 'frigid zone' with his vision of the North as the seat of future empire.

Left: Georgian troublemaker Josef Djugashvili, aka Stalin, 1913. Stalin was exiled to Kureika, high on the Yenisei River, in 1914, staying there until 1917, the year o the Russian revolution. As Soviet dictator, Stalin embarked on break neck industrialisation backed by ar attempt to transform the Arctic into the country's natural resource base. Grand infrastructure projects were advertised internationally as evidence of Soviet prowess. Less advertised were the Arctic prison camps accompanying them, far larger and more brutal than Stalin' own Siberian exile.

Left: Stalin embracing Valerii Chkalov, in 1936. In May 1937, Chkalov became the first pilot to land near the North Pole.

Right: The cover of *SSSR na stroike* (Soviet Union in Construction), a propaganda magazine published internationally, in May 1941, showing a convoy on the Arctic Northern Sea Route. In 1940, Soviet ships guided the German raider *Komet* through the route in a record twenty-one days. A year later – and a month after publication of this magazine – Nazi Germany invaded the Soviet Union.

Above: In the 1930s, Soviet activity in the Arctic entered popular culture through posters, banners, poems and books. This photo-montage – entitled *Vstrecha* (Return) – celebrates the 1934 airlift of 104 survivors from the *Cheliuskin*, a ship which sank in the Arctic Soviet Far East, six time zones east of Moscow. The seven pilots became the first recipients of the state's highest honour, Hero of the Soviet Union. The man seen greeting Stalin is Otto Schmidt, responsible for development of the Northern Sea Route. The backdrop is Red Square.

Above: Stalin believed the *Belomorkanal* (White Sea Canal), linking the Baltic to the White Sea, would transform the economic and strategic geography of the north. Its construction by free and prison labour was celebrated by Maxim Gorky as a redemptive project for prisoners working on it. Some 25,000 died. As shown here, the tools used were rudimentary. Having been built too shallow, and insufficiently maintained, the canal became useless after the fall of the Soviet Union.

Above: Sergei Prokudin-Gorskii's photograph of the Solovetsky Monastery, on the Solovetsky archipelago in the White Sea, in 1915, the second year of the First World War. The monastery was already nearly half a millennium old, powerful and prosperous. Two years later, the imperial regime would collapse, replaced by atheistic Soviet rule. The monastery was closed in 1920, only to reopen in 1990.

Right: In the Soviet period, the Solovetsky archipelago became better known for the prison camp which was set up there in 1923. Alexander Solzhenitsyn's famous book, *The Gulag Archipelago*, argued that it was here that the gulag system, which would eventually cover thousands of individual camps incarcerating millions of Soviet citizens, was "born and came into maturity". Over 40,000 died at Solovetsky alone.

Above: The Solovetsky monastery today. As the Orthodox church has rediscovered its historical role as a symbol of Russian nationhood under official state patronage, the monastery has been reborn. Some fear, however, that the memory of the monastery's use as a Soviet prison camp may suffer.

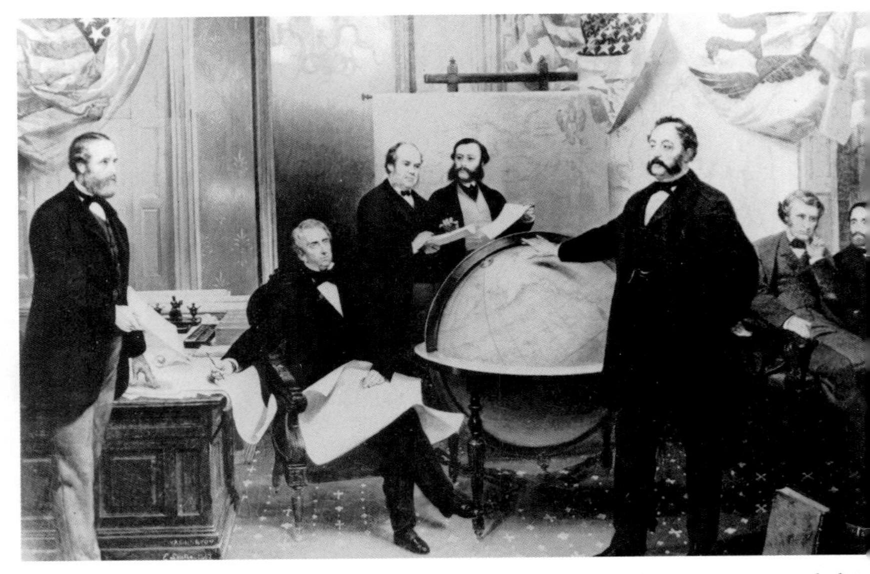

Above: In 1867, the United States became an Arctic power, purchasing Russian America (Alaska) This painting glorifies the scene as Eduard Stoeckl, Russia's minister in Washington D.C., stand by a globe, expounding the territories' benefits to US Secretary of State William H. Seward, seated at the desk on the left, the Stars and Stripes and the Russian imperial flag fluttering in the background. The artist took considerable licence – Senator Charles Sumner (who was not there is depicted seated at the back, while Alaska is presented far larger than its true size. Seward had little idea what he was buying. The Russians, meanwhile, having quite separately gained a firme hold on the western Pacific seaboard, were willing to offload Alaska (and its Inuit inhabitants) for $7.2 million (approximately $100 million in today's money).

a Soviet sea. At the middle of it all was the Soviet Union's heavily fortified Kola Peninsula, extending east from the Norwegian border at Kirkenes.

OSLO, NORWAY, 60° North: 'With the satellites, you could see the Kola bases,' John Skogan tells me, as we sit round the conference table at the Norwegian Institute for International Affairs (NUPI). The bases were impressive in scale and in number: nuclear submarine bases, bomber bases, seventy strategic air defence complexes, twenty-two airbases, nine major army bases, eighteen secondary airfields and enough pre-positioned material for a whole army.[37]

In the 1980s, Skogan's work at NUPI centred on Soviet naval strategy. Poring over scraps of information on the types of Soviet naval vessels being built, the pattern of their deployments and the military exercises in which they were engaged, Skogan tried to understand how the Soviet conception of sea power was changing and what this meant for global strategy and for Norway's place within it.[38]

The growing importance of the Soviet navy was manifest. Observers had noted an extraordinary build-up in naval forces since the 1960s, coupled with an increased willingness to operate outside traditional zones of Soviet influence.[39] The head of the Soviet navy, Admiral Gorshkov, echoed the great American geopolitical theorist Alfred Thayer Mahan, when he wrote in 1976 that 'all modern great powers are maritime countries'.[40]

By the late 1970s, Gorshkov's vision of a blue-water navy capable of projecting Soviet power across the world seemed close to realisation. The expansion of Soviet naval power observed by Skogan was accompanied by a growing militarisation of the Kola Peninsula, marked by a proliferation of naval bases and repair yards, and an upgrading of the land and air forces to protect them.

In Norway, American electronic intelligence technology had been used to track Soviet activities in the Arctic from the 1950s onwards. In the second half of the 1960s, Norwegian defence minister Otto Tidemand referred to the Soviet Arctic build-up in speeches to parliament. But the Norwegian public remained largely unaware of what this meant. By the 1970s, the importance of the Kola Peninsula to Soviet grand strategy – as home to a large portion of both the surface fleet and SSBN fleet – made it inevitable that the Arctic would be a key battlefield in any future war fought between NATO and the Soviet Union. In the early 1980s, Skogan's own conclusion was stark: 'If there was going to be some fighting in Europe, the chances of Norway being left out were nil.'

Around the same time, university researcher Johnny Skorve was using commercially available satellite images to study the geology of the Kola Peninsula, focusing on the copper-rich borderlands between north-eastern Norway and the Soviet Union. Looking at the photographs in more detail, however, Skorve was able to make out the roads, runways, airfields and barracks of the Soviet military. He was impressed. 'There was a lot of infrastructure,' Skorve remembers, some of it under construction. Taking his findings to John Skogan at NUPI, Skorve suggested that study of available satellite images of the Kola Peninsula could be used to draw a much better public picture of Soviet activities on Norway's doorstep.

By the time of Skorve's suggestion, the Cold War had entered a phase of acute tension. Soviet deployment of SS-20 tactical nuclear weapons had upset the nuclear balance in Europe, while the invasion of Afghanistan in 1979 raised fears of a newly assertive Soviet attitude to international affairs. In Moscow, faulty intelligence contributed to Soviet paranoia that a NATO attack was imminent, an obsession reinforced by a new American maritime strategy which rejected any 'Mare Sovieticum' in the Arctic, proposing a forward deployment of US attack submarines in the Barents Sea which would directly threaten the Soviet Union's strategic submarine 'bastion'.

A mood of confrontation and mutual suspicion had descended across the Arctic. US cruise missile tests in Arctic Canada – the only available terrain similar to northern Russia – provoked political crisis in Ottawa.[41] American military analysts argued that the Kremlin was 'probably . . . actually planning offensive operations against the Nordic countries'.[42] In Norway, one of Skogan's former NUPI colleagues, Arne Treholt, was convicted of passing state secrets to the Soviet Union in 1985.

It was in this context that John Skogan was able to raise funds to finance Johnny Skorve's satellite study of the Soviet Union's military complex on the Kola Peninsula. Despite opposition from parts of the intelligence community, they were able to go public with it in 1986.

To most Norwegians, the scale of the Soviet military presence, just across the border, was a revelation.

Challenged by the American forward strategy in the Arctic, in October 1987 the Soviet Union responded with a diplomatic overture. Mikhail Gorbachev chose to launch the initiative in Murmansk, the largest city above the Arctic Circle and home of the Soviet Northern Fleet. The occasion was a speech following the award of the Order of Lenin to the city, in recognition of its sacrifices in the Second World War.

Gorbachev told his audience – and those listening to his speech outside Murmansk and beyond the Soviet Union – that the United States was attempting to draw the USSR into a new arms race in the north, with the intention of undermining 'perestroika', domestic economic reforms introduced a few months previously. While Scandinavian countries – and the Soviet Union – had been seeking peace in the Arctic, the United States had been preparing for war.[43]

'One can feel here the freezing breath of the "Arctic Strategy" of the Pentagon,' he told his audience, picking apart the various elements of NATO policy in the north:

> A new radar station, one of the Star Wars elements, has been made operational in Greenland in violation of the ABM Treaty. US cruise missiles are being tested in the north of Canada. The Canadian government has recently developed a vast programme for a build-up of forces in the Arctic. The US and NATO military activity in areas adjoining the Soviet Polar Region is being stepped up. The level of NATO's military presence in Norway and Denmark is being built up.

Echoing President Reagan's challenge in Berlin several months previously to 'tear down this wall', Gorbachev called for a 'radical lowering of the level of military confrontation in the [Arctic] region'. The Soviet leader exhorted the world to 'let the North Pole be a pole of peace'.

Gorbachev proposed a 'nuclear-free zone in northern Europe . . . including possible measures applicable to Soviet territory'. He suggested consultations to reduce naval activities, including a 'limitation of rivalry in anti-submarine weapons'. As soon as the United States stopped nuclear weapons' tests, the USSR would cease its own testing on Novaya Zemlya. Beyond security, Gorbachev urged greater scientific and environmental cooperation between the Soviet Union and the other circumpolar states, including the development of Arctic energy resources. The Northern Sea Route could be opened to international shipping.

The Murmansk initiative was less generous than first appeared.[44] Given that the Northern Fleet carried the bulk of the Soviet Union's submarine-launched nuclear weapons, Gorbachev's offer to remove SSBNs from the Baltic Fleet was a political gesture, intended to draw Scandinavian states closer to the Soviet Union without making any real dent in Soviet strategic nuclear capability. Any agreed reduction in naval activities, particularly anti-submarine warfare activity, would benefit the Soviet Union. Applying such measures to the area that Gorbachev suggested – the Baltic, Northern, Norwegian and Greenland Seas – would severely constrain NATO's ability to counter the Soviet Union's submarine-based strategic nuclear force.

Proposed scientific and economic cooperation, and the offer to open the Northern Sea Route to international shipping, were more genuine. But such proposals cost the Soviet Union little and allowed the possibility that NATO allies could be split from the United States. Opening the Northern Sea Route – mooted in the past – would not be much of a concession. On the contrary, the Soviet Union would gain economically from any increase in northern shipping.

Gorbachev's Murmansk initiative was followed up in early 1988 by a Soviet diplomatic charm offensive, with premier Ryzkhov sent to Oslo and Stockholm. But the initiative ultimately failed. Norway rejected any attempt to split it from its NATO allies. And in any case, by 1988, the Cold War was almost over. Unlike his predecessors, Gorbachev did not send in the troops – or threaten to – when a series of revolutions swept across Eastern Europe the following year.[45] By the end of 1991, little more than four years after Gorbachev had stood in Murmansk and called for peace in the north, the Soviet Union had ceased to exist.

The end of the Cold War led to a steep decline in the Arctic's strategic significance. NATO's emphasis slowly shifted from territorial defence against the Soviet Union to conducting more flexible operations, such as peace enforcement in the Balkans, and more recently, out-of-area operations in Afghanistan. The United States maintained some of its Arctic infrastructure, such as the Thule airbase in Greenland, and kept up some of its military research programmes into Arctic survival and equipment, but other infrastructure was closed down. Much of the Distant Early Warning (DEW) Line was modernised, some of it was deactivated. In 2006, the United States finally pulled out of the Keflavik airbase on Iceland, ending over half a century of military presence.[46]

In Russia, the decline of military capability in the Arctic was both more severe and less controlled. The Northern Fleet, once the pride of the Soviet Union, was stripped bare. As the Soviet Union fell apart and the Russian economy collapsed, military salaries went unpaid and money available for upkeep and investment dried up. The final humiliation came in August 2000, when a Russian submarine, the *Kursk*, went down in the Barents Sea with the loss of 118 Russian lives. Everything about the sinking of the *Kursk* – from the initial disinformation spread by the Russian admiralty, to the slow response from Moscow, to the failure to accept offers of foreign assistance which might have saved lives – reeked of mismanagement.[47]

The threat of annihilation from Russian submarine-borne nuclear missiles was replaced in the 1990s by fears over the extent of nuclear pollution of the Arctic. After the collapse of the Soviet Union, evidence emerged that a number of Soviet nuclear reactors had been dumped in the Arctic, and that significant discharge from nuclear reprocessing and weapons' facilities had flowed down the Ob and Yenisei Rivers into the Arctic Ocean. On the Kola Peninsula, ageing nuclear power stations, inadequately stored radioactive material and an overhang of rusting Soviet nuclear submarines awaiting decommissioning have provided an enduring nuclear threat. In 1996, a report by the Bellona Foundation, a Russian–Norwegian environmental organisation, warned that radioactive 'waste is deposited haphazardly throughout the various navy yards and bases'.[48]

By 2001, there were still seventy-one submarines waiting to be decommissioned, with further waste at sites across the region, including at the facilities of Atomflot, the Russian atomic energy agency, near Murmansk.[49] The spent fuel from the remaining submarines at the Gremikha base is due to be unloaded by 2010 and taken to the Mayak processing plant in the southern Urals.[50] The storage facilities at Andreyeva Bay, where nuclear waste used to be stored in the open air, just thirty miles from the Norwegian border, are expected to be 'clean' by 2012. Until then, the risk of leakage remains.

Further north, on Novaya Zemlya, the extent of pollution caused by Soviet nuclear testing is not fully known. Complementing the Semipalatinsk range in a remote corner of Kazakhstan, approximately one fifth of Soviet nuclear tests took place in the Arctic. The most powerful nuclear weapon ever tested, nicknamed *Tsar Bomba*, was exploded in the air above Novaya Zemlya in late 1961. The shallow fjords around the islands became a dumping ground for a variety of nuclear waste over the years.

Rumours of the mismanagement of the Soviet Union's Arctic testing programme abound. By comparing aerial photographs of Novaya Zemlya taken by the German Luftwaffe during the Second World War with satellite photographs taken forty years later, Johnny Skorve exposed a 1973 nuclear test which he believes went badly wrong, displacing some 80 million cubic metres of rock.[51] Nonetheless, Russian government sources now report that Novaya Zemlya is clean. In the mid-1990s, the former boss of the nuclear testing site, then serving as President Yeltsin's minister for atomic energy, pointedly told Skorve's co-author that Novaya Zemlya is 'as clean as New York, even cleaner'.[52] In June 2009 Prime Minister Putin signed a decree designating some parts of the archipelago a national park.

Radioactive pollution in the Arctic is not just an inheritance of the military complex of the Soviet Union. The United States is thought to have lost a nuclear warhead near Greenland's Thule airbase in the 1960s, with serious health consequences for the local population. The United States, like the Soviet Union, considered the use of small nuclear devices for civilian purposes in the Arctic. (The Soviet Union actually used them on occasion, including during a botched attempt to cap a gas blowout in the Arctic Pechora region in 1980).[53] These days, both Norwegians and Russians express concern over nuclear pollution from the civilian nuclear plant at Sellafield, on Britain's west coast.

Meanwhile, concerns over radioactive pollution of the Arctic from military installations in the past have not prevented Russia from exploring how civilian nuclear technologies may aid the development of Arctic resources in the future. In 2007, at a conference in Anchorage, Alaska, the head of Russia's Kurchatov Institute, Evgeny Velikhov, suggested that nuclear-powered submarines might provide for sub-ice transport of liquid natural gas under the North Pole.[54] Russia's first floating nuclear power station is under construction, with suggestions that a fleet of such stations could eventually provide an important power source across the Russian Arctic. Things have come a long way since the voyage of the *USS Nautilus* in 1958.

The former head of Bellona's Murmansk office, Sergei Zhavoronkin, sees the decommissioning of the Soviet Union's Cold War nuclear infrastructure in the Arctic – much of which has been done with Western funding and technical assistance – as building the basis for international cooperation on wider Arctic development. There is a history, he points out, to European association with the Russian Arctic. Appealing to my nationality, Zhavoronkin reminds me that in Umba, on the southern part of the Kola Peninsula, there is English soil, dumped as ballast from English ships plying the White Sea trade routes in the nineteenth century.

'So, we have traded together,' Sergei tells me, 'we have fought together, we are decommissioning the Cold War together and, in the future, we will make great investments together for the development of the North.' It is a hopeful coda to a dark half-century of superpower competition in the Arctic. Partial remilitarisation of the Arctic is more likely.

In the Cold War, Arctic security policy was bound up with the single overwhelming threat of nuclear confrontation between the Soviet Union on the one hand, and the United States and its allies on the other. In the twenty-first century, that threat has largely gone. Instead, Arctic security policy in the future will be about a more fragmented set of challenges – many of them civilian – arising from the Arctic's growing economic importance and, partly as a result of climate change, its increased accessibility.

Elements of Cold War confrontation remain in the Arctic. It is still the main patrol area for Russia's submarine-based nuclear deterrent, and strategic bomber patrols have been resumed. But two crucial differences explain why the military–strategic future of the Arctic will not be the mirror of the recent past. First, the global context has changed, with the rigid bipolarity of the Cold War replaced by the more fluid geopolitics of the twenty-first century, including a much greater global interest in the economic opportunities of the Arctic. Second, the country with the greatest stake in the Arctic, Russia, is much diminished.

In the view of Rolf Tamnes, director of the Norwegian Institute for Defence Studies, 'It will take Russia a very long time to rebuild its armed forces; perhaps fifteen years to reach a decent great power status; to a French or a British level, say.' In 2008, Russia demonstrated its willingness to break international law in defence of its interests in the South Caucasus and, despite deep cuts announced in 2009, Russia will retain more nuclear warheads than any other country for the foreseeable future. Russia's ice-breaking and surveillance capabilities in the Arctic far outstrip those of any other Arctic country. But the force-projection capability of modern Russia does not approach that of the Soviet Union. Moreover, the domestic legitimacy of the Putin–Medvedev regime depends on economic prosperity more than military might. Barring the establishment of an ultra-nationalist regime in Moscow, a reprise of the Cold War in the Arctic is unlikely.

Nonetheless, the scope for geopolitical friction in the Arctic remains. Several potential flashpoints have returned to prominence in recent years, from the disputed sea border between Russia and Norway to the status of Svalbard. Until these issues are conclusively resolved, they provide an opportunity for friction in international relations. New potential flashpoints have emerged – not least those associated with the risk of overlapping claims to the Arctic seabed. Should Greenland ever achieve full independence from Denmark, it cannot be assumed that the island's current strategic alignment would continue. The Arctic's economic development – particularly shipping – will have security implications for Arctic coastal states.

Plans to protect vital energy infrastructure – over which NATO already has an area responsibility – will have to be updated as the number of oil and gas installations in the Arctic increases.[55] (Norway already has Special Forces trained to recapture offshore oil platforms – coming up from torpedo tubes in submarines, parachuting in from low altitude, or using fast boats and helicopters). The increasing accessibility of the Arctic to shipping will inevitably be accompanied by a reinforced naval and coastguard presence to protect the maritime 'global commons'. Ensuring maritime security (and keeping potential choke points open) has long been a task for the world's naval forces – from the preponderant British navy in the nineteenth century to the dominant US navy in the twentieth. If, in the twenty-first century, major trade routes develop in the Arctic as a result of climate change, naval forces from different countries – China and India as well as the United States and Russia – will pay as much attention to the Norwegian Sea and the Bering Strait as they currently do to the Panama Canal and the Straits of Malacca.

The security implications of a more accessible and more economically important Arctic will require a reconfiguration of the military and civilian resources of the Arctic states. In some countries, this reconfiguration is well under way; in others, it is only just beginning.

Outside of Russia, the country which is most advanced in terms of adaptation to the new security challenges of the Arctic is Norway. Arctic-capable frigates have been purchased to boost the country's naval strength, while the country's operational headquarters moved to Reitan, well above the Arctic Circle, in August 2009. Surveillance operations are being updated to allow for more intensive monitoring of the Norwegian and Barents Sea, an area which the head of the Norwegian defence ministry's policy planning department describes as 'as big as the whole Mediterranean'.[56] All this is expensive. Norwegian defence officials have been keen to raise the issue of High North security at the NATO level, reminding its partners that Oslo is shouldering a large share of the costs of securing the Arctic.[57] At an alliance level there is a serious question of who should do what.

Canada's military and civilian capacities to monitor and control its vast Arctic hinterland have not kept pace with rhetoric in Ottawa or the concerns of national security experts that the country is ill prepared for the challenges of a more accessible Arctic. There is a credibility gap between Canada's assertions of sovereignty over the Far North, and its ability to enforce them. Some conclude that Canada faces a security crunch in the north in the next few decades; others suggest the warning signs began long ago, when a Russian cargo plane stopped over without permission in northern Canada in 1998 or when an armed Chinese research vessel arrived unannounced in Tuktoyaktuk the following year.[58]

In May 1999 an Arctic Security Interdepartmental Working Group was convened on the initiative of Colonel Pierre Leblanc, one of the country's most outspoken advocates of the need for expanded Canadian military capability in the Arctic.[59] Since then, a series of joint military exercises – known as *Narwhal* and *Nanook* – have taken place with varying levels of success, involving the (brief) return of the Canadian navy to the Arctic after a gap of nearly fifteen years. The 2008 *Canada First Defence Strategy* confirmed plans to build a deep-water maritime facility at Nanisivik (to open in 2015) and six to eight Arctic/Offshore patrol ships (at a cost of at least $3 billion), though the first ship will not be delivered until 2014.[60] Twenty-four-hour surveillance of the Arctic by Canada's Radarsat-2 satellite is set to begin in 2010.[61] The Canadian government is investigating extending the Forward Operating Locations of the air force to Resolute Bay, at 74° North.

The United States already plays a significant role in Arctic security through its partnership with Canada in the joint North American Aerospace Defense Command (NORAD), and through the 2006 extension of the NORAD agreement to include a maritime warning element. But since the early 2000s it has been wars in Afghanistan and Iraq which have most occupied the attention of the Department of Defense and the army. In contrast, the navy has taken a considerable role in Arctic security. As early as 2001, at the same time that the Bush administration was casting doubt on global warming, the navy convened its first symposium on the impact of an ice-free Arctic on its operations.[62] In a reflection of growing concern about the challenges of future Arctic security, follow-up meetings were held in 2007 and 2009.

But it has been the United States Coast Guard (USCG) which has been most forceful of all in leading calls for an upgrade in America's military capabilities in the Arctic. In the summer of 2008, *Operation Salliq* revealed major shortcomings in the USCG's Arctic capabilities. In July 2009, Admiral Thad Allen, the most senior officer in the USCG, warned a Senate committee that the current American 'ice-breaker fleet is atrophying . . . we run the risk of losing that national capability'.[63] A 'minimum' capability would require at least three fully operational ice-breakers, at a cost of $700–900 million each.[64]

Finally, in June 2009, just three days after Greenlanders voted to accept an agreement granting the island enhanced autonomy from Copenhagen (but not including defence), the Danish parliament agreed defence plans for 2010–14 which will involve the establishment of a Danish Arctic Command (probably on Greenland) and an increase in the Danish military presence in the Arctic.

Remilitarisation of the Arctic is already under way.

The nature of twenty-first-century security challenges in the Arctic – less about war fighting and more about surveillance and control – is reflected in the scenarios of the future used by the American, Canadian, Danish, Norwegian and Russian armed forces to test their planning.

In 2007, a Canadian report offered four 'intrusion' scenarios: an emergency aircraft landing at Qausuittuq (Resolute Bay) carrying dangerous pathogens, the accidental explosion of a terrorist dirty bomb at the port of Churchill as part of an attempt to introduce the device from Russia into the United States, the unannounced transit of a French strategic nuclear submarine through the Northwest Passage and, finally, domestic sabotage of oil and gas infrastructure by the (non-existent) First Nations' Liberation Movement.[65] In one way or another, the response to any of these scenarios would involve Canada's armed forces; none would amount to a shooting war between the Arctic states.

In Norway, the head of policy planning at the Ministry of Defence offers a scenario in which Russian companies start to drill for oil in disputed areas of the Barents Sea. 'It's not a military attack,' Bård Bredrup Knudsen points out, but any diplomatic response would require a military element. It is this kind of complex scenario, rather than the straightforward military threat of the Cold War, which occupies the minds of the Arctic's strategic defence planners.

All-out interstate war in the Arctic is plausible, but unlikely. (One senior American coastguard official described interstate conflict arising from a dispute over maritime borders as the 'worst-case scenario').[66] But a proliferation of broader security challenges in an increasingly ice-free Arctic is unavoidable. Ultimately, the principal challenge for Arctic security in the climate-changed twenty-first century may not be too much surveillance and control, but not enough.

PART THREE
Nature

'A time will come in later years when the Ocean will unloose the bands of things, when the immeasurable earth will lie open, when seafarers will discover new countries, and Thule will no longer be the extreme point among the lands'

Seneca, Roman philosopher and dramatist

6
Signs: Nature's Front Line

'Soon it would be too hot. Looking out from the hotel balcony shortly after eight o'clock, Kerans watched the sun rise behind the dense groves of giant gymnosperms crowding over the roofs of the abandoned department stores four hundred yards away on the east side of the lagoon'

The Drowned World, J. G. Ballard, 1962

'Global warming connects us all. Use what is happening in the Arctic – the Inuit Story – as a vehicle to re-connect us all, so that we may understand that the planet and its people are one. The Inuit hunter who falls through the depleting and unpredictable sea-ice is connected to the cars we drive, the industries we rely upon, and the disposable world we have become'

Sheila Watt-Cloutier, former President, Inuit Circumpolar Council, Testimony to the US Senate, 2005

There are few straight lines in nature. Take almost any set of observations of natural phenomena over several years – the number of days over 100° Fahrenheit in a sequence of sweltering summers in New York, say, or perhaps the cumulative October rainfall in southern India between 1950 and 2010 – and you will get the same picture: dynamic, unpredictable, volatile. But set aside the 'noise' that the variability creates – the cloud of seemingly random fluctuations from the norm – and trends and patterns begin to emerge. Sometimes, those trends are difficult to discern – sometimes they are unmistakable.

Look at a graph of Arctic sea ice over the last few decades, showing the maximum and minimum extent of ice cover from year to year. The first impression is one of volatility, a stock market of results. Over the course of every year, the area of the Arctic Ocean which is covered in sea ice changes hugely, reaching its maximum extent in March and its minimum extent in September.[1] Beyond that annual cycle of expansion and contraction, explained for the most part by the seasonal rising and setting of the sun, there are large variations from year to year. In one year the extent of ice cover can fall below average, but then spike upwards the next. There is no year-on-year decline in Arctic ice cover.

What there is, instead, is an unmistakable trend. It is this trend, rather than the 'noise' of observations around it, which matters. Draw a line running through the midpoint of all those observations of maximums and minimums, over several decades, and it becomes clear that, on average, sea ice in the Arctic is diminishing. The 'bad' years for Arctic ice cover are getting steadily worse, and the 'good' years for Arctic ice cover just aren't as good as they once were. In 2005, the extent of Arctic ice cover reached 5.57 million kilometres (2.14 million square miles), the lowest on historical record. Then, in 2007, at the end of a particularly warm and cloudless Arctic summer, things fell off a cliff. At its lowest point, in September of that year, ice covered just 4.28 million square kilometres (1.65 million square miles) of the Arctic.[2] In 2008, the extent of Arctic sea ice recovered somewhat, leading some non-scientists to claim that the melting of the Arctic was a hoax. But such critics were making a fundamental mistake – confusing variations from the trend with the trend itself.

What actually drives the rate and process of melting and freezing of sea ice in the Arctic is enormously complex. It depends on a range of different factors, from oceanic currents to cloud cover and from air temperature to the pre-existing thickness of Arctic ice.[3] Some of the natural processes are well understood, others are not. Basic data, such as the thickness of the sea ice (and therefore its volume) rather than just its surface area, are incomplete.[4] Other factors, such as the level of overall global warming in the future, depend partly on human decisions whether to continue to increase emissions of greenhouse gases or to begin to reduce them. That the Arctic will warm more than the rest of the world – as a result of an effect known as Arctic amplification – is a given. Just how much more is unknown. Given the number, complexity and interrelationships of these variables, coming up with an accurate prediction of where Arctic sea ice will go from here is no easy task. It is not simply a matter of extending the trend line from the history of Arctic ice cover.

Opinions on the future of Arctic sea ice differ. At the Arctic and Antarctic Research Institute in St Petersburg, Ivan Frolov tells me that he believes that 'scientists will have to change their models again', once they realise that natural variability is about to send the process of Arctic warming into reverse.[5] 'For us it is absolutely clear cut,' Frolov tells me. 'Those scientists who totally connect climate change with CO_2 are making a big mistake.' Instead of an ice-free Arctic, Frolov argues that 'there will be an easier time' for a while, after which ice cover will extend once again. But this is a minority view. The report of the Intergovernmental Panel on Climate Change (IPCC) takes the line that 'the decline in arctic sea ice extent and its thinning appears to be largely, but not wholly, due to greenhouse gas forcing', suggesting that Arctic ice will continue to retreat as greenhouse gas concentrations continue to rise.[6] Many scientists believe the process may be accelerating.

In December 2007, Mark C. Serreze, a senior research scientist from the National Snow and Ice Data Center in Boulder, Colorado, stood up at a conference centre in San Francisco and told the fall conference of the American Geophysical Union why he thought that 2007 might turn out to have been a watershed. In his lecture Serreze identified three factors of particular importance. First, our models of Arctic climate change seem to have underestimated the rate of sea-ice loss. Second, there is a tipping point beyond which sea ice will no longer recover from year to year. Third, we may now be approaching – or may even have passed – that tipping point. As a result, the transition to an ice-free Arctic (in summer) may be abrupt.[7] Instead of the gentle slope of the trend line, Serreze argued, we may get change that is far steeper.

'The observed rate of decline is faster than any of our models,' Serreze told his audience. Worse, 'the ice is becoming younger and thinner'. The relevance of that observation, Serreze reminded the American Geophysical Union, is that younger and thinner ice tends to melt more rapidly in summer than thicker and multi-year ice – it is, in other words, more vulnerable. Serreze pointed to a study undertaken the previous year, looking at the risks of abrupt changes in Arctic ice, partly as a result of thinning ice.[8]

The direct cause of a shift from slow change to abrupt and irreversible change, the study showed, was a single warm summer. But the underlying cause was that once Arctic ice was sufficiently thin – and therefore vulnerable to rapid melt – a 'bad' summer could quickly become a disastrous one: 'Sea ice had thinned to this vulnerable state, and then came a kick from natural variability . . . and you kinda fell over the edge.' In the model developed by Marika Holland and her team at the National Center for Atmospheric Research the critical ice thickness was about 2.5 metres. And, as Serreze pointed out, 'We're there now.' 'It might be coincidence, blind coincidence,' he warned, 'but it's intriguing nonetheless – will 2007 be remembered as the tipping point?'

The idea of an ice-free Arctic Ocean – and the question of what an ice-free Arctic ocean would mean in terms of both local and global climate – is not new.[9] Scientists at the Lamont Geological Observatory in New York State have been trying to understand how an ice-free Arctic would affect both local and global climate since as far back as the 1950s. The principal aim of research by scientists such as Maurice Ewing and William Donn was to understand the processes behind natural cycles of climate change in the past: how previous ice ages began and how they ended.[10] Others wanted to understand what an ice-free Arctic might mean in the future and indeed whether there might be benefits to deliberate attempts to induce Arctic warming, either by blackening the Arctic ice pack – thereby increasing its absorption of heat – or even by exploding nuclear weapons in the Arctic Ocean.[11] But the overriding expectation of such scientists was that an ice-free Arctic Ocean was likely to be a distant phenomenon.

Today, an ice-free Arctic seems far more immediate. The question of exactly when the Arctic will become ice free in summer – no scientist is suggesting that the Arctic will become ice free in winter – has turned into a scientific guessing game. Some argue that it is as distant as 2100. A few, including Dr Wieslaw Maslowski, a Polish researcher at the Naval Postgraduate School in Monterrey, California, have suggested it could happen as early as 2013. Dr Mark C. Serreze himself has plumped for somewhere in the middle: 'I used to think 2050, 2070, maybe even 2100, but we're losing sea ice more rapidly than the models are projecting. My thinking is that 2030 is now not an unreasonable date to be thinking of'.

The most widely publicised consequence of an ice-free Arctic, at least in terms of its impact on the region's ecosystem, is the risk to the habitat of the region's polar bears. When it comes to advertising the impact of global warming in the Arctic, photographs of polar bears have become more or less standard, a visual cliché, for better or worse. Their fate is an emotive issue, used by some to raise awareness about the consequences of climate change and by others to imply that global warming is a matter of aesthetics.[12] Biologists who actually work on polar bear populations, such as Dr Steve Amstrup, a research scientist at the Alaska Science Center (part of the United States Geological Survey) have mixed views on the subject: 'I've been fortunate that the species I've been working on for so long has managed to capture people's imagination, but, for me, all parts of the ecosystem have their value.'

For Amstrup, who has been working with polar bears for nearly thirty years, there is no question but that the disappearance of Arctic ice, while not the only factor determining future polar bear populations, matters a great deal.[13] 'Sea ice is critical to the survival of polar bears,' he tells me, in his offices on the University of Alaska campus in Anchorage. It's simple mathematics: 'They [polar bears] lose one kilo a day [of bodyweight] onshore and they add two kilos offshore.' Whereas there is an abundance of low-quality food onshore, there is nothing onshore which can match the seal for its energy content. 'They're like these giant fat pills,' Amstrup explains.

The current picture of polar bear populations across the Arctic is mixed: in some areas polar bear populations are stable, in other areas they are falling already, and in a couple of areas they have risen over the last few years (albeit from a low starting point).[14] In the Russian Arctic, the data are simply too patchy to be able to tell. But if Amstrup and other polar bear scientists are right about the links between sea ice, polar bear habitat and population levels, then the future direction of polar bear populations will almost inevitably be down. Amstrup's own conclusion is stark: 'There may be some pockets where polar bears survive – maybe a few dozen, maybe a few hundred – but there will be nothing like the 25,000 you have today.'

The politics of polar bears is, however, a side issue. The local effects of Arctic climate change are bad in themselves, but the global consequences of Arctic climate change are potentially far more serious.

TROMSØ, NORWAY, 69° North: The plane banks hard to the right and turns up the fjord towards the airfield, ducking in and out of the cloud cover as we come in to land. The sea below is flecked with white tops. The land is covered in snow. When the plane hits the ground, battling against the crosswinds to stay on the runway, the strong headlamps on its wings illuminate thin streams of snow racing from right to left, like sand across a desert road. Welcome to the Arctic. By the following morning the wind has died down. The snowflakes are much larger now, floating down from the clouds like a million pieces of ice-cold confetti. I ask Kim Holmen, the Swedish research director of the Norwegian Polar Institute, how cold he thinks it is outside. 'Oh, judging from the snow I'd say it was about minus two degrees Celsius up in the clouds,' he says, as I wipe down my coat and let myself into his office.

Unlike most Scandinavians, Holmen didn't grow up with snow. Until the age of fourteen he had never seen a single snowflake, 'not even on a mountain top'. His childhood was spent in countries where snow was hard to come by – Egypt, India, Ethiopia, amongst others – following his father's job, installing telephone exchanges for Ericsson, the Swedish telecoms company. 'I read books, I dreamed of seeing snow as a young boy,' he remembers. But he sees plenty of it now. It's not just that Tromsø, a few degrees above the Arctic Circle, gets more annual snowfall than any other Norwegian town. It's that snow and ice, once considered rather unfashionable subjects in the natural sciences, have suddenly taken on tremendous importance for our understanding of Arctic climate change and, by extension, how much the melting of the Arctic is likely to influence patterns of global warming.[15] As the man largely responsible for deciding how the Norwegian Polar Institute spends its research funds, Holmen finds that snow and ice are taking up an increasing share of his time.

He swivels round in his chair and takes a thick volume off the bookcase behind him. 'You know this?' he asks. I nod. It's the *Arctic Climate Impact Assessment*, published in 2004, a massive scientific undertaking led by an American oceanographer, Robert Corell, assigned by the

Arctic Council to look into how global warming is changing the Arctic.[16] 'This book is a thousand pages long,' says Holmen, 'and it contains essentially one conclusion: surface processes in the Arctic are very poorly understood.' Filling that gap in the science – understanding how snow and ice interact with the environment, how they respond to climate change, and how they may alter the speed of future global warming – has become a major research priority for the Norwegian Polar Institute.

'The conclusions of our climate models are based on a certain description of the retreat of snow and ice and the albedo feedback mechanism, and so on,' he explains.[17] 'Unfortunately, understanding those snow and ice processes is one of the things which is weakest in our climate models – it's one of the largest uncertainties in the models,' he tells me.[18] This is hugely important. The predictions of scientists' climate models are the basis for the investments which will be required to mitigate climate change and, in some circumstances, to adapt to it. If the predictions turn out to be wrong, then the policy responses may prove to be wholly inadequate – or, alternatively, excessive. Reducing the areas of uncertainty in climate models is, therefore, the precondition for good climate policy. As far as Holmen is concerned, more research is necessary: 'I've engaged in a strategic action plan towards increasing our knowledge of snow and ice, and have convinced the government and the Prime Minister that this is something that we really need to work on.' It may take many years to fully understand these processes. 'But I'm a patient man by nature,' Holmen says.

The people at the Norwegian Polar Institute don't build the climate models themselves. In Norway, the 'number crunching' is done in Bergen, in the south of the country. But there are plenty of other climate institutes – in the United States, in Japan, in Britain – which have developed their own climate models. The Norwegian Polar Institute provides an understanding of the *processes* which feed into those climate models. 'You can't include all the known physics in these computer models, because the computers just aren't capable of solving all the equations,' Holmen tells me. 'So you have to make simplifications – called "parametrisation" in the jargon.'

The trick, of course, is to make sure that the simplified equations which are fed into the overall climate model correspond as closely to reality as possible. If you can't input everything, you have to make sure you have what is most essential. Holmen gives an example:

In the numeric climate model you might have only one number which describes the snow, but both you and I know that the snow can not only be deep, but it can be wet, it can be dry, it can be refrozen – all of those things have an influence on how rapidly it will melt. Of course, if you only describe it with one number – depth, say – then you are not capturing the other features which it is essential to capture. So you introduce another number that talks about, say, the thickness of the refrozen layer of the snow. But you're still not describing the whole thing. Our niche, and our goal, is to know these processes and live these processes so that we can provide the best inputs for the models, the closest fits to reality.

Holmen is most personally involved in investigating sea–air exchange, studying the heat flux between the ocean and the air. In the Arctic, the relationship between the two is changing rapidly. 'In the past, ice was a barrier,' Holmen explains. But things have changed. 'Take the ice away, and you have a totally different exchange of energy between the air and the ocean,' he says, 'and instead of having a local radiative balance, you have another flux, which can contribute to more rapid warming.' In other words, a reduction in Arctic ice cover tends to be self-perpetuating. Indeed, better understanding of the sea–air relationship in the Arctic is one reason why forecasts for when the region will be ice-free in summer have been coming down.

As Holmen sees it, the trends are accelerating. 'The world is moving faster than IPCC [Inter-Governmental Panel on Climate Change] and ACIA [Arctic Climate Impact Assessment] suggest,' Holmen warns. 'From being accused of being too alarmist and whatnot a few years back, we now think that the IPCC is too conservative.' Of course, there may yet be 'good' surprises – negative feedbacks that tend to return global climate to equilibrium. On the other hand, there may not. 'When I look back on my career there have been surprises all along – we have discovered new things,' he tells me, 'but very few of those discoveries have suggested that the greenhouse gas problem is smaller than feared . . . If it was just a random set of coincidences then half of them would be negative and half would be positive,' Holmen argues, 'but, to date, almost all have been negative,' suggesting that climate changes may happen more quickly than originally thought.

One of the most recent negative surprises has been evidence that Greenland's ice cap may be melting more rapidly than thought, with serious consequences for potential sea-level rise.[19] As the ice caps melt some water seeps through the ice sheets, lubricating their seaward progress. This, in turn, tends to speed up the process of melting as more ice comes into contact with warmer sea. 'Previously you were just calculating how rapidly the ice itself would melt – you heat it a bit, do the thermodynamics calculations and come to the conclusion that it will take a long time to melt the whole of Greenland,' Holmen explains. 'But if you lubricate the whole underside of the glacier then the whole movement changes, and then it is floating on water rather than on land, which of course has a direct impact on sea level.'

The idea that the progress of glaciers might be lubricated by melt-water is, in a sense, obvious. 'But no one suggested it,' Holmen says. The impact it may have on the length of time required to melt the Greenland ice cap is significant. 'The timescale with which large ice masses might alter sea level have changed,' I'm told, 'from several centuries to a handful of decades, perhaps.' No one is suggesting that all the ice on Greenland is going to melt by the end of the century. Nevertheless, given the huge volumes of water stored in the polar ice caps, the melting of even a small fraction of that ice will contribute noticeably to rising sea levels.

As a scientist, Holmen steers well clear of expressing a clear-cut position on the value that society should place on nature – a matter he understands as a question of public perception and personal values rather than scientific judgement. What people care about is not necessarily the same as what is most threatened: 'Everybody screams about the polar bear, but there are other examples of species which may disappear in the future – the ivory gull, for example.' And society's motivations for protecting nature can be as much about philosophy as anything else. 'Most people will never see a polar bear in the wild,' he says, 'but many people, at least in the West, will probably feel that the world is poorer without them.' Science, Holmen argues, has to engage with society and yet somehow remain above it. It has to be about facts, rather than values; it has to be approachable without being compromised.

That can sometimes be a fine line to tread, particularly when it comes to communicating the science of Arctic climate change to political and economic decision-makers – let alone the media – who will always prefer certainty and simplicity to uncertainty and complexity. Many scientists are uneasy about the purity of science being undermined by the dirtiness of politics. They fear that the critical distance required for good science doesn't translate into politics. But while Holmen recognises the challenge of communicating complex science, he argues that it is a vital input into political decision-making. Sure, when the issue at hand is as important and as complicated as the fate of the global environment the risks of distortion of science for political ends are greater than ever. But so are the risks of inaction.

Scientists like Holmen recognise instinctively that climate science – perhaps more than any other branch of science – must engage with politics. Holmen regularly hosts political leaders in Tromsø and at the scientific research station at Ny-Ålesund. Last year there were fourteen ministers, two European Union commissioners and 'hordes' of ambassadors, he tells me. This year, he's expecting to speak to a group that includes former President Jimmy Carter, former Secretary of State Madeleine Albright and media mogul Ted Turner, amongst others.

The spike in public and political interest in Arctic science is a good thing.[20] Not only does it drive a broader awareness of climate change, more specifically it helps research foundations to get funding for their work. But being in fashion has its drawbacks, too. 'You get a mixed bag of visitors,' Holmen tells me. 'Some come here for the thrill – there's a lot of those.' With heightened political interest in the Arctic comes the baggage of prestige politics. A strong element of national pride is involved in most countries' research programmes. 'Every nation worth its salt wants to be in the Arctic these days,' Holmen tells me, resulting in occasional problems of overlap and coordination. 'What if the UK wants to do its climate study of the Arctic and, fifty metres away, there are the Norwegians doing their climate study of the Arctic?' he asks, rhetorically. 'Globally, we need to come together, we need to be constructive,' he insists.

Having grown up in many parts of the developing world – albeit as the fortunate son of a Swedish telecoms engineer – Holmen is sensitive to the impact of Arctic climate change on places far away from wealthy Tromsø. The changing energy flux in the Arctic, Holmen points out, will have impacts on the intensity of the monsoon in southern Asia and, as a result, on the lives of millions of people.

'The redistribution of precipitation may be faster than the redistribution of population,' he worries. The melting of the Greenland ice cap – and its consequences for sea level – is of direct human relevance to low-lying cities and countries across the globe, from London and New York to Bangladesh and the Maldives.

Many of the major uncertainties in current global climate models relate to the nature and speed of change in the Arctic. Some of these uncertainties are offshore – such as the precise extent to which an ice-free Arctic Ocean will increase the amount of heat absorbed by the sea and how that will affect other global processes. But one of the most worrying uncertainties around how the Arctic will respond to a warming climate is onshore. As the Arctic warms, land which has been frozen solid for thousands of years – the permafrost – has begun to melt.[21] How far that process has gone, how quickly it is likely to continue and exactly what that means for the rest of the world is not yet clear.

FAIRBANKS, ALASKA, 65° North: About ten miles north of Fairbanks, as you drive up the Steese Highway towards Fox, there's a gravel track leading up to your right. In August the road-sign for the track is hard to spot, hidden behind the foliage of a warm central Alaskan summer. Up the gravel track a hundred metres or so is a clearing at the bottom of a steep hill, surrounded by a fence topped with rusty barbed wire. On the north side of the hill – known to the US Army Corps of Engineers as 'Hill 456' is a set of doors. Behind them, a tunnel stretches eighty metres into the frozen hillside. 'Kind of cool', says Dr Tom Douglas, a researcher working for the Cold Regions Research and Engineering Laboratory's (CRREL) office in Fairbanks. A heavy-duty freezer unit at the entrance to the tunnel, blowing out a constant stream of cold air, is intended to keep it that way.

The tunnel was built by the US army back in the 1960s, when one of the main reasons for interest in permafrost research was military: could nuclear missile silos or other facilities be hidden in the permafrost, for example? These days, the focus of CRREL's research – in Alaska and at CRREL's main facility in Hanover, New Hampshire – is largely civilian. The permafrost capital of the world, Fairbanks hosts research into not only the physical properties of permafrost – how it forms, how it melts and so on – but also into the geochemical and bacterial deposits which have been frozen into the permafrost for tens of thousands of years. 'It's like going back in time,' says Douglas.

Inside the tunnel, the combination of mould and bacteria makes it smell like the inside of a ripe blue cheese. Kept at a constant temperature of around -4° Celsius (25° Fahrenheit) the tunnel is the 'perfect place to maintain bacteria' says Dr Dan White, director of the Institute of Northern Engineering at the University of Alaska at Fairbanks. The side of the tunnel is bored with small holes where scientists have removed core samples from the permafrost, like biopsies from our living planet. Under conditions of global warming, some bacteria which have been trapped in ice will ultimately be released. 'There are some open questions here,' says White. 'What will come out? How will we respond to bacteria which have never been in contact with human beings?'

In the northern half of Alaska, evidence of permafrost is everywhere. The thousands of lakes which cover northern Alaska are formed by water which cannot seep beyond the first few centimetres of soil, because the ground below is an impervious mass of frozen earth. The strange geometry of the North Slope landscape – like the endlessly repeating forms of a geometric fractal – is the consequence of countless cycles of melting, freezing and cracking of the upper layer of the permafrost. The elevation of the Trans-Alaska Pipeline on stilts, and the installation of heat exchangers on the underside of the pipeline, are intended to prevent the permafrost beneath from melting.

Beyond Alaska, permafrost is a fact of life across the Arctic. Considerably larger and older areas of continuous permafrost crowd along the northern shores of Eurasia. Patches of permafrost dot many other parts of the northern hemisphere – as far south as the open plains of Mongolia and the mountains of the Himalayas. While researchers in Alaska may like to think of Fairbanks as the permafrost capital of the world, the greatest extent of permafrost – both in terms of coverage and depth – is across the Arctic Ocean in Russia. In parts of East Siberia permafrost can extend as deep as 1,500 metres into the earth.

The top permafrost scientist in Alaska, Dr Vladimir Romanovsky, is himself originally from Russia, having moved to the United States from Moscow State University in 1992. He's been working on permafrost issues since the mid-1970s. 'At that time we thought permafrost was more or less stable,' he tells me. 'We were thinking of changes over a glacial or interglacial time frame rather than anything quicker.' Given presumptions about the basic stability of permafrost – and the expense of data collection – the main focus was on modelling long-term change.

Things have changed since the 1970s. In terms of acquiring the raw data for permafrost research there has been a 'technological revolution'. 'Back in the seventies you had to physically be in all these places if you wanted to get continuous records,' Romanovksy explains. These days, equipment placed in boreholes can record data continuously and then be retrieved at a later date. The machines themselves are much simpler to use: 'School students can operate them.' The number of readings available to scientists has increased dramatically, as has their geographic extent.

But it's still not enough. 'It depends on the specific goal of the research,' he explains. Permafrost depends on so many different local factors that if you want to have a truly global view of the state of the world's permafrost then the number of data points required is large. 'We have gaps in Alaska even,' Romanovsky says. 'Even if we had two hundred boreholes it would not be enough.' It's a problem I hear again and again when talking to permafrost scientists – and a reminder of the extent to which so much of climate science is really about trying to reduce and manage uncertainties, and hoping that scientific understanding will keep pace with mounting evidence of climate change.

Beyond the engineering problems engendered by the thawing of permafrost – the fact that the foundations of many structures could become increasingly unstable and prone to collapse – the thaw could accelerate the processes of global warming. One risk is that vast quantities of methane which are currently stored in the deep permafrost as methane clathrates will be released into the atmosphere.[22] While most scientists consider the large-scale release of methane to be a long-term rather than short-term threat, the extreme potency of methane as a greenhouse gas has raised concerns that even a small additional contribution could push global climate towards a point of no return. A more immediate risk is that a warming permafrost will accelerate organic decomposition – a process which would speed the release of carbon dioxide and methane into the atmosphere. Both scenarios point to the potential for thawing permafrost to accelerate climate change.

Despite the potentially significant role that permafrost may play in the future of global climate, permafrost models are currently 'rudimentary'. Romanovsky warns that 'coarse' modelling of permafrost processes produce results that do not correspond to observable reality: 'You miss the latent heat effects and produce scenarios showing all the permafrost melting.' Reality is more complex. Thawing is not yet widespread, Romanovsky tells me. Where it is taking place, it tends to be slow. Deep permafrost is likely to remain frozen for a long time to come. Nearer the surface however, things are much more vulnerable to changes in atmospheric temperature. 'If climate continues to warm then a major degradation will begin in fifteen to twenty years, and peak in the next fifty,' says Romanovsky. There will be large-scale hydrological consequences, resulting in either increasing river run-off (and therefore more erosion in some river deltas) or an extension of wetlands, depending on local conditions. All of this has serious consequences for infrastructure – roads, railways, pipelines – currently constructed in areas of permafrost.

In the building across from where Romanovsky is looking at Arctic permafrost, Dr Glenn Juday is studying the relationship between global warming and trees. The two subjects are superficially very different. In fact, they are two sides of the same investigation of the past and future of the global carbon cycle and its consequences for climate. Permafrost is a major store of carbon below ground; boreal forests, which absorb carbon dioxide as they grow, are a major 'sink' for carbon above ground. Just as permafrost stores information on past climate over thousands of years, trees store information on changes in climate in the more recent past, from season to season and from year to year. The timescales may be different, but the principle is the same.

Juday has found himself increasingly concerned that the orthodox view of the relationship between trees and climate – that rising temperatures increase tree growth – is misleading. All other things being equal, trees *should* respond in the same way to a warmer climate as they do to a warmer summer, by growing more quickly. The existence of a treeline – above which it is too cold for trees to grow – demonstrates that temperature *can* be a limiting factor on the extent of tree growth. But Juday's work reveals a more complicated relationship between climate and trees than one of temperature alone.

Evidence gathered from Alaska's boreal forests shows that other factors may be just as important as temperature. 'The final nail in the coffin of the traditional view came in 2000,' Juday tells me, with the publication of research into the white spruce, the tree which covers the hills near Fairbanks.[23] Data collected by Juday and two of his colleagues showed that, on average, white spruces had not grown more quickly as temperatures had increased. They had actually grown more slowly. 'That's right,' Juday repeats. 'The warmer it was, the less the trees had grown.'[24] The reason was that temperature was not the only factor in play: higher temperatures tended to be associated with greater water stress. 'We had always thought that it was temperature which was the limiting factor on trees' growth, because that's what the treeline appeared to show,' he explains. But it turned out that moisture was still more critical.

Global warming may be contributing to two further threats to Alaska's boreal forests which, taken together, mean that the white spruce may be in a 'zone of probable species elimination' by 2030. In recent years, warmer summers and milder winters have increased the regularity and scale of outbreaks of the spruce bark beetle. Meanwhile dryer, hotter summers have increased the frequency of major forest fires. In 2004 there were flames from 'horizon to horizon'. For Juday, that was only the start: 'Within a period of decades every summer could be warmer than 2004.' And the growing problem of wildfires is unlikely to stop with boreal forests. In 2008 scientists published research suggesting that the frequency of wildfires on the Arctic tundra was likely to increase as well.[25]

The implications of the vulnerability and destruction of Alaska's boreal forests are as widespread and as worrying as the implications of the melting of permafrost. What is true for the boreal forests and tundra of North America is likely to be equally true in Russia. Given the role of boreal forests in the regulation of global climate, their disappearance is a cause for global concern. The idea that global warming might accelerate tree growth, thereby increasing carbon dioxide uptake from the atmosphere and helping to balance out anthropogenic emissions, now seems rather naive. The picture emerging from current research suggests that changes in the Arctic are likely to accelerate the pace of global warming rather than slow it down.

Given the complexity of our climate system – with so many inputs, variables and processes in play – global policy on climate change is bound to be made under conditions of some uncertainty. But reducing the bounds of that uncertainty should allow for a more informed balance to be struck among the range of options – mitigation, adaptation, preparation, migration – which the world will have to adopt to manage climate change in the future.

These are not abstract questions: get it right and we will save money and lives, get it wrong and we will waste money and lose lives. The case for supporting Arctic scientific research, therefore, is unanswerable. And the case for supporting its internationalisation – to allow for the best global talent to emerge, and to allow for a common scientific understanding of climate change to be forged – is compelling.

NY-ÅLESUND, NORWAY, 79° North: The air tastes clean at the Zeppelin atmospheric station, perched on a mountain-top at 78°54′ North, on the north-east side of Svalbard. On three sides of the Zeppelin station lies a succession of jagged ridges poking through fields of snow. On the fourth side, down at sea level, 470 metres below, lie the scattered buildings of the Ny-Ålesund base itself, the northernmost permanently inhabited settlement on earth. I draw the cold Arctic air into my lungs and hold it there for a few seconds before exhaling. 'People don't get sick here,' says Dorothea Schulze, the German scientist who rides the cable car up the mountain every day to make sure that the sensing equipment – from Switzerland, Sweden and South Korea – is functioning smoothly. Today Dorothea has some additional work: replacing a couple of the air filters which capture and record variations in atmospheric pollutants.

Inside the Zeppelin there's equipment to measure almost every conceivable atmospheric variable. Stockholm University has installed equipment to look at carbon dioxide concentrations, which past data show to have risen more or less continuously – although with natural seasonal variation over each twelve-month period – since the Zeppelin station opened in 1989.[26] The Norwegian Institute for Air Research (NILU) is taking measurements of methane concentrations in the air above Ny-Ålesund every fifteen minutes. Other equipment measures changes in levels of atmospheric pollutants, covering everything from mercury to the toxic cocktail of man made pollutants known principally by their acronyms: POPs (Persistent Organic Pollutants), PCBs (Polychlorinated Biphenyls), PCDDs (Polychlorinated Dibenzo-p-Dioxins) and, lest we forget, PCDFs (Polychlorinated Dibenzofurans).

For atmospheric scientists, Ny-Ålesund's remoteness is a virtue. Here, thousands of miles from the great industrial heartlands of Europe, Asia and North America, scientists are measuring overall changes in the composition of global atmosphere which would be masked by local variation if the same measurements were taken further south. The relatively pristine air of the Arctic is helping scientists understand the pollution of the mega-cities of the south – in New York, Mumbai and Beijing. For other scientists, it is the combination of Arctic conditions with the accessibility of up-to-date experimental facilities that make Ny-Ålesund such an extraordinary place to do science. Glaciologists and permafrost experts only have to hike a few miles away from Ny-Ålesund itself to find a suitable area for study. Those studying how marine life operates in the fragile conditions of the Arctic – and how it may be affected by global warming – have the use of the King's Bay Marine Laboratory, opened in 2005, and partly funded by ConocoPhillips, an oil company.

And then, of course, there is the unquantifiable benefit of different types of scientist from different countries and scientific backgrounds meeting in one place.[27] If scientific discoveries are partly brought about by unplanned chance encounters, then surely it is a good thing that there is such a range of nationalities at Ny-Ålesund and that fields of research vary from measuring levels of Persistent Organic Pollutants (POPs) in the atmosphere to looking at how the blood of spider crabs is adapting to changes in ambient sea temperature.[28]

The current internationalisation of Ny-Ålesund, with national scientific stations from ten countries – Britain, China, France, Germany, India, Italy, Japan, Korea, the Netherlands and Norway – is principally a reflection of a growing understanding of the global usefulness of Arctic science. But it is also a reflection of the desire of many countries to make sure they have a national presence in the Arctic.

That the main presence on Ny-Ålesund is Norwegian is no surprise. For a start, Spitsbergen falls under the sovereignty of Norway – albeit within limits set out in an international treaty signed in 1920. The Arctic is a key part of Norwegian national identity. 'We have a long tradition in the North,' says Trond Svenøe, manager of Norway's Sverdrup research station, looking out of his window at a bust of Roald Amundsen, the great explorer. 'When we got independence in 1905 we were a small and poor nation,' he tells me, 'so we had to show what we could do.' Ny-Ålesund itself used to be a Norwegian coal-mining settlement. Most of the buildings – and a short stretch of a railway line used to carry coal from the mines down to the shore – date from that time. But since a major mining accident forced the mines to close in the early 1960s – toppling the Norwegian government of the time as well – Norway found a new use for its northernmost outpost as a scientific research establishment. The King's Bay Company, the original owner of the coal mines, has been transformed under public ownership into the organisation responsible for the overall logistics and management of Ny-Ålesund. Much of the oversight for the science done at Ny-Ålesund, however, comes from the Norwegian Polar Institute's Sverdrup station.

Beyond the Norwegians, most of the longest-established research stations at Ny-Ålesund are controlled by the 'old' Arctic research powers, in particular Britain, France, Italy and Germany.[29] There are large variations in the scale and type of scientific research that those countries are conducting in Ny-Ålesund. Much is determined by the priorities of national funding bodies and the efforts and interests of particular individuals. The biggest European station, combining French and German efforts in a symbolic gesture of Rhenian cooperation, is open year-round.[30] The British station, in contrast, is only open for the summer. The Italian station has suffered from research cutbacks in recent years – and it's closed when I visit.

The Dutch scientific research station, really just a couple of small huts crammed with clogs and scientific papers, is largely the work of one man: Dr Marten Loonen, from Groningen. 'First time I came up here I just bought a ticket to Longyearbyen [the main town on Spitsbergen, half an hour's flight south] and they put me up in the old school house, without water,' he remembers. Returning almost every year through the 1990s, by 2003 Loonen had managed to scrape together the money to officially open the Dutch research station. This year he's brought up his young son to help him for a few weeks in his research on barnacle geese, a pest for Dutch farmers and a potential carrier of the H5N1 virus.

So much for the European presence at Ny-Ålesund, the legacy of Europe and America's historic predominance in Arctic science. It is the growing presence of Asian scientists at Ny-Ålesund which may point to the future. In the past few years China, India, Japan and South Korea have all established or expanded their presence on Spitsbergen. In the process, Ny-Ålesund's research community has been broadened and enriched.[31]

China's presence at Ny-Ålesund is symbolic on many levels, representing the entry of the world's most populous country into an elite polar club previously dominated by the West and further indicating that China's scientific and political interests span the globe. The opening of the 'Yellow River' station is part of a much broader Chinese drive in both Arctic and Antarctic science and politics. Politically, China has beefed up its role in the Arctic by becoming an observer member of the Arctic Council; commercially China has taken a growing interest in gaining access to Siberian oil and gas supplies and the potential of a trans-Polar maritime route. Scientifically, China has invested in research at both the North and South Poles, funding a series of expensive Arctic and Antarctic cruises and developing its Antarctic research infrastructure. As one European scientist puts it, 'Money is clearly no object.' According to the acting director of the 'Yellow River' station, China's commitment to polar research is only likely to increase. 'The government is putting a lot of money into infrastructure to support the scientists,' he tells me, handing me a glossy brochure on the recent achievements and future plans of the CAA (Chinese Arctic and Antarctic Administration).

Sitting around the dining table at 'Yellow River', one of China's Arctic researchers, Wei Luo, explains her country's interest in the Arctic in terms of improving understanding of how China may suffer as a consequence of global warming. 'It's sure that climate change will have an impact on us in China,' she tells me. 'Lots of scientists in Tibet are looking at the glaciers which feed into our rivers and lakes,' she says, 'and day by day the administration is getting to know how climate is changing – more typhoons around the coastline and that kind of thing.'[32] The Arctic might seem far removed from China – but, increasingly, China doesn't see it that way.

The Indian 'Himdari' station at Ny-Ålesund is representative of India's wider scientific ambitions to be a leading research nation in its own right, rather than sending so many of its brightest and best research scientists to study and work in the United States or the United Kingdom. National symbols are important: while the doorway to China's station is flanked by two sculptures of Chinese dragons, the entry to India's station is marked with a bas-relief of Ganesh. But 'Himdari' is not just about prestige. Using Arctic science to understand India's own environmental vulnerabilities is a key motivation.

My host at 'Himdari' is the affable Dr C. G. Deshpande, a senior scientist from the Institute of Tropical Meteorology in Pune, near Mumbai. His colleagues at Ny-Ålesund come from the National Centre for Antarctic and Ocean Research in Goa and the universities of New Delhi and Lucknow.

Deshpande's work involves monitoring the electrical charges of Arctic air, using components he devised and built himself, standing in the springy tundra behind Sverdrup research station. Dr B. C. Arya is looking at carbon monoxide release from the Arctic snow pack and at ozone concentrations, which may ultimately improve understanding of atmospheric chemistry, an issue of major importance in some of India's more polluted cities. Rakesh Mishra and Dr Druv Sen Singh are working on the glaciers around Ny-Ålesund and comparing their dynamics with those of glaciers that Mishra has worked on in the Himalayas, up near Manali, north of Chandigarth. 'Compared to the Himalayas, this is nothing,' he tells me.

Warming faster than other parts of the world, and hosting highly vulnerable ecosystems which are already suffering the consequences of global climate change, the Arctic is nature's front line. It is also nature's storehouse – a mine of biological substances and processes which we barely understand, and which may yet be of tremendous industrial, pharmaceutical and medical use. The fact that the Arctic is relatively unknown – like the Amazon or the deep sea – makes it all the more likely that we will find the unexpected.

The hunt for active biological substances and processes in the Arctic is known as bioprospecting, inviting comparison with the search for gold or oil. It has yet to garner the attention won by climate science, but bioprospecting is a growing element of the practical day-to-day science conducted in the Arctic region. The United Nations has already identified forty-three commercial enterprises with bioprospecting operations in the Arctic.[33]

Most interest in Arctic bioprospecting lies in enzymes, microorganisms and proteins which operate at low Arctic temperatures. Arctic micro-organisms may be useful in bioremediation of polluted land and water – speeding up the biodegradation of an oil spill, for example. Naturally occurring Arctic antifreeze proteins which allow Arctic marine life to survive at temperatures near or below 0° Celsius (32° Fahrenheit), may be useful to the commercial food industry – Unilever is developing ice creams containing antifreeze proteins from Arctic fish, which will allow the fat content of the ice cream to be reduced. The same proteins might also have medical applications in the cryo-preservation of cells, bodies and blood. In the future, it may become possible to use Arctic marine bacteria to manufacture long-chain polyunsaturated fatty acids – normally known as fish oils – thus reducing their cost and increasing their availability. Study of Arctic hibernation, meanwhile, may improve our ability to reduce damage caused by heart attacks or strokes.

To Trond Svenøe, manager of Norway's Sverdrup station at Ny-Ålesund, the notion of commercialising Arctic research is anathema. For him and for others, Arctic science should be held to higher values. Nansen's old dream of the Arctic as a source of redemption for the modern world fits uneasily with the idea of individual profit. In the end, however, attitudes towards Arctic science will have to strike a balance between the image of Arctic science as a non-commercial symbol of international cooperation, and the reality that science in the Arctic is becoming more and more like science in the rest of the world, subject to the same demands, pressures and conflicts.

Aside from the possibilities offered by bioprospecting, the Arctic offers up unique conditions for preservation. Since 2008, seeds from all over the world have been shipped to Spitsbergen, and placed in a specially designed vault more than 100 metres inside a mountain, maintained at between -18° and -20° Celsius, behind a set of steel doors and two airlocks. Designed to contain samples of up to 4 million crop varieties, the idea of the Global Seed Vault is to provide a backup system for the world's genebanks, needed to ensure that global agricultural biodiversity is maintained. It is, in essence, a manmade Arctic refuge.

In addition to the obvious security benefits of placing such an important facility inside a frozen mountainside, several hundred miles above the Arctic Circle, there's a natural benefit too: even under the worst-case scenarios for global climate change, the vault will remain naturally frozen for at least 200 years. 'It's the best-insulated freezer in the world,' says Cary Fowler, the American who heads the Global Crop Diversity Trust, one of the organisations jointly responsible for the Global Seed Vault.

The location of Fowler's offices – in the white marble headquarters of the Food and Agriculture Organisation, in Rome – is a reminder that maintaining global agricultural biodiversity is ultimately about maintaining the ability of the world to feed itself. The range of challenges is daunting: 'population growth, increasing global affluence [which tends to mean people have more meat-rich and therefore agriculturally intensive diets], water stress and then those traditional problems of disease and pests'. Then comes climate change. Research from Stanford University and the University of Washington suggests that global warming will have huge impacts on crop yields all over the world.[34] Disastrously hot summers – previously experienced as freak occurrences – are likely to become the norm.

It's not just a matter of allowing crops to migrate north to cooler growing areas: 'Most of the crops are sensitive to photo periods depending on daylight – so if you move crops north to find the right temperature you'll be altering other things in the crop's environment which determine when it germinates, when it grows – everything will fall out of sync.' 'There's no quick fix,' says Fowler. 'We've got an agricultural system that's evolved over ten thousand years,' he argues, 'so to think that we can just rearrange it quickly and cleanly over the next fifty years is truly naive.' The purpose of the Global Seed Vault on Spitsbergen is to make sure that the adaptive capacity contained in the hundreds of thousands of naturally occurring varieties of crops is maintained. By working out which crops grow where now, and then matching those characteristics against projections of future climate in different regions of the world, scientists can work out what will need to be done for agriculture to adapt.

Although the popularisation of the Global Seed Vault as a 'doomsday vault' has been helpful in raising the profile of crop diversity, in a sense that misses the point. 'We're not some doomsday cult standing around saying, "Repent, the end of the world is near,"' Fowler says. When it comes to maintaining crop diversity, the true threats are far more immediate and more mundane than a doomsday scenario would suggest. 'For crop diversity, doomsday comes every day,' Fowler tells me. As it stands, many countries have gene banks – which the Global Seed Vault is planned to complement, not replace – but many of the most important gene banks are vulnerable to physical disruption: in Colombia, Peru, the Philippines, Syria. Fowler shows me a photo of one of the world's major regional seed banks after a flood a few years back. 'They lost a lot of diversity that day – crop varieties which are now as dead as the dinosaurs,' he says. 'This happened in September 2006. But do you remember doomsday happening then? I don't.'

In the mid-1980s, Norway had proposed internationalisation of its own crop diversity backup system, deep in a Spitsbergen coal mine. Suspicions about ownership of the material, and technical problems with the facility, prevented the proposal from taking off.[35] But the idea of some kind of failsafe backup remained. When the idea resurfaced in the early 2000s, Spitsbergen remained the obvious choice – not just for environmental reasons but because the reputation of Norway's government as stable and transparent would be helpful in persuading other countries to send samples of their valuable biological natural resources to an international facility. 'If we'd tried to do this in another country,' Fowler says, 'we would have found it far harder to get the agreement of both developing and developed states.'

Strangely enough it was Hurricane Katrina which, for Fowler, provided part of the final impetus to move from thinking about a Global Seed Vault to actually building one. 'There were a lot of recriminations after Katrina,' he remembers:

> Everyone knew that, sooner or later, a force 5 hurricane hitting New Orleans was inevitable and that, if it did, the levees wouldn't hold. The experts knew all this. And what did they do about it? So, Henry [Shands] and I were talking and we thought, sooner or later, won't we be in the same situation with crop diversity. Sooner or later, we're going to lose an entire gene bank – maybe in a regional conflict we'd lose several. And then there are accidents which take place every day – mismanagement, equipment failures. Silently

we're losing diversity every day, in ways that don't ever hit the headlines. So, what are we going to do about it?

In 2006, drilling began and a tunnel was bored into a mountainside near Longyearbyen. On 26 February 2008, the Global Seed Vault opened for business.

Fowler is understandably proud of what has been achieved. 'It's the first time in my lifetime that countries have got together to do something this far-sighted,' he tells me. As one Arctic storehouse of biodiversity is being raided in order to provide short cuts for industrial, pharmaceutical and medical science, another Arctic storehouse with the potential to help global agriculture adapt to a different, climate-changed world, is being accrued.

In itself, fluctuation in climate – in the Arctic and elsewhere – is nothing new. Millions of years ago, parts of the Arctic which cannot now support anything more than the barest of vegetation were covered in swamps and forests. Within the last few millennia, the rhythm of spread and retreat of human populations in the Arctic seems to have followed a pattern of cooling and warming periods in Arctic history. Within the span of a single lifetime, conditions in parts of the Arctic have warmed and cooled and warmed again.

What is different this time around is the human element, not only in terms of manmade influences on the process of climate change but also in terms of the human consequences of climate change. Natural variations in climate are overlain by human influences – emissions of greenhouse gases and deforestation (which tends to reduce the capacity of the global environment to absorb rising concentrations of carbon dioxide). Just how important anthropogenic factors are compared to natural factors in contributing to observed climate change is still open to discussion, but the overwhelming majority of climate scientists believe they are significant. If this view is correct, the long-term trajectory of climate depends to a significant extent on whether concerted global action is taken to reduce future greenhouse gas emissions.

Discussions of global warming have begun to shift from questions of causality in the natural world to questions of the consequences of global warming on human society. What does global warming mean for patterns of global trade or for the global balance of power? What balance will political leaders strike between mitigation of climate change and adaptation to it – if, that is, a global deal on such issues can be struck at all?[36] What countries are likely to bear the greatest costs of climate change, and – more controversially – are there individuals, businesses or countries which may actually benefit from climate change in some way? To paraphrase a famous British politician, 'the era of procrastination' is being replaced by 'a period of consequences'.[37]

Globally, the consequences of climate change – whether the changes themselves are principally manmade or not – represent a costly disruption to human society, threatening the lives and livelihoods of millions, if not billions, of people. Overall, the prognosis is bad. If the projections of scientists are correct, then nowhere will be unaffected by global warming. Low-lying areas of the world will be vulnerable to rising sea levels of a metre or more over the century. Some water-stressed areas of the world are likely to face more frequent and more serious water shortages, accentuating the risks of domestic unrest and cross-border tension.[38] Forest fires may grow more frequent and less controllable. Agricultural areas will be forced to adapt to changing weather patterns by either changing the crops that they grow or, in some cases, by abandoning their lands altogether. Fishing communities will be forced to adapt to the changing location, size and type of fish stocks. As temperatures increase globally, tropical diseases are likely to become more prevalent, moving into cities and towns once invulnerable to them by virtue of latitude or altitude. Tropical storms will become more frequent and more powerful.

For many Arctic communities, climate change is a death warrant. Rising sea levels, increased coastal erosion, diminishing sea ice and melting permafrost will undoubtedly damage Arctic infrastructure and threaten traditional 'life ways' which rely on specific environmental conditions, such as sea ice from which to hunt. For others, however, the prospect of an ice-free Arctic offers the opportunity for shipping routes, economic development and Arctic prosperity.

Ultimately, of course, it is only possible to make a judgement on the overall costs and opportunities of climate change in the Arctic by taking a view on how those costs and opportunities should be valued. Any such view is necessarily subjective. How can we compare, for example, the risk of losing the traditional Inupiat whale hunt – something of incalculable cultural value to a small number of people – with the possibility for cheaper global trade routes through an ice-free Arctic Ocean? Putting value judgements aside, the consequences of global warming in the Arctic will involve both tragedy and success, destruction and innovation, risk and opportunity and, for better or worse, losers and winners. What is certain is that a static vision of the Arctic is unsustainable in an era of rapid change and shifting climate.

7 Consequences: Reworking Geography

'Ah, for just one time I would take the Northwest Passage
To find the hand of Franklin reaching for the Beaufort Sea
Tracing one warm line through a land so wide and savage
And make a northwest passage to the sea'

'Northwest Passage', Stan Rogers, 1981

'Our government is considering the issue of opening the Northeast Passage – results will be conveyed to all parties very soon'

Arthur Chilingarov, Russian Arctic expert, 1990

'An old dream of establishing new shipping routes in the North is coming true'

Report to the Icelandic Foreign Ministry, 2005

In climatological terms, the consequences of a warming Arctic are global. But in human terms, the direct impacts of Arctic warming are, at least initially, mostly local. Many of the communities in the Arctic are particularly vulnerable to global warming because their cultural traditions and 'life ways' depend on a relatively unchanging set of environmental circumstances. Others, more prosaically, are vulnerable because they are built in areas which will be physically altered by climate change. The foundations of structures built on permafrost will become unstable. In some places, low-lying coastal areas may be exposed to rising sea levels or, as the protection of sea ice disappears for much of the year, to higher levels of coastal erosion.[1] In others, increasing river discharge may cause inshore areas to become clogged up with silt.

The US Army Corps of Engineers estimates that, as a result of coastal erosion, the largely Inuit communities of Newtok, Shishmaref and Kivalina will all need to be moved over the next ten years, at a cost of up to $455 million.[2] Although that seems a lot of money for three settlements with a total population of around little more than a thousand, for those who have lost their homes, it's hardly enough. In some respects the costs of moving Newtok, Shishmaref and Kivalina are the tip of the iceberg: they are nothing compared to the much greater costs which will be incurred as a result of climate change in the Arctic region as a whole, not just to coastal communities, but potentially to inland roads, railways and pipelines constructed across newly vulnerable permafrost. In Alaska alone, the state has identified hundreds of communities at additional risk as a result of global warming. A Canadian government report on the consequences of climate change in the Canadian Arctic warned of wide-ranging challenges for infrastructure in the future, alongside potential gains from improved maritime access.[3]

Calculating the potential, direct economic costs of climate change in the Arctic – damaged infrastructure, destroyed homes, and so on – is difficult. It means combining two layers of uncertainty: uncertainty about the speed and extent of future global warming, and uncertainty about the extent of vulnerability of buildings and communities to a given set of physical changes in the natural environment. In some ways, it's a quixotic task too – after all, we'll only find out the true economic value of lost housing or lost infrastructure when we have to replace it. Indeed, in many parts of the Arctic, that's exactly the approach taken – dealing with the consequences when they appear, rather than trying to anticipate them and prepare for them. As long as the direct consequences of climate change are relatively small and manageable, that is, perhaps, an understandable approach. But what about the future? Getting a sense of the scale of the problem a decade or more from now is key to understanding what our policy options are for dealing with it.

'You could call your chapter "structural uncertainty,"' puns Pete Larsen, a senior researcher at the Nature Conservancy in Alaska. We're sitting in a restaurant in downtown Anchorage on a Friday night, 5 p.m., and the bar is beginning to fill up with the after-work crowd. There's a warm buzz to the place, complemented by country music from what Alaskans call the 'lower 48', the rest of the United States.

In 2007, Larsen and a group of researchers at the University of Alaska in Anchorage published a report on how climate modelling and economic modelling could be used to estimate the costs of climate change to Alaska's public infrastructure, at risk from erosion, permafrost melt or other changes in the physical environment. Providing sufficient data for the model was hard enough, requiring 16,000 pieces of public infrastructure to be catalogued and (approximately) valued. Actually getting a result, using a range of different climate change scenarios and 100 statistically simulated runs of the model, was harder. 'It took something like 28 hours to run 100 Monte Carlo simulations up to 2080,' Larsen tells me. 'We burned out two processing boards.'

The final estimates that Larsen and the team came up with – between $3.6 billion and $6.1 billion up to 2030 (with the potential for costs to be 'billions of dollars higher' without adaptation) – suggested that the state could be saddled with around $5,000 to $10,000 of additional costs per inhabitant.[4] Spread over a number of years, that doesn't seem an overwhelming expense, even taken in addition to the $32 billion which the team estimated would have to be spent whether there was climate change or not. But, as Larsen is keen to point out to me, those numbers only covered public infrastructure, not privately owned homes and infrastructure, which is of far greater value. Moreover, some of the valuations and assumptions made in the model were, Larsen now thinks, too conservative. 'You could easily double or triple that number and publish a new paper now,' he says. And, of course, you'd have to multiply that number several times again to get anywhere close to the potential costs of damage caused by climate change in the Arctic as a whole.

In some ways, this is frontier economics. Given the combination of uncertainties involved, and the difficulty of getting the right data, no one is expecting a clear-cut, single-figure, wholly reliable result. But then, that's not the point. Building an economic model estimating the direct costs of climate change in Alaska is not just about coming up with a final number, it's about starting a debate. In that sense, the report has been a success. In September 2007, Governor Sarah Palin set up a climate change subcabinet to develop a strategy for how the state manages the local consequences of global warming. Large amounts of federal money may be at stake: in Washington, Alaska's Senators and Congressmen have been trying hard to make sure that Alaska receives a generous share – tens of billions of dollars – of any federal appropriations for adaptation measures.

Interest in the direct economic consequences of Arctic climate change is not limited to Alaska and it's not limited to government. Mining companies and oil companies across the Arctic – in Canada, Greenland, Norway and Russia as well as the United States – want to understand the possible impacts of climate change on both their existing and planned infrastructure. In some places, those impacts may actually be helpful – most obviously in terms of improving sea access. In others, however, the impacts may be costly. What happens if low-lying areas such as the Yamal Peninsula, a region of the Russian Arctic touted as a major source of natural gas in the very near future, begin to flood? What happens if ice-road access to mines and communities in Arctic Canada becomes less reliable? What happens if Russia's Siberian pipeline network, much of which is already in a poor state of repair, is destabilised by thawing permafrost?

In Norway, Grete Hovelsrud, an Arctic anthropologist at the Center for International Climate and Environmental Research Oslo (CICERO) takes a community-focused approach to understanding the consequences of climate change in the Arctic, leading an international project known by the acronym CAVIAR (Community Adaptation and Vulnerability in Arctic Regions), due to produce its final report in 2010.[5] 'We start by going to the communities and asking them what we should be looking at, rather than sitting in an office somewhere,' Hovelsrud explains. To some natural scientists, the social scientific approach, focused on what communities themselves view as their vulnerabilities, is hard to grasp. But to Hovelsrud it makes total sense. It is, she argues, the only way of matching up the natural science of climate change with the reality of how people will actually experience its consequences.

Not all the consequences of climate change which Hovelsrud is looking at in Norway – particularly for the city of Hammerfest, on Norway's Arctic northern coast – are as dramatic as those hitting Newtok, Shishmaref and Kivalina. More days when the temperature is just below freezing is likely to lead to heavier snow, clogging up Hammerfest's roads and increasing the risk of mini-avalanches on the hills around the city. Higher snowfall may not bring death and destruction to Hammerfest – and the problem is hardly insuperable – but managing it will require investment, changes in regulation and new building designs. These might be relatively small costs, but they are small costs which add up – particularly if they are replicated in the much more populous and potentially far more vulnerable Russian Arctic.

There are other possible consequences of Arctic warming identified by Hovelsrud which have more obviously negative impacts. Climate change may make Arctic storms more frequent and more powerful, causing problems for the airport, the port and, importantly, the most expensive piece of infrastructure ever built north of the Arctic Circle: the Melkøya liquefied natural gas (LNG) plant owned by StatoilHydro, Norway's largest oil company and one of Hammerfest's largest employers.

'When you speak to Statoil they say they have no problems,' Hovelsrud tells me. But a photo of Melkøya draped in snow, looking more like an ice palace than a multi-billion-dollar LNG plant, tells a different story: This is the facility after a big storm, from the direction they expect to be the prevailing one in the future.' 'Half of the plant didn't work,' Hovelsrud says. Even a temporary shutdown costs money and discourages future investment.

For many of the Inuit communities of Alaska, Canada and Greenland and the European Arctic, the consequences of climate change pose an existential threat to homes, livelihoods and culture. The locations of Inuit settlements, and the means to sustain them, are a function of the natural environment. As global warming changes the Arctic, it has impacts on almost all aspects of Inuit life – from the structural security of housing across the Arctic, to more difficult grazing conditions for reindeer in the European Arctic and changes in migration patterns of caribou in the American Arctic. For the Inupiat community of northern Alaska, whale hunting, traditionally conducted from solid winter pack ice, has become more dangerous. As Arctic pack ice becomes thinner and more frequent stretches of open water – known as 'leads' – have opened up, the difficulty and risks of traditional whaling increase. The Inupiat were amongst the first to warn of changes in Arctic sea ice, back in the 1990s.[6] As the sea ice retreats they are amongst the first to suffer the consequences.

There's a distinctly patronising element to many southern ideas of the fate of these communities, reinforced by the traditional romantic view of the Inuit as a once-noble society, now helpless, perhaps irretrievably spoilt by its brushes with modernity. But despite (or perhaps because of) that tragically inclined image of the Inuit – laden with a sense of cultural distance – there's also the possibility that Inuit voices, once completely ignored, may at last be heard. Some hope that the experience of Inuit communities in the Arctic – and the moral authority which derives from those experiences – may draw attention to the immediate challenges of adaptation, and galvanise global measures to mitigate climate change.

Since 2004 a political alliance has begun to emerge between two sets of regions and states which are particularly vulnerable to climate change: Inuit communities in the Arctic and a group of nations, mostly tropical islands, known as the SIDS (Small Island Developing States). The link between them is fundamental: 'As we [in the Arctic] melt, the small developing island states sink,' as one Inuit leader has put it.[7] Together, the SIDS and the Inuit have greater political influence than they ever could separately. Suitably enough, their alliance is called 'Many Strong Voices'. John Crump, the Ottawa-based polar issues coordinator of the United Nations Environment Programme tells me that, nonetheless, 'It's hard to be heard, so we have to be strategic.'

Crump has been working on Arctic issues since he moved to the Yukon in the 1980s. Then, the Inuit were only beginning to emerge into the global political limelight, largely as a result of land settlement issues in the United States and Canada throughout the 1970s. In 1999, the Inuit communities of Arctic Canada won the separation of Nunavut from the Northwest Territories, leading to a much greater degree of Inuit self-government within Canada. Greenland, which has enjoyed Home Rule (within Denmark) since 1979, took a step closer to full independence in 2009 when a new, enhanced Self-Rule regime entered into force. As global attention has turned to the Arctic in recent years,

Crump sees an opportunity for the difficult adaptation issues facing northern communities to help get across a broader, global message: '450 ppm [concentration of atmospheric carbon dioxide] is too high.'

Though the total population of affected Inuit communities is small in global terms, Crump still believes in the power of 'moral suasion' to mobilise political action to mitigate global warming and to help local communities adapt to its consequences. 'It's happened before,' he tells me, citing the role of indigenous communities in getting international agreement to manage (and eventually eliminate) Persistent Organic Pollutants (known as POPs).[8]

But, for some, the impacts of climate change on indigenous communities – while unquestionably taking place, and while undoubtedly of great concern – can deflect attention from a broader picture of Arctic changes. 'Another thing which is totally oversold is the issue of indigenous peoples,' says Olav Orheim, former director of the Norwegian Polar Institute. 'You get good pictures,' he tells me, 'and you intuitively understand that they will be affected by climate change, and you intuitively understand that that is bad – which is correct, I'm not saying that it's not – but it's only a small part of the total story.' For people such as Orheim, the vulnerabilities of the existing local population need to be set against other, grander, global consequences – both negative and positive – which may arise from Arctic warming.

In some respects, Orheim's right. Beyond the direct and local impacts of Arctic warming – both quantifiable and unquantifiable – there is a range of indirect impacts – on fisheries, on agriculture, and on shipping – which may have yet more wide-ranging consequences for the future history of the Arctic.

Global warming should theoretically increase rates of biological marine productivity. That doesn't mean that the fishermen of the North are automatically going to become rich as a result of climate change. The health and size of fisheries are already subject to tremendous natural variability. Climate change is an additional complicating factor. Complex maritime ecosystems will not respond linearly to changes in sea temperature. Warmer seas make it far more likely that non-native invasive species alter the internal dynamics of the marine ecosystem. Ocean acidification – an observed byproduct of higher atmospheric carbon dioxide concentrations – will affect some Arctic species, just as it affects maritime species around the world. The retreat of sea ice will tend to harm species that depend on sea ice, and strengthen sub-Arctic species that don't.[9]

While some species will adapt to a warmer Arctic, many won't. And although some fishing communities may gain from changes in the location and type of fisheries, most will need to be highly adaptive to survive. Climate change will lead to the migration of some fisheries across national maritime borders, raising the issue from domestic politics to international diplomacy. (Norwegian experts have already expressed concern that fishing grounds may be moving east, into Russian waters.) In the United States, the increased physical accessibility of the Arctic has led for calls for additional restrictions on Arctic fishing, so as to protect existing stocks for both local and outside fleets.

The possibility of additional biological productivity in the sea is matched by the possibility of additional biological productivity on land. The dream of northern prairies being opened up by global warming is not entirely far-fetched. Research conducted for the Canadian government suggests that global warming could bring a significant northward extension of land available for arable farming in Canada (though the study stopped short of the Arctic Circle itself, only looking at provinces below 60° North).[10] The study projected a 50 per cent gain in land suitable for continual cropping in Alberta, with much lower gains in Saskatchewan and Manitoba. Further north, in the Arctic Yukon and Northwest Territories, global warming may yet improve the currently dismal economics of Arctic farming.

If the world needs to produce more food, the Arctic is not the first place to look. There are other places where agricultural productivity could be more easily improved (not least in Africa). But with suitable crops, sufficient water, the right market incentives and the right investment, some Arctic and sub-Arctic areas of the world could profitably provide an additional source of food for global supply.

Perhaps the most important indirect consequence of Arctic climate change, profoundly reworking the economic geography of the North, is the increased accessibility of the Arctic to global shipping. As the natural environment of the North begins to change out of all recognition, the idea of the Arctic as an impenetrable physical barrier will be overturned.

According to the legend, an English trading ship called the *Octavius* began her voyage home from China in 1762. Attempting to cut short the return journey from Asia to Europe, the captain of the ship decided to sail north, rather than south around the coast of South-east Asia. He would attempt, like so many before him, a navigable Northwest Passage across the top of Arctic North America.

At first, things went well. Passing through the Aleutian Islands – a necklace of small islands draped across the mouth of the Bering Strait – the ship sailed up the west coast of Alaska, then a Russian colony. But somewhere off Alaska, as summer turned to autumn and as the Chukchi and Beaufort Seas began to freeze up, the *Octavius* became trapped in ice. A dream of wealth and fame turned to despair. One by one, the crew of the *Octavius* died of cold or starvation. But the ship survived, carried by the pack ice from east to west until, thirteen years later, in 1775, its remnants were discovered by a whaling ship, the *Herald*, off the west coast of Greenland.

The story is yet another tall tale inspired by the trials and tribulations of Arctic navigation. But in a sense, the doubtful veracity of the story of the *Octavius* serves to emphasise the enduring mystique and romance of Arctic shipping. The history of attempts to cut short the shipping route from Europe to Asia by sailing either across the top of America (the Northwest Passage) or across the top of Asia (the Northeast Passage, or Northern Sea Route) is littered with greed, impatience, loss, and defeat in the face of insurmountable natural obstacles.[11]

Nonetheless, the potential gains from a shortened commercial passage between Europe and Asia – particularly before the construction of the Suez Canal in the nineteenth century and the Panama Canal in the twentieth – meant that the idea of an open Arctic route was never entirely abandoned. In the twentieth century, the only country to successfully pursue the project was the Soviet Union. While the Northwest Passage undoubtedly retained its romance – the result of the heroic failures of the nineteenth century – it was the Northeast Passage which was prised open first.

From the 1920s onwards, Moscow invested millions of rubles and thousands of lives in the attempt to turn the north coast of Eurasia into a major shipping lane. Over the seventy-year span of Soviet history, the volume of freight shipped along the Northern Sea Route grew to over 100,000 tons in the early 1930s, to over 1 million tons in the early 1960s and to over 6 million tons by the late 1980s. Some freight was carried from the western ports of Murmansk and Arkhangelsk all the way east to Magadan and Vladivostok, Soviet outposts on the shores of the Pacific. But the vast majority of Arctic shipping in the Soviet Union was not trans-oceanic but regional – connecting the shores of the Barents Sea, the White Sea and the Kara Sea.[12]

Support from the Soviet Union's nuclear ice-breakers – the pride of Communist engineering – was key.[13] In 1977, the most powerful ice-breaker ever built, the *Arktika*, crushed its way through mile after mile of pack ice to become the first surface ship to reach the North Pole. It was ships like the *Arktika* – expensive, slow but hugely powerful – which made extensive shipping in the Soviet Arctic possible. The costs to the state were obscured by the confusion of targets and creative accounting which characterised the Soviet economy. In Moscow, Soviet leaders occasionally considered reducing their overheads by opening the Northern Sea Route to foreign shipping.

In the early 1990s, with the collapse of the Soviet Union and the implosion of the Russian economy, the rationale for much of the shipping along the Northern Sea Route vanished. At the Russian Central Marine Research and Design Institute (CNIIMF) in St Petersburg, Vladimir Vasilyev brings out a slide showing the number of transits along the Northern Sea Route between Europe and Asia over the last twenty years or so. In 1985, there were five; in 1993 there were thirty. Then, the numbers fell drastically. In 2001 there were just two transits along the Northern Sea Route. 'Now, there are basically none at all,' Vasilyev tells me. Despite a Japanese-funded project to look at the feasibility of commercial shipping along the Northern Sea Route in the 1990s, large-scale use of the Northeast Passage has been put on hold.

In the twenty-first century, things are beginning to look up for Arctic shipping, not least because one of the expected consequences of global climate change is that the Arctic will become relatively ice free (for part of the year) within a matter of years. Long before it is totally ice free in summer, the ice may have reduced sufficiently to make an ice-free shipping passage between the Pacific and the Atlantic. With ice-free summers, sea ice will, by definition, never be more than one year old – much easier to break than the multi-year ice which still covers a large area of the Arctic Ocean.

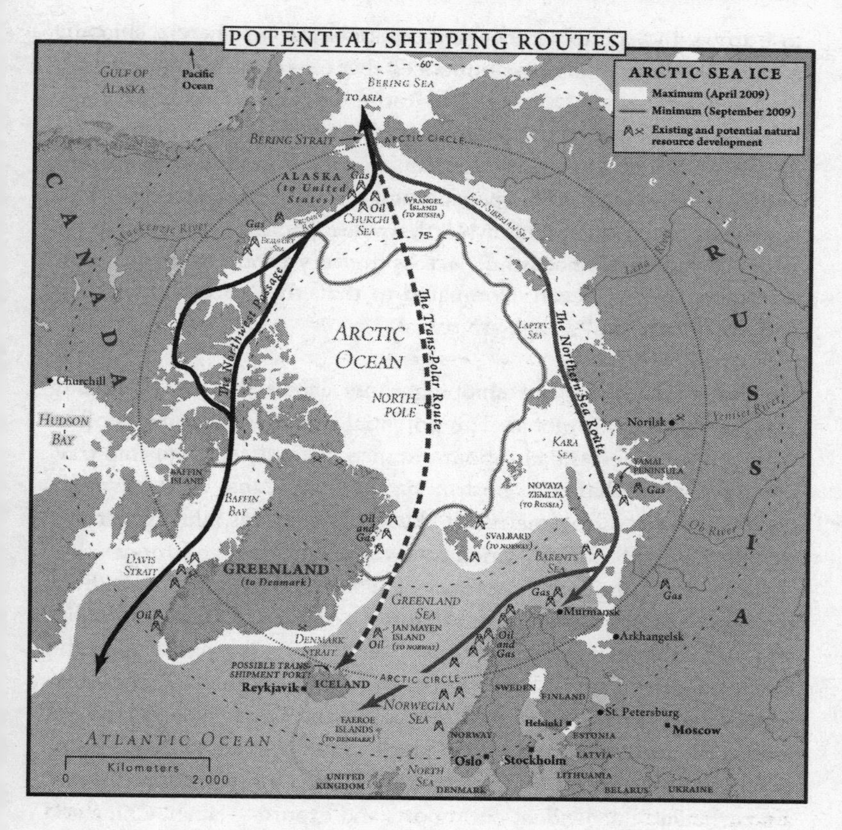

As in the past, most public attention in the West has focused on a Northwest Passage between the Pacific and the Atlantic, despite the obvious practical problems and environmental concerns arising from sending large ships through the narrow passageways and shallow waters of the Canadian Arctic. Though improved ice conditions and general economic development of the North may lead to increased shipping in the North American Arctic, the Northwest Passage is not the most likely candidate for a major trans-Arctic shipping route.

It's more probable that a Northeast Passage will assume that role, despite the physical drawbacks of a number of narrow straits and shallow seas (at least on the 'inner' Northern Sea Route). Vasilyev estimates that the current capacity of the Northern Sea Route is 5 to 6 million tons from Europe to Asia and 2 to 3 million tons from Asia to Europe. In September 2009, a German-based commercial shipping company, Beluga Shipping, announced that two ships had successfully transited the Russian Northern Sea Route, carrying an industrial cargo from South Korea to a customer in the Netherlands.[14] Beyond the 'inner' Northern Sea Route, an even shorter route between Europe and Asia may eventually open up as a direct conseqence of Arctic warming: the so-called Outer Northern Sea Route, running far to the north of the Russian coastline, across the very centre of what is now the frozen Arctic Ocean. Compared to that, the Northwest Passage and the Northeast Passage are trivial.

Excitement over a major shipping route across the Arctic, linking Europe and Asia, is global. The potential commercial benefits of an Arctic route are essentially about distance. Whereas a cargo ship travelling from Yokohama to Boston via the Suez Canal must currently cover some 12,865 kilometres of the world's oceans, an Arctic route would shave off 3,000 kilometres, saving time, fuel and money. But there are also considerable security benefits from an Arctic route. Both the Suez and the Panama canals are in zones of political instability and vulnerable to terrorism. They will require major capacity upgrades in the coming years. A route across the Arctic, particularly the deep-water Outer Northern Sea Route, would avoid those choke points and reduce the vulnerability of global trade to disruptions – intentional or otherwise – in central America or the Middle East. For countries that are particularly dependent on imports and exports – Japan, China and South Korea – Arctic shipping is in the national interest.

Countries lying along a prospective Arctic route between the Atlantic and the Pacific could also gain, both economically and geopolitically. In the Russian port of Murmansk, hopes are high that global warming will transform the importance of Russia's northernmost warm-water port, making it not just a port for prospective oil and gas developments in the Arctic, but also a shipping hub for trans-Arctic goods trade. At the other end of the so-called 'Arctic Bridge', the Canadian port of Churchill is touting itself as a North American terminal for trans-Arctic trade, connected directly to the Canadian railway system. OmniTRAX Incorporated, the Denver-based company which owns the port, has tapped Lloyd Axworthy, Canada's former foreign minister, to chair the Churchill Development Corporation, charged with promoting the port's expansion. In Iceland, meanwhile, a number of locations have been identified as possible sites for an Arctic transshipment port, along the lines of a (much smaller) Arctic Singapore.[15]

Improved Arctic shipping – without continuous ice-breaker support – is not simply a matter of waiting for the ice to disappear.[16] There are numerous other challenges. A warmer Arctic may be less icy, for example, but it may well also be more stormy.[17] Even when there's no ice in the sea, Arctic shipping will remain vulnerable to ice forming on deck, with potentially dangerous consequences. Taken together, these challenges explain why, when a former US Coast Guard captain and Arctic shipping expert was asked to brief Panama on how Arctic warming would impact their business, he told them not to worry too much – yet. It may only be a few years before high-value, time-sensitive cargos regularly transit the Arctic. But until the cost of dealing with Arctic conditions comes down, the bulk of low-value shipments will continue to use traditional routes.

As for the sea ice itself, while the general trend in Arctic sea ice may be towards a reducing extent and diminishing thickness, variability will still occur from year to year. Even once the Arctic Ocean is regularly ice free in summer, climate change won't banish ice from the Arctic in all places and for all time. Most of the Arctic will still freeze up in winter, and the dates on which it does so will vary from year to year. The distribution of sea ice will be different from one year to the next, just as it has always been.

Variability is the sworn enemy of the kind of regular, predictable, cost-efficient logistics on which the transport industry relies. From the point of view of industry, variability is a cost. Ice variability is another reason why, time-sensitive and high-value cargos aside, large-scale trans-Arctic shipping is a decade or more away.

However, smaller-scale Arctic shipping within the Arctic will become much more frequent, much earlier. In some cases there is simply no alternative. Any increase in Arctic oil and gas development, for example, implies an increase in Arctic shipping. Shipments in the Barents Sea and Kara Sea – particularly shipments of crude oil – will expand far faster and far earlier than a trans-Arctic trade route between Europe and Asia. The value of the potential oil and gas reserves justifies the expense associated with investing in ice-breaker support or ice-capable tanker ships.

Improved marine technology, spurred on by the prospects of opening up the Arctic, is making things much easier. The development of ice-capable container ships which do not require expensive breaker support could transform the economics of shipping in the Arctic during periods of the year when sea ice will remain. Several independent ice-capable container ships have already been built. Several more are on the drawing board. Climate change will not turn the Arctic into a year-round Mediterranean, but combined with advances in ship design, it may do just enough to make the Arctic a viable commercial sea.

HELSINKI, FINLAND, 60° North: When it comes to shipping across ice-filled seas, no one does it quite like the Finns. Although Finland has no coastline in the Arctic north – having lost its access to the Barents Sea as part of a peace deal with the Soviet Union in 1944 – not a single one of Finland's Baltic ports is totally free of ice in a hard winter. The shallow and relatively unsalty waters of the Baltic freeze far more readily than do the harbours of the Arctic Barents Sea to the north. Given the inadequate capacity of Finland's road and rail links to its immediate neighbours – Russia, Norway and Sweden – open access to the sea is a matter of economic and strategic survival. Without its ice-breakers and ice-capable ships, Finland would shut down in winter. The Finnish company, Aker Arctic, has been involved in either the design, testing or construction of some 60 per cent of the world's ice-breakers.

Aker Arctic's headquarters are at Vuosaari – a Helsinki suburb twenty minutes' cab ride from the centre of town made up of brightly coloured shopping centres and brand new office blocks. The company's offices are by the Gulf of Finland, between a thick pine forest and the sea. From the outside, Aker's offices look like the laboratories of a high-end pharmaceutical company, but inside everything is about shipping. A huge map of the Arctic, centred on the North Pole, dominates the entrance hall. More than a decoration, the map is a declaration of ambition – to open the Arctic for commercial shipping.

Göran Wilkman, Aker Arctic's Vice-President for Technology, ushers me into a room on the first floor. He apologises that the boardroom is occupied today: 'It's full of Russians,' he explains. A few minutes later a lady pops in with a large flask of weak coffee and a plate of pepparkakor, the spicy biscuits you get all over Scandinavia.

Things have changed since Wilkman started working in the Finnish shipping industry as a student in the mid 1970s. The industry has been through several cycles. Each bust has brought a wave of mergers, acquisitions and bankruptcies. An ever-increasing share of the actual process of building ships has moved east, principally to Korea and China. The European shipbuilding industry has become smaller and much more specialised.

These days, Aker Arctic itself does not build ships at all.[18] Instead, it acts as a design and consultancy company, generating advanced blueprints for ice-capable vessels. Test facilities at Vuosaari allow it to simulate a range of different ice conditions, and to see how various model hull designs and propulsion systems work in ice. 'Our main thing,' Wilkman explains, 'is the concepts at one end and then the testing at the other.'

It is a global business. Aker Arctic is helping to supervise and advise on projects from Japan, on the other side of the world, to the new ice-breaker being built for Estonia, on the other side of the Gulf of Finland. But most of Aker Arctic's business is with Korean companies – including Aker's largest shareholder, the STX shipbuilding company – and with the Russians. Finnish companies have long experience of dealing with the complexities of doing business in Russia. In the past, that was a matter of necessity, given the strong economic ties between the Soviet Union and Finland. Now it is a matter of commercial choice, particularly as Russian mineral companies with operations in the Arctic – from the oil and gas companies to Norilsk Nickel, the mining company – consider how best to get the riches of the Russian north to market. 'Everybody wants to go north,' says Wilkman.

The key, in his view, to expanding Arctic shipping operations lies in transport ships that can handle ice conditions without ice-breaker support. In general, building ice-capable container ships has meant compromising on speed in open water. The main innovation in Aker Arctic's approach is a propulsion system which allows their ships to run astern in ice, rather than forwards, using what amounts to a second bow, specifically designed for the job. 'An ice-breaking bow is one thing,' says Wilkman, 'but with a real double-acting ship with the same power you can get through two metres of ice and still be efficient in open waters.'

Wilkman shows me a photo of one of these new ships in action.

'The ice was sixty centimetres thick, but this ship could run astern at eight knots,' he says. Norilsk Nickel's *Arctic Express*, a 170-metre long ice-capable container ship designed by Aker Arctic, is one of five built for the Russian mining company which has made it independent of the ice-breaking services of the Murmansk Shipping Company (MSC). Instead of having to rely on MSC's expensive nuclear ice-breakers, Norilsk Nickel is now able to ship nickel direct from Dudinka – on the Yenisei River, flowing north from Siberia into the Arctic Kara Sea – to the port of Murmansk. Aker Arctic estimates that the new ships are saving Norilsk Nickel up to €80 million ($100 million) a year.

In addition to the double-acting boats, Aker-designed shuttle tankers, each weighing in at 70,000 tons, have been built for the Arctic operations of two of Russia's largest oil and gas companies: Gazprom and Lukoil. 'And now we're talking about trans-polar containerships,' says Wilkman, producing a cross-section of a ship that would be able to sail across the Arctic even in the kind of ice conditions which will continue to exist in some parts of the Arctic Ocean (and at some times of the year) long after ice-free summers have become a regular occurrence. (He tells me that, if an order were placed today, such a ship could be built within three or four years.) Finally, Wilkman pulls out designs for an Arctic liquefied natural gas (LNG) tanker – which could, at some point in the future, carry LNG from the vast gas reserves of the Arctic to markets in Europe, Asia and the United States.

Further development of Arctic shipping is a double-edged sword. The most obvious risk associated with any expansion in Arctic shipping is pollution. Oil spills are bound to rise. The ballast water of ships from outside the Arctic region threatens to introduce foreign species, upsetting the delicate ecosystems of the North. For the Inuit, Arctic shipping, along with the more direct consequences of global warming, will be disruptive – welcomed by some, and rejected by others. And while some governments may rejoice at the prospect of increased shipping in or near their territorial waters, overturning the equation of climate change with economic disaster, any major expansion of commercial shipping will be accompanied by the expense of improved surveillance and additional support vessels. The governance regime for Arctic shipping will have to evolve.[19]

The prize, however, is a realignment of the world's commercial geography, boosting the Arctic's economic and geo-strategic importance, and confronting the cultural myth of the Arctic's marginality head on. The key factor in making that realignment possible is climate change. Diminishing Arctic ice and increasing Arctic accessibility are the prerequisites for an Arctic shipping route alongside its more southerly competitors. But how this transition occurs (and how soon) depends on politics, power and, above all, economics.[20]

PART FOUR
Riches

'Let us be wise'

Patricia Cochran, Chair of the Inuit Circumpolar Council,
October 2007

8
The (Slow) Rush for Northern Resources

'I wanted the gold, and I sought it;
I scrabbled and mucked like a slave.
Was it famine or scurvy – I fought it;
I hurled my youth into a grave.
I wanted the gold and I got it –
Came out with a fortune last fall, –
Yet somehow life's not what I thought it,
And somehow the gold isn't all'
'The Spell of the Yukon', Robert Service

'This is the fifth time that the world is said to be running out of oil. Each time . . . technology and the opening of new frontier areas has banished the specter of decline'
Dan Yergin, Chairman of Cambridge Energy Research Associates, 2006

In the late 1840s the gold rush hit California, transforming San Francisco from a Pacific outpost into a boom town. In the early 1850s the southern hemisphere was mobbed with prospectors when gold was discovered at the Australian settlements of Ballarat, Beechworth and Bendigo. The nearby town of Melbourne grew into the British Empire's second largest city, laid out with grand buildings expressing all of Australia's confidence in its future. The 1860s saw a flood of interest in the crown colony of British Columbia, as thousands of Americans went north in search of a quick fortune. Finally, in 1897, just as a financial panic hit the world economy, news filtered into the United States of a gold find in the high north of Canada, just a degree below the Arctic Circle. In a time of economic crisis, the Klondike became a symbol of hope.

The first to make their fortunes on the Klondike were the professional prospectors of the upper Yukon, staking their claim to the shores of Eldorado and Bonanza Creek in 1896, months before news of the finds had found its way to the United States. But by summer of the following year the national newspapers were awash with stories of men such as Clarence Berry, a poor miner who had struck gold in the Klondike and returned to Fresno, California with $130,000. When the *Excelsior* steamship docked in Seattle in July 1897, the *New York Times* quoted the Captain as saying, 'Out of sixty-eight passengers there is hardly a man on board who has less than $5,000. One or two have over $100,000'.[1]

In 1897 and 1898, thousands followed the promise of Arctic riches and headed north, travelling by steamship from San Francisco or Seattle to the Alaskan port of Skagway, and then hauling themselves and their mining equipment over the Chilkoot Pass into Canada. Those who arrived in winter set up camp on the shores of Lake Bennett or Lake Lindemann, waiting until the spring melt to travel down the Yukon River to Dawson City, the boom town on the Klondike. (Embracing the spirit of American enterprise, a New Jersey resident and former soldier in the imperial German army, Charles A. Kuenzel, planned to avoid the logistical difficulties of land and sea transport by travelling to the Klondike in an airship of his own design.[2]) Veterans of previous gold rushes mixed with first-timers out to try their luck.

Most prospectors were disappointed when they got there. The best claims had already been staked, leaving only the more marginal areas, where the chances of finding gold were much lower. Some made their fortunes; most spent what little they had to survive. Food shortages in Dawson City led prices to soar. Salt was worth its weight in gold.[3] In winter 1897, conditions had deteriorated to the extent that US President McKinley considered sending a humanitarian relief mission to the Canadian Yukon. By 1899, just a year after construction began on a railway designed to open up the Yukon to further exploration, the Klondike rush was over. As prospectors' attention turned to more accessible regions of the North – the coast of western Alaska in particular – the plans were shelved. It would be another sixty years before Dawson City was connected by rail to the rest of the world.

Of all of the gold rushes of the nineteenth century, the Klondike rush of the late 1890s has left the strongest imprint on popular culture. Set against the dramatic background of the Yukon boom, the novels of Jack London were hugely popular not just in North America, but in the Soviet Union, where they served as parables of man's struggle against nature – a very Soviet theme – and as first-hand evidence of capitalism's tendency to crisis and collapse.

Even today, a folk memory of the Klondike still lingers over the North, inspiring popular fantasies of an Arctic resource bonanza. But in most respects, the resource booms of the twenty-first century are very different from the gold rushes of the nineteenth.

Whereas the nineteenth century gold rushes were chaotic and individualistic enterprises that relied on crude technologies affordable to anyone, today's rush for resources is an organised, corporate affair, involving technologies both highly complex and hugely expensive. The main players in the rush for northern resources in the twenty-first century are not individuals but companies – ranging from the established oil majors and minerals conglomerates to the smaller companies on the high-risk frontier of mineral development, gambling on the discovery of a single large deposit to make their owners rich. Only companies (both private and state owned) have the scale and expertise to extract mineral resources from the Arctic at a profit. Where individuals still pan for gold – as in parts of Alaska – they represent an attachment to a heroic past rather than a realistic expectation of striking it rich.[4]

In the nineteenth century, governments tended to view gold rushes as destabilising – encouraging unwanted migration, undermining public health and, on occasion, provoking a breakdown of law and order. With good reason. The history of resource booms is punctuated with violence: from the Eureka stockade at Ballarat in 1854 to the revolutionary activities of Josef Stalin in the oilfields of Azerbaijan in 1904–05. In the Klondike rush, the Canadian government scrambled to give sufficient resources to the North West Mounted Police, to ensure that the gold rush did not turn into humanitarian disaster or worse, a bridgehead for American annexation.

Today, governments are much more supportive of the development of Arctic resources, seeing them as a cornerstone of future prosperity and security. Arctic infrastructure projects – from pipelines to storage terminals – often require the financial and political support of the state. Ottawa is spending over $100 million on programmes to map the mineral resources of the Canadian Arctic, with a view to accelerating their development. The Kremlin has identified the extraction of Arctic oil and gas as key to Russia's continued status as an energy superpower. Greenland's Self-Rule government promotes exploitation of its Arctic minerals as an economic precondition for independence from Denmark. Only in the United States, torn between the competing political demands of environmental protection and energy independence, is central government support for Arctic mineral development unsure. (In Alaska, of course, backing for development, both from the state government and from the citizens of the state, is much higher.)

And whereas the Klondike rush was about gold – an inert store of value – the Arctic resources most coveted in the twenty-first century are the fuel of modern industry and wealth creation: oil, gas, nickel, and zinc.[5] Strategic necessity and the search for commercial profit have replaced individual acquisitiveness as the driving force behind the rush for Arctic resources.[6]

In one respect, however, the Arctic has changed little since the Klondike. For all the advances in northern engineering and geological mapping, the Arctic remains a relatively unknown corner of the planet. (As one American geologist explained, it will take fifty years for Alaska to be as well mapped as Vermont or New York, while the subsea geology of the oil-rich Gulf of Mexico is understood far better than the Arctic Beaufort Sea).[7] The Arctic remains a province on the geological frontier, more than a century after the Klondike rush and more than 200 years after the discovery of Arctic oil.

The first Arctic oil came not from the land, but from the sea. Whale oil, used for lighting in the dark towns of northern Europe and North America, was the first globally traded Arctic commodity.

At the turn of the seventeenth century, the 'Spitsbergen oil rush' led to one of the first waves of southern interest in the Far North.[8] The Dutch took an early position on the Arctic archipelago, establishing the settlement of 'Smeerenburg' – literally, 'blubber town' – in 1619. In the eighteenth and nineteenth centuries, following both the demand for whale oil and the supply of whales, the United States became the world's leading whaler nation, bringing an industrial scale to the enterprise and nearly wiping out the Arctic whale population in the process.[9] On the shores of the American Arctic, the family names of New England whalers are still carried by the dynasties of Inupiat hunters.

The existence of geologically formed oil in the Arctic was known as far back as the eighteenth century, with limited geological exploration throughout the nineteenth and early twentieth centuries. Alexander Mackenzie noted its presence in northern Canada as early as 1789. In 1888, a report to the Canadian Senate concluded the area around the Mackenzie River was the site of possibly the 'most extensive petroleum field in America, if not the world'.[10] Across the border in Alaska, American surveyors reported oil in the south of the newly purchased territory in the 1870s and referred to earlier Russian finds.[11] In 1923, fearful of an impending global scarcity of oil and its consequences for the mobility of American power, President Harding set aside a large area of resource-rich northern Alaska for future exploitation by the navy.[12] Danish geologists recorded seepages of oil on Greenland in the nineteenth century. In the Russian Arctic, some thirty shallow wells were drilled near Ukhta between 1869 and 1917.

Commercial production of Arctic oil took longer to develop.[13] On 24 August 1920 a drill team led by American Theodore August Link hit oil on the Mackenzie River in Canada's Northwest Territories, a few hundred miles from the site of the Klondike gold rush:

> At 783 feet a strong flow of oil was struck. For ten minutes, a column of oil spouted from the 6-inch casing to a height of 75 feet above the derrick floor, after which the well was capped. The flow on that occasion probably exceeded 600 barrels of oil. On two subsequent occasions the valve was opened, with similar results.[14]

One Canadian newspaper breathlessly reported Link as saying that the so-called Norman Wells find was 'the biggest oilfield in the world', stretching from the Great Slave Lake in the south, to the mouth of the Mackenzie River in the north, and opening into the Arctic Beaufort Sea. The oilfields of Peru and Mexico – then among the world's biggest oil-producing countries – were viewed as 'miniature' in comparison. In an echo of the Klondike rush, some prospectors travelled north, claiming land along the Mackenzie River.

But the company that had financed Link's wildcatting expedition provided a more sober assessment. As M. B. Green, a manager for Imperial Oil, a subsidiary of the American giant Standard Oil of New Jersey (now ExxonMobil), told the *Calgary Herald*:

> As for Fort Norman, the oil there is not and will not be worth one red cent, for it is doubtful if a railroad, the only means by which the oil could be transported on a commercial basis, would ever be built to that far northern outpost; certainly not unless oil should be found there in overwhelming quantities . . . The only value of oil at the present time . . . is purely scientific. It is now known that oil is there . . . [but] whether it is there in sufficient quantity to warrant the spending of millions of dollars to attempt to get it out is not known yet by any means.

For most of the next fifty years, Green's assessment of the economics of Norman Wells applied to the Arctic as a whole. After ninety years of production, Norman Wells remains the northernmost oilfield in Canada.

Men like Theodore Link never gave up on the promise of Arctic oil. But only in the Soviet Union, where strategic factors outweighed economic rationales and labour was artificially cheap, would the production of Arctic oil be possible on any scale before the 1960s. As part of Stalin's drive to exploit the mineral wealth of the Soviet north – making forced labour available for the task – the Chib'yu field in the Timan-Pechora basin came into production in 1930. The Soviet Union's Arctic oil and gas development was driven more by orders from Moscow than by economic logic.

In Canada, Green's argument was harder to counter. Even if, as Norman Wells confirmed, there was oil in significant quantities in the Arctic, it would remain in the ground until lower costs of production and transport made it profitable on the open market. Apart from near the wellhead itself, Arctic oil was commercially unviable. While gold could be taken from the Klondike in suitcases and still make its owners rich as soon as they docked in San Francisco, oil was different. Without the infrastructure to transport large volumes of oil to consumers, there would be no Arctic oil rush.

After the initial excitement of the find at Norman Wells wore off, the economic realities set in. By the end of 1920, Norman Wells was limited to producing just 100 barrels of oil a day. There was no incentive to build a railway or pipeline to southern Canada or the United States. Even as demand for gasoline for motor transport rose across North America – oil consumption rose from 1.03 million barrels of oil per day in 1920 to 2.58 million barrels by 1929 – it was more cheaply available from a string of newly discovered fields in the American west than from the farthest reaches of the Canadian north.[15] As fears of scarcity gave way to a glut, the oil price slumped.[16]

The Discovery Well at Norman Wells and a basic refinery built to convert its crude into usable petroleum were closed in 1925. Production restarted in 1932 when a radium mine opened at Great Slave Lake and a limited local market emerged. But output remained at a very low level – just 1,000 barrels over the whole of 1932, rising to around 23,000 barrels a year by the late 1930s.

For a moment in the Second World War it looked as if Norman Wells might be transformed from an isolated oil well supplying the local Arctic mining industry into an asset of strategic military significance. In 1942, the Japanese occupation of Attu and Kiska – two of the Aleutian Island chain off Alaska's west coast – sparked American concerns that maritime oil shipments to Alaska might be vulnerable to attack. To prevent Alaska being cut off entirely, the United States army built the so-called Canol ('Canadian Oil') pipeline. Norman Wells was connected to a newly built refinery in Whitehorse, 600 miles away. Three onward pipelines were built to take refined products to Skagway and Fairbanks in Alaska, and to Watson Lake in the Canadian Yukon.

Theodore August Link, the man who had first struck oil at Norman Wells in 1920, returned to the Mackenzie region as the Canol project's chief geologist, with a mandate to boost its oil production. A series of seventeen wildcat wells drilled up and down the Mackenzie Valley came up dry. But further wells drilled at Norman Wells itself were more successful.[17] By 1944, annual production had risen to 1 million barrels.

The expansion was short-lived. Almost as soon as the Canol pipeline was operational, it was deemed a waste of money. While the army viewed it as a necessity of war, a US Senate committee determined that it had no peacetime future.[18] In 1945, against the wishes of the military, Canol was duly closed down. Many of the oil wells drilled at Norman Wells between 1942 and 1944 were capped. The one producing oilfield of the Western Arctic returned quietly to its pre-war role as a supplier of oil to the limited local market of Canada's Northwest Territories.

The long post-war expansion of the global economy was fuelled by an abundance of cheap oil. Rising demand was met by a boom in production, rising from just over 10 million barrels per day in 1950, to just under 50 million barrels per day in 1970. As a result, oil prices were kept low and remarkably stable. The price of crude oil rose by just nine cents in twenty years – from $1.71 to $1.80 per barrel. With the global market mechanism apparently working well, and with prices so low, producing oil in the Arctic was of marginal economic interest for much of that period.

The five-fold increase in global supply – and particularly the flood of oil onto the international market – was dominated by the rise and rise of the Middle East as an oil-producing region, buoyed by costs as low as 11 cents per barrel. Growth in Middle Eastern production was complemented by the return of Russian oil to the international market in the late 1950s, rising exports from traditional producers such as Venezuela and the emergence of entirely new oil provinces in Africa. (In Libya, production grew from zero to 3.3 million barrels per day between 1959 and 1970.)

But throughout these years, the world's largest producer of oil, and the bedrock of the system's stability, was America. Not only was US production far greater than that of any other country, but its significant surplus production capacity – the happy unintended consequence of policies introduced in the 1920s to prevent oversupply from driving prices too low – gave it the unique ability to boost global supply on a massive scale in the face of any oil shortage.[19] In 1956, when Middle Eastern oil supplies to Europe were interrupted by the Suez Crisis, the United States was able to act as a 'swing producer', alleviating European shortages through a trans-Atlantic 'oil lift'. It did much the same in 1967, following the Six Days' War. The Texas Railroad Commission, the administrative body which set the level of American oil production, was considered far more powerful than the presidents of Latin American producing countries, or the princes of the Middle East.

The American age of oil – based on massive domestic production and surplus capacity – could not last for ever. But in 1959, President Eisenhower reluctantly made a decision which accelerated its demise. Under pressure from domestic American producers who feared a flood of cheap Middle Eastern oil would drive them out of business, Eisenhower introduced quotas on imported foreign oil (with exemptions for Canadian and Mexican producers).[20]

Those in favour of protection for domestic producers had argued that it would boost American oil production and thereby benefit national security. Critics of the scheme were more prescient. They retorted that while quotas on foreign imports might boost American production in the short term they would, by the same token, bring forward the peak in domestic production. America's surplus capacity would be squeezed, and eventually it would disappear. Instead of prolonging America's pre-eminent role in the world of oil, import quotas would, eventually, destroy it. If achieving national security were the true objective, it would be better to conserve America's oil in the ground for use when global supply was scarce, rather than providing incentives to use it up when global supply was plentiful.

As the 1960s wore on, the validity of that criticism became more and more evident. Even with the incentives to domestic oil production provided by import quotas, the number of drill rigs in the United States fell into steep decline – falling by two thirds between 1955 and 1971.[21] Growth in production slowed down and, in 1970, production peaked.[22]

One consequence of the slackening of new production in the continental United States, and of the incentive that import quotas provided for American and Canadian production, was that exploration began to move further north. The American Arctic, long considered a potential source of oil, was slowly put into play.

Initial results were not encouraging. In Arctic Canada, the Norman Wells field was still producing a steady trickle of oil, with output rising from 471,000 barrels in 1950 to 852,000 barrels in 1970, a little over 2,000 barrels of oil a day. But ambitions to turn the wider Canadian north into a major oil-producing province came to little, despite periodic exploration efforts in the Mackenzie Delta and elsewhere. In the mid-1950s, the Geological Survey of Canada undertook *Operation Franklin*, demonstrating the geological potential for oil on Canada's Arctic islands. The Conservative government of John Diefenbaker, promising an Arctic boom, backed the issue of exploration permits across the Canadian north. But such grand visions of Arctic promise were easily derailed by the enormous cost of Arctic exploration and by geological bad luck. Dome Petroleum drilled the first commercial test well on Canada's Arctic Islands – Winter Harbour #1 – in the winter of 1961–62.[23] Both it and two subsequent wells – on Cornwallis and Bathurst Islands – were dry.

In Alaska, the focus of commercial exploration and production in the 1950s was in the area around Cook Inlet, in the sub-Arctic south. In the Arctic North, exploration was initially led by the government. Between 1944 and 1953, the United States Geological Survey surveyed Naval Petroleum Reserve-4, drilling thirty-six test wells. In 1958, a comprehensive report produced by John C. Reed, the US navy commander who led the survey, confirmed the existence of oil deposits, while emphasising the logistical and engineering challenges to production.[24]

In the late 1950s and early 1960s there were only sporadic commercial exploration efforts in Arctic Alaska. British Petroleum (BP) was one of the first, entering into an exploration agreement with Sinclair Oil in the late 1950s in the hope of diversifying the company's oil resources away from the Middle East. But after six dry holes on the North Slope – the flat plain of northern Alaska which slips down to the Beaufort Sea – the company abandoned its efforts. Standard Oil of New Jersey – the core of what would later become ExxonMobil – began an Arctic exploration push in 1964, using its Humble Oil affiliate to partner with Atlantic Richfield (Arco), an independent producer from California. But as with the BP/Sinclair partnership, exploration results for Arco/Humble were poor. By 1967, Humble Oil was ready to pull out of the Alaskan North. For all its apparent geological promise, the Arctic seemed destined to remain a footnote in the history of oil.

The year 1968 was one of social and political upheaval across the world, from Prague and Paris, to Saigon and Memphis, Tennessee. In the Arctic, it was the beginning of an economic revolution.

In March that year, after months of rumours, Arco/Humble annonced a major oil discovery at Prudhoe Bay, on the North Slope of Alaska. Nearly half a century after Norman Wells, oil had been struck in the American Arctic. De Golyer & MacNaughton, a reservoir engineering firm retained by Arco/Humble, initially estimated that the Prudhoe Bay field held up to 10 billion barrels of oil, making it the largest field ever discovered in North America. In fact, the original reserves in place in Prudhoe Bay were even greater.

Although Arco/Humble had discovered the field, one of the major winners from the discovery was British Petroleum (BP). Until 1967, BP had adopted a wait-and-see policy, preferring to let others make the initial investments required for North Slope exploration. Outbid by Arco/Humble for the central crest formation of Prudhoe Bay – the geological rise to which any oil on the North Slope would tend to migrate – BP had settled for the flanks of the area, securing a total of 96,396 acres. When the size of the Prudhoe oil discovery became apparent, that investment suddenly looked very astute. For little outlay, BP had secured a major stake in the future of the North Slope.

Following its successful strike, Arco/Humble offered to buy out BP's untested Prudhoe Bay acreage. BP rejected the approach. Instead, a drill rig from the Cook Inlet area of southern Alaska was shipped up to the North Slope and reassembled on the banks of the Putuligayak River, just a few miles from Arco/Humble's original site. BP commenced drilling in November 1968. Five months later it confirmed oil. While Prudhoe Bay catapulted Atlantic Richfield into the major league of domestic US producers and reconfirmed the status of Standard Oil of New Jersey, the find allowed BP to diversify away from its historical focus on Iran and thus establish itself as a global oil company.[25]

After the false starts at Norman Wells in the 1920s and 1940s, and the disappointments of Dome Petroleum's wells on Canada's Arctic islands in the early 1960s, the discovery of Prudhoe Bay in 1968 marked the beginning of the first sustained Arctic oil rush.

Initially, the scramble for North Slope real estate was frenetic. Alaska's 1969 sale of North Slope leases broke all previous records and surpassed all expectations. Instead of the $12 million raised by the three previous lease sales, the 1969 sale raised $900 million (around $5.3 billion in today's money).[26] Whereas in 1965 BP had paid just $17.80 an acre for land in Prudhoe Bay, four years later prices had soared. Amerada / Getty paid $28,233.25 per acre for a 2,560-acre tract.[27]

Interest in the promise of Arctic oil was not limited to Alaska's North Slope. Across the border in Canada, the government had backed the establishment of a national Arctic exploration company, Panarctic Oils Ltd, even before the Prudhoe Bay find was announced.[28] After 1968, its exploration programme accelerated. American-owned Imperial Oil, the other main player in northern Canada, ramped up its own search for Arctic oil and gas over the same years.

Both companies were successful. Imperial's efforts were focused on the coastal areas near the Mackenzie River delta, striking oil at Atkinson Point in 1970 and natural gas near the Inuit settlement of Tuktoyaktuk the following year. Panarctic's finds were still further north, on Canada's Arctic islands. A major gas field was found at Drake Point on Melville Island in 1969, followed by a second on King Christian Island the next year. In 1974, Panarctic discovered the Bent Horn oilfield on Cameron Island, several hundred miles north-east of Prudhoe Bay, at 77° North.

In time, oil companies' attention turned to the offshore Arctic. Geologically, it seemed likely that the onshore oil-bearing structures of the North Slope and the Mackenzie Delta would continue out under the shallow Beaufort Sea. Identifying and developing those resources, while difficult, did not seem insuperable. Offshore drilling had been transformed by the experience of developing the Gulf of Mexico and, later, the North Sea. Whereas the first offshore rigs of the late nineteenth century had been, in effect, standard onshore rigs mounted in shallow water at the end of pontoons, by the early 1970s the offshore industry had developed its own technology and its own expertise.

This technology could not be applied to the Arctic without modification. Whereas the main natural threat to offshore production in the Gulf of Mexico was hurricanes, and the main problem in the North Sea was rough seas, the principal complication facing offshore drilling in the Arctic would be ice. But the industry had demonstrated it could cope with icebergs, having developed techniques of handling them both in southern Alaska and off the Canadian coast of Labrador and Newfoundland. (One technique was to use thick nylon lassoes to pull icebergs away from rigs.[29])

Pack ice presented a more specific Arctic challenge, threatening to crush drill rigs and pull them off location. But here as well, the technical challenge could be overcome by the use of drill ships – which could move off before the pack ice reached them – or by the construction of artificial islands in shallow waters – which would allow the use of standard onshore exploration equipment. Some viewed the ice pack – and the expectation in the early 1970s that global cooling in the future would extend and thicken it – more positively.[30] A geologist who worked in the Canadian Sverdrup basin in the 1970s argued that 'ice was our friend', providing more stable conditions for exploration than might exist under twenty-first-century conditions of global warming.[31]

Many concluded that, despite the costs of exploration and the difficulties of low-temperature operations in the Arctic, its future as a major oil-producing province had arrived.

This future, however, depended on a solution being found to the problem raised by Green at Norman Wells in 1920: oil in the ground was one thing, but getting it to market quite another. The future of oil and gas development in the Arctic would not hinge on the existence of the resource, or even on the technical challenges and costs associated with extraction, but on the infrastructure to deliver it to market. For several years in the late 1960s and early 1970s it was not clear whether the challenge of transporting Arctic oil and gas would be met, or what geographic route it would follow if it were.

In Alaska, several options were on the table. Governor Walter J. Hickel – nominated US Secretary of the Interior by President Nixon in January 1969 – wanted to use the railroad, echoing the parallel development of oil and rail which had been so vital to the early establishment of John D. Rockefeller's Standard Oil.[32] (The Department of Transportation estimated that Hickel's proposed extension of the existing Alaska railroad from Fairbanks to Prudhoe Bay would cost $300 million.[33]) Another option was to ship oil from the north shores of the American continent through the Northwest Passage. Humble spent an estimated $40 million on an eighty-day test voyage of the 1,005-foot *SS Manhattan* ice-breaking oil tanker, completing a round trip from the east coast to Prudhoe Bay and back in the late summer and autumn of 1969. A final option was a pipeline. In February 1969, a group of companies – including Humble – announced a plan to build an all-Alaska 500,000-barrel-per-day pipeline from Prudhoe Bay to the southern port of Valdez.[34]

It was this third option which triumphed. The railroad had found too few advocates, and no support from the oil companies. The passage of the *SS Manhattan*, while it had demonstrated the technical feasibility of shipping oil in the Arctic summer, led Humble to conclude that the cost of marine transport was too high, adding between 90 cents and 1 dollar to the cost of a barrel of oil. (Diplomatically, meanwhile, the fallout from the voyage of the *SS Manhattan* hardened American–Canadian antagonism on the legal status of the Northwest Passage).[35] Construction of a pipeline seemed both technically feasible and widely supported. Japanese steelmakers delivered the first $100 million of steel for the pipeline in September 1969, just six months after the project had been announced. It was expected that oil from Prudhoe Bay could flow south as early as 1972.

This initial phase of Prudhoe Bay's development, from 1968 to 1969, was hectic. The quicker Prudhoe Bay could be brought on-stream, the sooner oil companies' investments in the leases would be paid off, and the sooner the State of Alaska would start receiving royalties (as early as 1972). Any pause or delay would weaken momentum for development. Cost and environmental damage were secondary considerations.

The construction of a road from Fairbanks to Prudhoe Bay, completed in March 1969, was a case in point.[36] Once built, the road provided all-important additional capacity to meet the demand for the transport of heavy equipment and supplies to the North Slope. But only at tremendous environmental and economic cost. One estimate placed the cost of transporting one ton of material by road from Fairbanks to Prudhoe Bay at $200 – 25 per cent more expensive than air freight.[37] And with the emphasis on speed of construction, the engineering lessons which Commander Reed had carefully explained in 1958 were forgotten. Less than two months after the road was completed, it was abandoned as impassable when spring meltwater turned it into a river.

By March 1970, plans for the Alaska pipeline had run up against a series of difficult political and legal obstacles: native land claims on territory over which any pipeline would have to pass, lawsuits claiming that possible environmental damage caused by the pipeline had not been adequately assessed, and requirements for other pipeline routes to be considered. The initial, hectic phase of development ended; a slower, deliberative and more uncertain phase began, with its decisive episodes played out in Washington and the US federal courts more than in the Arctic. As much as economics, politics was proving itself to be a key factor in determining whether Arctic oil or gas development would take off.

The first obstacle to the pipeline project was resolved by the passage of the Alaska Native Claims Act, extinguishing native claims in return for money and a one-off transfer of land. Signed into law by President Nixon in December 1971, the Act handed over $962.5 million (approximately $5 billion in today's money) and 40 million acres of land.[38] The second obstacle – the use of the 1969 National Environmental Policy Act (NEPA) to launch a legal challenge to the permitting of the Alaska pipeline – would take longer to overcome, requiring an entirely new (and much longer) environmental impact statement, a complex series of court cases and a major redesign of the pipeline itself.[39] (Computer modelling demonstrated that sending hot oil through an underground pipeline in the Arctic would melt the surrounding permafrost, causing the pipeline to twist and buckle. To avoid this, the pipeline would be built above ground, raised on stilts, for half its length.[40])

The third obstacle – the requirement under NEPA for alternative pipeline routes to be taken into account – led to consideration of a trans-Canadian route along the Mackenzie River valley, in place of the trans-Alaska option. Had the Canadian route been adopted, taking oil from Alaska through Canada to the Midwest, the development of the Canadian Arctic might have been accelerated by forty years or more.

At first, Ottawa supported the principle of a pipeline through Canadian territory. Not only would such a route prevent oil from being shipped along the coastline of British Columbia – which many saw as the greater environmental risk to Canada – but it would encourage the development of oil and gas resources along the pipeline route.[41] With the heavy transport infrastructure in place, the prospects for oil and gas production in the Canadian Arctic would be transformed.[42]

For the production companies on the North Slope, the trans-Canadian option was a distraction. Compelling business reasons to support an all-Alaskan option abounded. It would allow companies to deal with only one set of regulators. It would allow for the possibility of exporting crude oil by tanker to East Asia – thus skirting the restrictions of the Jones Act and preventing the lock-in to the US domestic market that would surely result from a pipeline to the Midwest (via Canada).[43] Above all, opting for an all-Alaska route would mean that delayed construction work could begin right away, rather than waiting for another round of planning and permitting from Canadian authorities.[44]

The risk of additional delay – inherent in the Canadian option – was ultimately viewed as a national security problem. The US Department of Interior's security analysis concluded that:

> From a national security point of view, it is important to get North Slope oil to the lower 48 states as soon as possible . . . The Alaska pipeline gives promise of bringing in a significant quantity of North Slope oil to the lower 48 states by 1975 . . .[45]

A Canadian pipeline, in contrast, would not be able to deliver oil to the United States before 1981 at the earliest.[46]

Support swung behind the trans-Alaska option in Congress, with attempts to legislate away the remaining legal obstacles to the pipeline's construction. But by the autumn of 1973, with Washington intensely focused on the Watergate scandal, the final permits for the construction of the pipeline had still not been given.[47]

In the end, it was another set of events, thousands of miles from the Arctic, which provided the final impetus for work on the Trans-Alaska Pipeline to begin.

On 6 October 1973, still smarting from the embarrassment of defeat in the 1967 war, Egypt and Syria launched an attack on Israel. The threat of war in the Middle East had been rising all year, but when the attack came, on the Jewish Day of Atonement (Yom Kippur), the Israeli Defence Force was unprepared. Its losses were unexpectedly high. Fearing that defeat of Israel would significantly strengthen the Soviet Union's hand in the Middle East via its Arab proxies the United States intervened, sending supplies to Israel by sea and air. Days later, the oil-producing states of the Middle East decided to punish the

United States, first by cutting oil production and then by placing an embargo on exports to the US and the Netherlands.

In a matter of days, the spot price of oil tripled. Long after a cease-fire had been reached in the Yom Kippur War – the war lasted for only a few weeks – the Arab oil embargo continued. The average price of oil in 1973 had been $3.29 a barrel; in 1974 it was $11.58. The oil price would never again be as low as it had been in early 1973.[48] (In 1979–80, a second oil shock, caused by the Iranian revolution and fears over the consequences of the Soviet invasion of Afghanistan, pushed oil prices up again, from $14.02 a barrel in 1978 to $36.83 in 1980.[49])

The 1973 crisis was a 'shock' in terms of price, but the underlying power relationship between consumer and producer countries – and the mostly Western international oil companies which dominated global production – had begun to shift years earlier.

The process began not in the Middle East, but in Latin America, with nationalisation of the Mexican oil industry in 1938 and a bench-mark 'fifty-fifty' deal struck in Venezuela in 1943, introducing the then revolutionary idea that income from oil should be split equally between the company and the host state.[50] In 1950, the 'fifty-fifty' formula was applied to Saudi Arabia, ending a situation in which Aramco (the American producing consortium) paid more to the US Treasury in tax than it paid to Saudi Arabia in fees and royalties.[51] The US State Department was broadly supportive of such rebalancing, anticipating that a more equitable arrangement would cement America's alliances and secure foreign oil for the long term. In 1960, the power of producer states – minus the United States and the Soviet Union – was further enhanced by their decision to band themselves into the Organisation of Petroleum Exporting Countries (OPEC).

But the truly critical shift in power was geological. In 1970, US oil production peaked. Although America would remain the largest producer of oil for years to come, the US would never pump more crude oil out of the ground than the 4.1 billion barrels it pumped in 1970, at an average of 11,297,000 every day. America's ability to act as a global swing producer – increasing production in times of need, and reducing it when prices fell – was compromised. OPEC looked set to replace the Texas Railroad Commission at the core of the global oil market.

The end of the age of American oil dominance was more sudden than most experts had predicted. In 1970, a taskforce assembled by President Nixon predicted that US oil imports from the eastern hemisphere would not reach 10 per cent of national consumption before the mid-1980s. They could not have been more wrong. Eastern hemisphere imports had reached fully 15 per cent of consumption as early as 1972, accelerating America's path towards import dependency.[52] In 1973, James E. Akins, a future American ambassador to Saudi Arabia, warned that 'the wolf is here' on the pages of *Foreign Affairs*.[53] The result was a paradigm shift in global oil, from a market mechanism stacked in favour of consumer countries and their corporate representatives to one in which producer countries had far more control.

The impacts of this shift were deep and wide-ranging. First, the International Oil Companies (IOCs), which had thought themselves masters of the global oil market, now found themselves as supplicants. (Even before the oil shock itself, Aramco was forced to accept Saudi participation for the first time.) Second, the Arab oil embargo showed importing countries to be highly vulnerable to decisions on oil production taken in Riyadh and Baghdad. While European states were used to their vulnerability to Middle Eastern politics, in the United States this came as a jolt.[54] Third, the embargo increased the financial strength of exporter states and demonstrated their geopolitical importance. After 1973, American foreign policy would become ever more inextricably bound up with American energy policy. The Middle East would become the focus of that intersection.[55]

For the Arctic, the impacts of these larger shifts within the global oil market were enormous. Immediately following the 1973 crisis, the last political obstacles barring construction of the Trans-Alaska Pipeline were overcome. Just as in 1959, when national security arguments were used in support of quotas on foreign oil, so concerns over America's energy security were now adduced in support of the speedy development of Arctic oil. On 13 November 1973, just five weeks after the outbreak of the Yom Kippur war, the Senate passed the Trans-Alaska Pipeline Authorization Bill by eighty votes to five. President Nixon signed it into law three days later. Finally, in 1974, work on the long-awaited pipeline began once more.

The oil shock of 1973 – and its successor in 1979–80 – provided a renewed impetus to Arctic oil and gas development.

Arctic oil was seen as costly, but secure. As investments in the Arctic continued to pile up in the late 1970s and early 1980s, money flowed in from Japan, a country with no hydrocarbon resources of its own, and therefore highly vulnerable to any shut-off in supplies from the Middle East. In the United States, the navy initiated a drilling programme in the NPR-4 area of northern Alaska, spending $1 billion between 1976 and 1982 to evaluate the strategic reserve set aside by President Warren G. Harding fifty years previously.

For the IOCs, the American Arctic came to be seen as a safe haven for investment as much as a frontier for exploration. Summarily ejected from their pre-eminent positions in the Middle East, the relative political stability of the United States and Canada was attractive. More importantly, with oil prices suddenly three or four times higher, developing Arctic resources was more commercially viable than ever.

In Canada, government support for Arctic development and the prospect of a gas line from the Mackenzie Delta, set an expectation that Canada's Arctic hinterland would become a major source of oil and gas in years to come. Offshore, exploration in the Beaufort Sea expanded massively. Dome Petroleum – one of the major players in Canada's Arctic oil rush – predicted that $40 billion (nearly $100 billion in today's money) would be spent on Beaufort Sea development in the 1980s.[56]

Outside North America, higher oil prices provoked a surge of interest in offshore North Sea development, launching both the United Kingdom and Norway on the path to becoming major oil exporters. (Decades later, as the North Sea began to decline, Norway would move its sights to its Arctic back yard: the Barents Sea.)

For the Soviet Union, the hike in global oil prices offered an opportunity to boost its own oil revenues and secure its place on the global market.[57] Between 1972 and 1982 Soviet oil production increased by half. The day when Russia, having turned first south and then east for the bulk of its hydrocarbon production, would be forced to turn to the Arctic North, loomed ever nearer.

The oil shocks of the 1970s elevated the Arctic from the status of marginal oil province to a key part of IOCs' reserve portfolios and a crucial element in the energy strategies of Canada, Russia, and the United States. In 1983, ten years after the first oil shock and fifteen years after the discovery of the Prudhoe Bay oilfield in Alaska, the president of Cambridge Energy Research Associates argued that 'the Arctic is absolutely crucial to the world supply picture'. Interior Secretary James Watt told one journalist, 'The big issue in energy is Arctic oil.'[58]

But such confident pronouncements on the future of Arctic oil and gas were premature. As in the past, the bright expectations of booming Arctic production were upset by events both inside the Arctic and, more importantly, beyond it.

In Canada, prospects for the expansion of Arctic oil and gas production were undermined in 1977 when the final report of an inquiry into the consequences of a natural gas pipeline from the Mackenzie Delta recommended a ten-year moratorium on the project. Not unreasonably, Justice Thomas R. Berger warned that a gas pipeline would likely be followed by other oil and gas developments along the Mackenzie River, creating an 'energy corridor' in the Canadian Arctic. Given the risks of environmental damage and, more importantly, the opposition of the local Inuit population, Berger warned against the idea, embarrassing the Canadian government in the process.[59]

But the damage caused by Berger's report was not immediately fatal. In the early 1980s, Ottawa continued to support Arctic exploration through the tax system, and private investment held up, running into billions of dollars.[60] Dome Petroleum remained keen, predicting in 1981 that it would commence major commercial production from the Beaufort Sea within five years, using ice-breaking tankers to transport Arctic oil south. In 1985, Panarctic shipped the first 100,000 barrels of Canadian Arctic oil from Bent Horn to Montreal. Gulf followed with a shipment of 316,000 barrels from Amauligak to Japan. But the politics and economics of Canadian Arctic development were turning. In 1985, Canada's National Energy Program was cancelled. With falling oil prices and without a permanent pipeline infrastructure, oil and gas exploration in the Canadian Arctic returned to the deep freeze.[61]

In the United States, the dynamics were different. The opening of the Trans-Alaska Pipeline in June 1977 – a month after Justice Berger's report had come out in Canada – meant that Alaska had the basic transportation infrastructure which its neighbour lacked. The cost barriers to developing new Arctic fields were lower as a result. In the early 1980s there was no sign of let-up in the North Slope boom. In 1982, oil companies paid $2.07 billion (around $4.6 billion in current dollars) for the right to drill in the Beaufort Sea. But from the mid-1980s, as the price of oil fell, the momentum of exploration and production began to slow. In 1988, less than ten years after the first barrel of oil had been pumped from Prudhoe Bay to Valdez, output from the North Slope peaked.

The underlying cause of the end of the first Arctic hydrocarbon boom was not a lack of natural resources, or even a lack of technology to develop them, but a combination of politics and price. In 1985, the price of oil stood at $27.56. A year later, it had fallen to $14.43.[62] The reason? After trying (and largely failing) to support the global price of oil by gradually reducing its own production Saudi Arabia shifted course.[63] In 1986 alone, Saudi production increased by 40 per cent. Prices fell precipitously and fears of scarcity were forgotten. Once again, the Arctic was pushed to one side. It would remain there until prices began to rise again over twenty years later: from $28.83 in 2003 to $72.39 in 2007, and to $139 in 2008.

At the beginning of the twenty-first century, global exploration and production budgets exploded, and the Arctic became a target of investment once again. Many of the projects on the table in the American Arctic are variations of plans from the 1970s and 1980s which never made it off the drawing board. Whether those schemes now materialise or not will ultimately depend on the same combination of politics and price which determined the fate of development projects several decades ago.

In Canada, the idea of a Mackenzie 'energy corridor' has been revived by plans for a 1,220 km, $16.2 billion gas-line, strongly supported by Ottawa, running from the Mackenzie Delta to Alberta (where natural gas could be used as an energy source to develop the oil resources of the Athabasca tar sands). A billion dollars has already been spent on exploration for gas to fill the pipeline and on feasibility studies for its construction.[64] A long-running environmental review has cost more than three times the original estimate of $6 million. Whereas in 1977 Justice Thomas R. Berger scuppered an earlier plan for a Mackenzie gas line partly on the basis of opposition from Inuit groups, this time around Inuit groups were heavily involved from the start, in the form of the Aboriginal Pipeline Group. Regulatory approval of the pipeline is expected in 2010.

In Alaska, two competing projects for a gas line from the North

Slope – both of which would which would terminate in Canada – are also gaining traction. (A proposal backed by the Chinese oil company Sinopec, which would have involved a gas line to southern Alaska and an LNG terminal to liquify gas for export to China, was rejected.) State support has been crucial, with some $500 million of Alaska state funds made available for an option led by TransCanada and ExxonMobil (the rival Denali gas line is sponsored by BP and ConocoPhillips). Both sides expect to spend over $100 million on an 'open season' in summer 2010. Both projects – of which only one is likely to ultimately be built – carry price tags of $25–$30 billion, making them amongst the biggest construction projects ever undertaken in North America. Neither would be ready before 2018.

Meanwhile, oil companies have scrambled to bring back long-retired drillers with Arctic experience, tempting them back with generous consulting contracts. Specially designed Arctic drill ships, abandoned ten or twenty years previously, have been refitted and put back in operation. Data acquired by oil companies in the exploration boom of the 1980s have been revisited. Old arguments – about the importance of the Arctic to America's energy security and about the environmental risks associated with Arctic development – have meanwhile received a new lease of life.

At first glance, the global oil picture at the beginning of the twenty-first century bears a striking similarity to that of the 1970s. Then, as now, oil had become deeply politicised, with resources concentrated in regions more politically challenging for the West than had previously been the case. Then, as now, the idea of a world running out of oil led to questions about the sustainability of the global supply, and the sustainability of oil as a key component of overall energy requirements. Then, as now, the role of IOCs seemed in flux, with fears that the business model of Western oil majors no longer reflected the geological and geopolitical realities of global oil.

In some ways, the world's energy outlook is far less severe than in the post-oil-shock world. The energy intensiveness – and particularly the hydrocarbon intensiveness – of global GDP has declined. In some developed economies, demand for oil may soon peak, even as aggregate global demand continues to rise.[65] Natural gas has largely replaced oil for electricity generation, and the prospects for its wider use are good.

Oil is simply not as central to economic growth as it was thirty or forty years ago, except perhaps in countries such as China and India, where its use is growing rapidly. Finally, the alternatives to oil as an energy source, restricted by prohibitive cost or rudimentary technology thirty years ago, are now far closer to being competitive. Transition away from oil and gas is more plausible now than in 1973.[66]

On the exploration and production side, things have also markedly improved. Three-dimensional seismic technology has dramatically improved what geologists can 'see' underground, making it easier to identify new resources. Companies can operate in areas that would have been inaccessible thirty years ago, particularly in terms of deep water offshore, and extract a far higher proportion of reserves. Directional drilling and remote drilling have led to an extraordinary reduction in the surface footprint of a major field. Technology has meanwhile reduced, but not eliminated, the risk of environmental disasters.

And yet, in some ways, things are worse. The urgency of the need to remake our energy system is greater. A shift from hydrocarbons to alternative sources of energy – or the ability to use hydrocarbons in such a way that carbon dioxide can be captured and sequestered – is pressing. The fear in the 1970s was pollution; in the twenty-first century the fear is irreversible climate change. The geopolitics of energy are more pronounced. The concentration of remaining global oil and gas reserves in the Middle East is greater now than thirty years ago, partly because policies designed to boost production in some parts of the world – the United States, for example – have accelerated their depletion.

Worse, many argue that global oil output is close to a peak beyond which it will inevitably plateau and fall.[67] The notion that the world is physically running out of oil – and a host of other minerals, such as nickel, platinum and zinc – is contested.[68] But the idea that the world's 'easy oil' has now been produced – leaving us with deposits that are either more technically difficult to develop (such as the Alberta tar sands) or more physically inaccessible (such as deep-water oil) or both (such as Arctic oil) – is commonplace amongst oil company executives.[69] The challenge to maintaining an adequate future supply of oil is accentuated by rising demand in China and Asia, and chronic underinvestment in exploration and production in the past. That is why the International Energy Agency predicts oil prices averaging $100 per barrel in the period up to 2015, rising to $120 by 2030 (or $200 per barrel in nominal terms).[70]

From the perspective of the IOCs – currently undergoing a 'serious identity crisis' – the outlook is still more complex.[71] Most IOCs are producing less oil than in the past, and there are fewer and fewer places where new oil reserves – the key to IOCs' market valuation – are available on good terms. The balance of power between producer states and Western companies has become steadily less favourable in recent years. There was a brief moment in the 1990s when the opening of oil provinces in the Former Soviet Union – previously off limits to IOCs – seemed to point the way to a market-oriented future for the global oil and gas industry. But the last five years have seen a rising tide of resource nationalism, with producer countries and state-owned national oil companies (NOCs) increasingly dominant.[72]

All these global factors play into the Arctic. In Russia, developing the Arctic is a national imperative, reinforced by the global energy context: relatively high oil prices, growing resource scarcity and the potential for control of hydrocarbons to boost geopolitical leverage. In the United States, the corporate logic of Arctic development is complemented by a national security discourse on the importance of domestic oil and gas production, particularly in a global context of high prices and uncertain future supply. In Canada and Norway, while exploitation of Arctic hydrocarbons serves broad national objectives, it is a globally set price for oil which makes development attractive. For the Western oil industry meanwhile, unable to replace reserves elsewhere, the Arctic is simply the next best place to look.

Though no one really knows how much oil and gas there is in the Arctic, it is not for want of estimates, projections or guesses.

The most reported – and misreported – estimates of the Arctic's reserves are from the United States Geological Survey (USGS). They estimate that around 240 billion barrels of oil and oil equivalents, mostly natural gas, have already been found above the Arctic Circle, nearly as much as the entire proven oil reserves of Saudi Arabia.[73] They estimate that an additional 412.2 billion barrels of oil and oil equivalents remain undiscovered in the Arctic.[74] By any reckoning, these are huge numbers. The estimate for 'undiscovered' Arctic hydrocarbon resources represents 20 per cent of the global total.

But while the headline figures are astounding, so are the levels of uncertainty surrounding them. After being burned by publishing single estimates for reserves which proved to be wildly wrong the USGS now produces probabilistic assessments that vary widely from the mean. For West Siberia, for example, the USGS calculated a 5 per cent probability that there were 795 million barrels of oil remaining undiscovered at one end of the distribution, and a 5 per cent probability of some 9 billion barrels of oil remaining undiscovered at the other.[75] In another Arctic area, the West Laptev Grabens Assessment Unit (off the north coast of East Siberia), two different geological scenarios resulted in an even more disparate set of estimates. One provided a mean of just 1.1 billion barrels of oil; the other provided an estimate of over 15.8 billion barrels of oil, more than the entire proven oil reserves of Mexico, one of the largest producers of crude oil in the world.[76]

Many are highly critical of the USGS approach, and sceptical of its results. The French geologist Jean Laherrère, with experience of working everywhere from Algeria to northern Canada from the 1950s on, and one of the strongest advocates of 'peak oil', argues that the USGS estimates are a triumph of statistics over geology. He suggests that the actual technical field data from historical exploration is far less positive than the picture drawn by the USGS, though the historical approach has its own problems in a province like the Arctic.[77] Laherrère suggested to me that an earlier, and much-quoted, estimate for Arctic resource potential was, more or less, 'pulled out of a hat'.[78] Personally, he is convinced that 'we'll find nothing more than small fields in the Arctic'. His own estimates are 50 billion barrels of 'undiscovered' oil and 28,000 billion cubic metres of natural gas, many times below the USGS estimates.[79] Other estimates, however – from Russia in particular – are far higher than the USGS number.[80]

Everyone agrees that much of the undiscovered hydrocarbon resource of the Arctic will probably be natural gas rather than oil. Of the USGS's total estimate of 412.2 billion barrels of undiscovered oil equivalent in the Arctic, some 80 per cent is thought to be natural gas – globally more abundant than oil, and both harder and more expensive to transport.[81] The resource base of the Arctic is more like

Qatar than Saudi Arabia. And whatever resources there are in the Arctic will not be spread evenly. Projected ratios of oil to gas vary widely between different Arctic hydrocarbon provinces.[82] While the USGS projects that the vast majority of the undiscovered oil in the Arctic is in or near Alaska, the lion's share of Arctic natural gas is expected in the West Siberian and East Barents basins, stretching from the Russian Arctic to the Norwegian Arctic.

There is another hydrocarbon resource of the Arctic which scientists and companies are only beginning to seriously consider as an energy source of the future. Methane hydrates – essentially methane trapped in ice-like water structures under great pressure or at low temperatures – are found all over the world.[83] But they are thought to be most accessible from Arctic permafrost.[84] Methane hydrates were first brought to global attention in the 1960s by Soviet scientists, who discovered their natural occurrence in the gas fields of West Siberia. In 1972, they were found in Alaska. Since the 1990s, Japanese and Canadian teams have been studying the possibilities of methane hydrate production in the Canadian Mackenzie Valley. Estimates of methane hydrates on the North Slope alone range from 3,200,000 billion cubic metres to as high as 19,000,000 billion cubic metres.

As with Arctic oil and gas, the mere fact of methane hydrates' discovery has not resulted in production. But if oil and gas prices do eventually return to the highs of 2008 – as many analysts expect – methane hydrates may well be seen as an attractive source of 'unconventional gas'. If offshore development is restricted or even banned in some parts of the Arctic – a distinct possibility in the offshore American Arctic – gas from onshore permafrost may become viable. In Alaska, both ConocoPhillips and BP are involved in projects with the National Energy Technology Laboratory to determine the commercial viability of methane hydrate production, due to report in 2010.[85] The ConocoPhillips project, which aims to extract methane from hydrates by a process of carbon dioxide replacement, has the potential to produce energy and sequester carbon at the same time.

But there is a catch. Methane is a powerful greenhouse gas, some twenty times more potent than carbon dioxide. Large-scale production of methane hydrates would inevitably result in the release of at least some methane into the atmosphere. How significant this issue is determined to be, along with the price of oil and gas, will decide whether or not the Arctic becomes a production site for methane.

The Klondike gold rush of 1897 became a cultural phenomenon. It inspired poems, books and songs. It provided the backdrop for a Charlie Chaplin silent movie and, in the early 1930s, for a Mickey Mouse cartoon.[86] In 1936 Mae West wrote and starred in *Klondike Annie*, a feature-length film promising that its heroine 'made the frozen north . . . red hot!'

Though its future path is just as uncertain, the rush for the Arctic resources of the twenty-first century is somehow less romantic. Its promises seem more distant and more intangible: energy security rather than sudden wealth, corporate profits rather than self-made millionaires.

But the stakes are much higher in the twenty-first century than they were in 1897. The development of Arctic hydrocarbons is of global economical, environmental and geopolitical significance. The cost of today's potential investment in Arctic infrastructure runs into hundreds of billions of dollars. The environmental standards to which oil and gas development in the Arctic are held, meanwhile, will help define a global balance between resource exploitation and environmental protection. In Greenland, Arctic resource exploitation may offer a pathway to full independence. In Russia, development of Arctic oil and gas resources will help determine a superpower's fate.

9
Russia's Arctic Dilemma

> 'The Soviet Union attaches much importance to peaceful cooperation in developing the resources of the North, the Arctic . . . According to existing data, reserves of such energy sources as oil and gas are truly boundless'
>
> General Secretary Mikhail Gorbachev, October 1987

> 'I never saw a bureaucracy produce a single barrel of oil'
>
> Rex Tillerson, CEO of ExxonMobil, June 2008

> 'Our first and main task is to turn the Arctic into a resource base for Russia in the twenty-first century'
>
> President Dmitri Medvedev, September 2008

ST PETERSBURG, RUSSIA, 59° North: It's summer 2008, and Russia feels rich. Oil is at $139 per barrel and Russia is once again the world's largest producer.

At the St Petersburg International Economic Forum, everything is calculated to impress. From the faux classical music played when each minister strides onto the auditorium stage, to the scale of each company's stand in the conference hall outside, no expense has been spared to convey a single message: Russia is back.

Look around and you can still see the Soviet Union. It's in the name of the conference centre, *Lenexpo* – a throwback to when St Petersburg was Leningrad, and Soviet trade fairs were a byword for Eastern Bloc carousing, an opportunity for Soviet citizens to meet foreigners – albeit mostly from Communist countries – and to pretend to sell the Soviet goods which no one in the West would buy.

It is not just in the name of the place. It is in the ill-fitting suits that most Russian businessmen wear, just a little too tight around the shoulders and just a little too short in the leg. It's in the statue of Cheburashka outside; the Soviet Union's most popular cartoon character. Above all, it's in the inescapable feeling that this smoothly run conference – like many things in Russia – is the result of massive expense rather than basic efficiency. As with any functioning element in Russia, there is the constant Soviet sense of narrowly averted disaster. And just as the Soviet Union would go into overkill when it wanted to show off to the world – at the US-boycotted 1980 Olympics or any given May Day parade on Red Square – so the feeling of overcompensation hangs over this conference like a pall.

Old habits die hard, perhaps. The swagger of the Soviet Union relied on military power, oil production and the waning hold of history – all of which were either unsustainable or in decline by the 1980s. Today, Russia's undisguised confidence is still based on elements of that Soviet inheritance. But there is also a sense of real achievement in recent years, under-appreciated outside the country. Westerners tend to view the 1990s as a decade of Russian success – democratisation, the peaceful secession of former Soviet states and the appearance of political freedoms unimaginable even during the period of Perestroika and Glasnost. But for most Russians, the 1990s were the years when their country touched the void. In 1998, Russia essentially went bankrupt. Ten years later, the St Petersburg International Economic Forum is holding discussions on the investments of Russia's multi-billion-dollar sovereign wealth fund. It is a remarkable turnaround, and one of which many Russians are justifiably proud.

There are few illusions about liberal democracy here. To many, it failed them in the 1990s, just as the West failed them by preying on their weakness to deprive Russia of her historical sphere of influence. The term 'sovereign democracy' is more popular these days, amounting to a warning to the West not to interfere in whatever domestic political system the Kremlin decides to institute. Though President Medvedev, a former lawyer, talks encouragingly about the rule of law and making the bureaucracy more responsive to the population – Soviet leaders such as Kosygin and Gorbachev did the same – the fundamental intention is to better manage the country, not to break the state's hold on power.

The President is late, making his audience wait for forty-five minutes in their seats before finally appearing to a fanfare of trumpets. But then this is the political theatre of the new Russian regime, at once authoritarian and liberal, speaking of reform while maintaining the delicate political balance between the clans which rule the Russian state. Medvedev must appear as both the first servant of the people and as first director of the nation, Prime Minister Putin aside. 'No one would have respected him if he had been on time,' the Russian next to me confides.

For most Russians, the dual nature of the regime is accepted without question. Indeed, for many, it is the price of success. Whether Putin was skilled or lucky – he took over Russia just as global commodity prices were beginning to increase – he is credited with a transformation. Ten years ago, people did not get paid. Now they do.

Inside the trade pavilion is a large model of the SuperJet100, Russia's attempt to produce a mid-size passenger jet to compete with the likes of Embraer of Brazil and Bombardier of Canada. Behind it are models of Russia's more well-known aviation successes, the Sukhoi military jets, part of a fleet of Russian military aircraft that, even now, are superior in many respects to their American and European counterparts. Off to one side, ancient Russian scientists are discussing the government's decision to invest in home-grown nano-technology. It is hardly Silicon Valley, but Russians insist that they now have the financial means to compete on the international level.

These are the industrial symbols of Russia's attempt to diversify away from oil and gas. It is a reflection of the past research priorities of the Soviet state, rather than Russian intentions for the future, that so many of these technologies have their origins in the military–industrial complex. (Although the SuperJet100 is a civilian aircraft, it will be made by a state company that until recently specialised in military jets.) But the urgency of turning Russia's swords into ploughshares is a matter of practicality in the global economy. Russia's political and economic leadership wants to avoid the over-dependency on oil and gas exports which ultimately caused the Soviet economy to collapse.

But apart from oil and gas, Russia has little the world wants. Oil and gas represent a greater share of Russia's national exports than for any other Arctic country.

Perhaps that is why the largest stand in the pavilion, yet somehow also the most discreet, is that of Gazprom, the biggest Russian company of them all and the world's largest producer of natural gas. It occupies a plain two-storey glass structure. At its back there is a wrought-iron bridge recalling the elegant bridges across St Petersburg's many canals. In front are two model LNG tankers – a reference to Gazprom's future ambitions to produce and export LNG from Arctic Russia. Compared to the other stands, Gazprom's is a model of sobriety, and unlike the others, it is closed off, guarded by a Russian heavy wearing an earpiece. The glass is opaque; entry is by invitation only. But then, unlike other companies in the pavilion, there is no need for Gazprom to advertise. Both at home and abroad, Gazprom is the acme of Russia's attempt to combine statecraft with business.

On the one hand, the political links of Gazprom are obvious. David Victor, a leading expert on gas geopolitics, has called it 'a political colossus which happens to be in the gas business'.[1] The company is majority owned by the state, and the current President once served as the company's chairman. As is perhaps natural for a company which emerged as the successor to the USSR gas ministry, many of the company's employees have links to the state.[2] More disconcerting is that so many of those links are to the FSB, the successor to the KGB. Gazprom's potential to interrupt gas supplies to other countries provides a powerful instrument for state policy towards Russia's neighbours, though both the frequency and effectiveness of its use are much disputed.[3] While oil exports may provide Russia with money, it is Gazprom's natural gas exports that provide Russia's political leverage.

On the other hand, the company is seeking to be treated like any other commercial enterprise, attempting to buy up downstream companies in Western Europe, expressing interest in development in Alaska and advertising itself as the only gas company capable of working effectively in Arctic conditions. In 2007, the company's deputy chief executive, Alexander Medvedev, announced Gazprom's intention to be valued at $1 trillion by 2017, making it the world's largest company by far.

With control of the world's largest reserves of natural gas, a commercial ambition to be the world's largest company, and the political support of one of the world's most powerful states, Gazprom can afford for its stand to be less showy than the rest. In Russia, transparency is for those who need to be transparent and advertising is for those who need to advertise. Gazprom needs neither.

Back in the conference hall, Gazprom's Chief Executive, Alexei Miller, is master of ceremonies at a session on Russia's oil and gas future, hosting an audience of top oil and gas executives, all of them jostling for a share of the Russian future. Present are a Who's Who of the international hydrocarbon industry: the CEOs of no fewer than eight of the world's largest oil and gas companies. The total sales of the companies that they represent amounted to $1,883 billion in 2007, more than the GDP of Russia.[4] As outside the hall, the tone inside – at least from Russian participants – is one of self-congratulation.

Vagit Alekperov, CEO of Lukoil, shows a publicity film on the Varandey oil terminal, developed with ConocoPhillips in the Pechora Sea, high in the Arctic. Designed to allow ice-breaking tankers to load oil from the Arctic Timan-Pechora province at any time of the year, the $4 billion terminal is offered as evidence of Russian companies' ability to work in Arctic conditions just as well as their Western rivals. ConocoPhillips provided some of the technology from its Alaskan operations, and some was brought in from Finnish contractors, but Lukoil is keen to present this as a predominantly Russian operation. The first tankerload of crude oil left Varandey in June 2008, bound for the refinery of Come by Chance, on the coast of Newfoundland.

Miller himself talks confidently of the planned establishment of an exchange that will trade gas in Russian rubles rather than US dollars, undermining one of the key pillars of the dollar's position as the global reserve currency. Promising stability of gas supply for the future, he boasts that 'the size of reserves allows us to confidently say we can meet any solvent consumer's demand'. Gazprom, he insists, is 'investing in Europe's energy security'. Indeed, opposition within Europe to Gazprom is qualified as 'astounding'. 'One gets the impression some European officials cannot decide which they fear most: energy shortages or a fictitious Russian threat,' he complains, alluding to the position of some European countries – including the United Kingdom – that Gazprom should not be able to buy downstream

European firms. It is a bravura performance, and one which echoes the argument of Russia's government, that the West engages in a policy towards Russia consistent only in its double standards.

But there is dissent from the Gazprom view of the world here, too. There are whispered doubts in the room about Russia's ability to continue producing oil and gas at current rates. And despite claims of the scale of Gazprom's investments and Lukoil's know-how, whether Russian industry can develop the prize of Arctic resources – particularly offshore – is still an open question.

Most Western oil executives choose to talk around the subject, wrapping their analysis of Russia's hydrocarbon future in the success of specific projects or the global nature of the oil market. The CEO of Chevron, David O' Reilly, cites the need for hydrocarbon investments on a global scale, taking refuge in the statistics of the International Energy Agency suggesting a global investment shortfall of $140 billion in 2008. Jim Mulva of ConocoPhillips prefers to talk about his company's successful partnership with Lukoil – in which ConocoPhillips holds a 20 per cent stake – suggesting that $500 million has already been saved by the application of American technology to Russian fields.

Meanwhile, Thierry Demarest, CEO of the French oil major Total, and partner in the first phase of development of Russia's Arctic Shtokman field, limits himself to encouragement of domestic energy efficiency, a crucial issue if Russia is to export an increasing share of its production, but not a controversial one. In the audience, a delegation from ENI, the Italian oil and gas company so accommodating in its support of Gazprom's grand strategy for control of the European gas market, is silent.

Perhaps because they feel they have less to lose, or perhaps because they have already been burned, the CEOs of Shell, BP and ExxonMobil are more critical in their analysis of doing business with Russia.

Jeroen van der Veer, the tall, professorial head of Shell, the world's second-largest international oil company, renowned for his genial sense of humour and his Dutch plain-spokenness, suggests that investment decisions will be far harder 'in countries where the sanctity of contracts is under pressure'. In part, van der Veer's comments are a reflection of Shell's own problems with the multi-billion-dollar Sakhalin project in Russia's sub-Arctic Far East.[5]

This summer it is Tony Hayward, the British CEO of BP, who is on the hot seat. Caught in the middle of a boardroom struggle over control of TNK-BP, a fifty-fifty partnership between BP and Russian investors, Hayward is in an extremely delicate position in St Petersburg. TNK's BP-anointed CEO, Robert Dudley, has had difficulties with immigration officials. He may be forced to leave the country. BP's ambitions in Russia – both for TNK-BP, and for some kind of partnership with Rosneft on the development of Arctic oil and gas assets – seem on the point of blowing up.[6]

All eyes in the room are on Hayward – CEO for only a few months since the departure of his illustrious predecessor Lord Browne.[7] If he is too critical of Russia, BP will burn its chances to acquire 'bookable' Russian oil and gas, the holy grail of the industry and the main element of stock-market valuation. Too accommodating, and BP will signal its willingness to be permanently consigned to the second tier of oil companies, accepting the role of service and technology provider with no ownership of the oil and gas itself.

Hayward comes out fighting. His message, loosely translated, is simple: You need us. 'As everyone knows, production [of Russian oil] has stalled this year,' he tells the audience. His solution – an improved fiscal structure, application of Western technology and project management skills, creation of effective partnerships with Western companies – is standard. But the implied criticism of the current arrangement of the oil and gas industry is powerful, and for Russians who see oil and gas as the basis of the country's power, the suggestion that it might be best served by welcoming Western companies is a paradox. Zero-sum games, rather than win-win agreements, are still the dominant mental models of many in Russian government.

It is an old story in the Russian oil and gas industry, which stretches back to the Baku boom of the 1890s and the concessions awarded to foreign oil companies in the Soviet Union in the 1920s: a constant tug between the scale of Russian hydrocarbon resources in the ground and the Western technology and finance needed to get them out. In the 1890s it was the Rothschilds and the Nobels; in the twenty-first century it is Shell, Total, BP and the rest of the international oil companies (IOCs). The dilemma for Russia – a country where fear over losing control of its natural resources lies deep in the national DNA – is to what degree it is prepared to sacrifice national control for efficient production of resources. In the Russian Arctic, home to Russia's greatest future production prospects, where the technological and financial barriers to its realisation are greatest, the dilemma is at its most acute.

In St Petersburg, it falls to Rex Tillerson, CEO of ExxonMobil, to spell the message out, serving up his industry home truths with a generous dose of Texan charm. 'I never saw a bureaucracy produce a single barrel of oil,' he tells his audience, good-humouredly. Later, the kicker is more potent: 'There is no confidence in the rule of law in Russia today.'

In the evening, there is a reception in the grounds of one of St Petersburg's many palaces, hosted by the governor of the region, an up-and-coming figure keen to maximise the political benefits of a major economic conference in her city. Kamchatka crab from Russia's Far East is served with champagne from France. Smoked salmon is stacked up on my plate as if all of Russia's rivers have been destocked for this one night. In the gardens, actors dressed in eighteenth-century period costume stalk around, reminding Russians of the elegant glories of St Petersburg's past. On the lake, Russian gondoliers steer passengers around a candlelit raft where round-faced girls belt out famous classical arias.

Back on shore, a Russian businessman, a little unsteady on his feet, boasts about his nation's future. 'The country has lots of money, see,' he tells me, pointing around him as the party enters full swing. But how long will it last?

In June 1997, a senior government official from Moscow submitted a 218-page dissertation to the St Petersburg Mining Institute on the role of mineral resources in the Russian economy. The dissertation, for an advanced degree in economics, was probably begun a year earlier. Perhaps the government official who submitted it, having lost his job in St Petersburg in 1996 following the defeat of Mayor Anatoly Sobchak in that year's elections, thought that an academic qualification would improve his chances, at the age of forty-five, of finding a challenging new job.

But by the time Vladimir Vladimirovich Putin had submitted his dissertation, he had already landed a new job: working for the administration of President Yeltsin in Moscow. Eighteen months later, after a lightning ascent, he would assume the role of acting President of the Russian Federation following Yeltsin's surprise resignation on 31 December 1999.

The text of Vladimir Vladimirovich Putin's thesis was subsequently taken out of public circulation and classified. Some have argued that this was more to protect him against allegations of plagiarism from an American textbook – translated into Russian by the KGB in the early 1990s – than to protect state secrets.[8] Whatever the detail of the dissertation itself, a subsequent article in the inhouse journal of the Mining Institute, apparently reprising and expanding its themes, provided a road map for the role of energy under Putin's presidency and its de facto continuation under President Medvedev.[9]

The core thesis of Putin's 1999 article – 'Mineral Resources and the Strategic Development of the Russian Economy' – was that Russia's future as a great power depended on the rational management of natural resources, valued at $28 trillion.[10] To him, 'furthering the geopolitical interests and maintaining the national security of Russia' should be a key objective of energy policy. But in order to realise the potential of Russia's natural resource endowments – and to use those endowments to support the development of Russia's non-resource-based economy – the relationship between the industry and the state would have to be radically revised. After the break-up of the oil industry in the 1990s – the gas industry largely avoided Balkanisation – and the asset stripping of entities more interested in exhausting depreciating assets than investing in Russia's future production, the state would have to reassert itself.

This was not about a return to the Soviet model – which Putin suggested was responsible for a 'low level of effectiveness and lack of capacity in world markets'. What he envisaged instead was the creation of vertically integrated oil and gas companies, capable of competing with Western rivals and attracting foreign investment. Instead of Soviet-style autarky, Putin argued that Russia simply needed to set 'the conditions for the sustainable entry of Russia into the world economy'. The state did not necessarily need to own everything, as long as its right to determine the strategic direction of the industry was unquestioned.

The Byzantine history of the Russian oil and gas industry over the last ten years has been, at heart, the realisation of the ideas laid out by Vladimir Vladimirovich Putin in his 1997 dissertation and his 1999 article. Chief amongst them are the twin principles of the state's strategic control of the domestic industry, and the use of that control to enhance the country's power. For both producers and consumers of Russian oil and gas, all roads lead to the Kremlin.

For the most part, Putin's aims have been met. Since 2000, Russia's oil and gas industry has been rationalised, state power has been reasserted and potential political rivals to the President have been forced out.[11] Mikhail Khodorkovsky, an ambitious (and allegedly unscrupulous) oil magnate has been imprisoned. Yukos, the company which he ran, has effectively been expropriated. By October 2003, when Khodorkovsky was arrested, he was reportedly on the point of selling a large part of Yukos to US companies ExxonMobil and Chevron. With Khodorkovsky's arrest the hopes for a strategic energy alliance between the United States and Russia – much heralded by the Bush administration in 2001 and much questioned by experts – have fallen apart.[12]

The main winners of the Kremlin's reassertion of state control, and its focus on Russification of the industry, have been Gazprom and Rosneft – both public companies in which the state holds a controlling share, and both closely tied to political interests deep in the fabric of the Russian state. In the offshore Arctic, it is Rosneft and Gazprom who have been selected to develop Russia's resources. In July 2008, Igor Sechin, former KGB agent and Russian Deputy Prime Minister, announced that only companies with five years' prior experience of working on the Russian continental shelf – most of which lies in the Arctic – would be eligible to bid for their oil and gas reserves. Handily, only Gazprom and Rosneft – the oil company of which Sechin is also Chairman – fit that criterion.

Foreign oil and gas companies, never quite sure of where they stand with Russia, have been forced to endure uncertainty, harassment and, in some cases, indirect expropriation. Production-sharing agreements (PSAs) agreed on in the 1990s – referred to by Putin as 'colonial' – have been repudiated. Nonetheless, humbled, embarrassed or both, foreign oil companies have kept returning to Russia. Driven by the need to maintain their reserves, they have had little choice.

The external part of the Kremlin's strategy, to use the country's natural resources to promote Russian power and influence abroad, has focused on Europe. It is in Europe, and particularly in the pipeline politics around Europe's fringes, that the Kremlin's controlling instincts have been most in evidence.

Russian influence in Europe flows partly from Europe's declining domestic production of natural gas, and Russia's growing role as the continent's supplier.[13] Without any realistic prospect of Europe's main non-Russian suppliers – Algeria and Norway – making up for Europe's falling production, dependency on Russia is forecast to rise. (The European Commission has warned that European dependence on imported gas – most of it Russian – could rise from 60 per cent to 77 per cent as early as 2020).[14] Both in Brussels and in Washington, Europe's growing dependence on Russian gas is cause for concern, with fears that much-needed Russian gas will come with political strings attached.[15]

Russia's hand is already strong. But it has been critically strengthened by European disarray. In the absence of a single integrated European gas market, Russia can pursue bilateral deals with favoured European countries, while excluding those it views as difficult – notably the Baltic states, Poland, and the Ukraine.[16] And as long as Russia controls the physical infrastructure of European gas lines – and can physically circumvent countries opposed to its ambitions – Russia can hope to divide its European neighbours against themselves, and boost its influence as a result.

In the North, Russia's pipeline strategy has focused on the Nord-Stream gas line to Germany. A 1,200-kilometre route along the floor of the Baltic Sea, skirting the Baltic states and Poland, should allow Russia to provide its largest market with 55 billion cubic metres (bcm) of long-term natural gas – much of it planned to come from the Russian Arctic – without transiting its former Warsaw Pact allies, now amongst its most vociferous critics.[17] In the south, Russia has revived a historic energy alliance with Italy and played off Austria – the first western European country to import Soviet gas, beginning in the late 1960s – against Hungary.[18] The aim has been to ensure (possibly without success) that the Russian-controlled South-Stream pipeline is built, instead of a US-backed alternative which would allow Europe to import Central Asian (or even Iranian) gas, without transiting Russia.[19]

As a grand strategy for Russia's resurgence as an energy superpower – asserting state control over Russia's natural resources at home while asserting Russian control over supply routes abroad – the Putin system has been remarkably successful over the last decade. Its future, however, is more in doubt.

The sustainability of Putin's model of Russia's energy superpower status depends on three factors: control, price and production.

On control, Putin was skilful: putting the Kremlin at the core of a reorganised national oil and gas business, and maintaining Russia's new-found energy leverage internationally through a series of blocking moves, bilateral deals and diplomatic manoeuvres. On price, Putin was lucky: the run-up in world oil prices began a matter of months before Putin became President. On production, Putin took advantage of a Russian oil industry on the rebound, boosting production from its mid-1990s lows. Over the last few years, Russia's oil and gas production has been consistently high.

But there are indications that these factors may not hold in the future as they have over the last decade. President Medvedev may not be able to make the Putin system work as well as it did for his predecessor.

The fall in the price of oil and gas from the stratospheric heights of 2008 may not be the true long-term challenge for Medvedev. Though the decline has been precipitous since summer 2008, oil prices – and the gas prices which are largely calculated in reference to them – remain high in historical terms, and most expect them to stay there (or recover further) over the next few years.

Of greater concern is the long-term sustainability of Russia's production. (Indeed, Kremlin control over the domestic oil and gas industry may be exacerbating the problem.) If Russia's energy superpower status is to be sustained, maintaining Russia's faltering production is imperative. Production from the Arctic – particularly of natural gas – is the Kremlin's last best chance of avoiding a devastating production crunch. In the eyes of the Kremlin, producing Russia's Arctic resources is not a choice, it is a strategic necessity.

Were Russian oil production to peak, the direct consequences would be mostly economic: a further hit to Russia's solvency, a further weakening of the ruble, and a further reduction in the ability of the Russian government to buy its way to economic diversification. But a shortfall in Russia's production of natural gas would be altogether more serious, spelling disaster for Russia's geopolitical ambitions and undermining

Russia's leverage over its European consumers. A Russian failure to honour contracts with its major European consumers would be so calamitous that one analyst has argued that 'Russia may even prefer to lose money on importing [and then re-exporting] Central Asian gas rather than default on export contracts'.[20] Another way of preventing a shortfall in exportable natural gas would be to bring further Arctic resources – the largest remaining reserves of Russian natural gas – into production.

The problem for both oil and gas is one of production, not of reserves. Russia has some 79.4 billion barrels of proven oil reserves, more than any other country outside the Middle East bar Venezuela, and nearly three times the proven reserves of the United States. In natural gas, Russia's reserves are still more impressive, amounting to more than one quarter of the world total. At 44.6 trillion cubic metres Russia's reserves of natural gas exceed those of its nearest rival, Iran, by more than two thirds; they are nine times as great as the proven natural gas reserves of the United States.

The recent history of Russian oil and gas production is not as positive as the scale of its reserves might suggest. For much of the 1990s, the narrative was one of a country running down its assets, not developing them. Even now, Russian oil production is substantially lower than in the final years of the Soviet Union, having fallen from 11.5 million barrels per day in 1987 to 6.1 million in 1996 and then having risen to just over 10 million bpd in 2007.[21]

According to the Russian government, future oil production will peak at 10.7 million bpd in 2020 before entering a terminal decline.[22] The International Energy Agency predicts a lower plateau – some 10.4 million bpd in 2015 falling to 9.5 million bpd in 2030.[23] But others argue that both those figures are far too high, and that the peak has already been passed, arguing that the underlying pattern of production has been one of decline since 2006.[24]

For several years after the collapse of the Soviet Union, geological work virtually ceased. Discovery and development of new Russian oilfields have been slow. In 1999, Putin pointed to the need to 'halt the lagging growth in reserves relative to the volume of mineral resources extracted', suggesting that Russia should urgently 'expand the study and exploitation of the resources of the [Arctic] continental shelf and open seas'. Now, the Medvedev–Putin regime is doing just that. But progress is slow: as of early 2007 there were only fifty-eight wells in the Russian Arctic, compared to 1,500 in the Norwegian Arctic.[25] Most oil production still comes from fields discovered by the Soviet Union. Russia's offshore Arctic oil prospects are virtually untouched. The country's first Arctic offshore field, Prirazlomnoye, is only expected to enter production of 120,500 bpd in 2011, after years of cost overruns and delays.

The recent history of Russia's natural gas production is far less dramatic than that for oil. From 431 billion cubic metres (bcm) produced in 1985 the figure rose to just under 600 bcm in the final year of the Soviet Union, dipping in the 1990s, but reaching a new high of 612 bcm in 2006.

The relative stability of Russia's natural gas production after the collapse of the Soviet Union is partly explained by the structure of the gas industry: whereas the oil industry was split up and sold off, causing widespread disruption, the Russian natural gas industry held together. The last Soviet gas minister became the first chairman of Gazprom.[26] But while the stability of the gas industry's structure may have prevented a slump in production in the 1990s, the same factor may hinder attempts to boost production in the future. The dominance of Russian gas production by Gazprom – a political behemoth as much as a commercial enterprise – does not inspire confidence in its ability to develop new gas fields. One American official I spoke to went as far as to argue that 'Gazprom is more interested in kick-backs than in pumping more gas'.

In 2007, Russia's aggregate natural gas production fell by 0.8 per cent, while Gazprom's output dropped 1.3 per cent, to 548 bcm. Publicly, the company blamed the fall on a lack of demand from Europe, enjoying an unusually warm summer.[27] In reality, the decline stemmed from the terminal decline of Gazprom's main sources of gas, the super-giant fields of the Nadym Pur Taz region of West Siberia.[28] As a result, Gazprom was only able to achieve its 2008 contractual obligations in terms of European supply by the transhipment of gas from Turkmenistan, a former republic of the Soviet Union.[29]

Such an expedient is not a long-term solution. For that, Gazprom needs Arctic gas to replace its West Siberian reserves. Only then can Gazprom prevent a temporary dip in overall production from escalating into something more serious.

By 2020, Gazprom expects half its natural gas to come from the north-west Russian Arctic: the low-lying Yamal peninsula, protruding into the frozen Kara Sea and the offshore Shtokman field, 600 kilometres into the Barents Sea.[30] Yamal alone is forecast to provide 75–115 bcm of Russian gas as early as 2015, rising to 310–360 bcm by 2030.

But both projects are technically complex and massively expensive. Yamal will require some $80 billion of investment while development of Shtokman will require a further $20–30 billion. Many doubt the gas fields can be opened on time – indeed, the start dates for Yamal and Shtokman have been put back several times already.[31] In the context of a global economic crisis, finding the money to develop them may be harder than ever.[32] But if the projects are delayed further, European consumers may find that their problem with Moscow is not too much Russian leverage, but too little Russian production.

Arctic development is essential to Russia's continued status as an energy superpower. Without Arctic oil, Russia's output will stagnate and decline. Without Arctic gas, a manageable shortfall in production will turn into an unsustainable deficit in its ability to export. It is with good reason that a Gazprom publication advertises, 'There is no alternative to Yamal!'[33] The Kremlin agrees.

As a result, the real question for Russia's leaders is not whether or not to develop the Arctic, but how. There is a lot of national prestige riding on the answer. Development of technically challenging Arctic oil and gas reserves offers Russian companies the opportunity to shake off a reputation for poor management and inefficient production practices. It offers the state the opportunity to prove that the 'national champion' model works, and that foreign expertise and finance are no longer necessary. To succeed, however, Russia will need to overcome the history and legacy of the Soviet oil and gas industry.

The Soviet industry was renowned for its heroic feats of production. During what Lev Tchurilov, the last Soviet oil minister, described as the 'golden age' of Soviet oil, production increased four times over. In 1960, the USSR produced 2.9 million bpd; in 1980, it pumped 12.1 million bpd, more than the United States and Canada combined.[34] But much of this growth, relying on basic technologies and trial-and-error tactics, was inefficient and wasteful. 'We like to create dreadful problems, then solve them heroically,' as Tchurilov put it.

The culture of the Soviet oil and gas industry was determined by the structure of incentives set by Moscow. Frequently, the result was absurdity: wholesale flaring of gas produced as a byproduct of oil production, not only because of low gas prices, but because the oil ministry had no incentive to pay for its delivery to the gas ministry; drilling of many shallow holes rather than fewer deep ones, partly because the quality of Soviet steel was not sufficient to allow for drilling through harder ground, but also because drillstaff were awarded bonuses on metres drilled rather than production found; over-speedy production of oil and gas fields such that the eventual output was much lower than would have been technically feasible had the fields been developed more carefully.

The first victim of Soviet productivism was the environment. Local concerns about environmental damage caused by development were always trumped by Moscow's demand for output. Tchurilov describes the brutal 'Siberian technology' employed:

> We didn't always succeed in treating nature with proper care . . . Trees were cut down, and if there was no time or equipment available, we would just use explosives to remove stumps . . . The mess is still there now . . . [People] turned a blind eye to such barbarism. Anyone who objected was fobbed off with the explanation that the faster we worked, the more oil we would produce.[35]

The consequence of Soviet productivism was the development of an industry with no culture of conservation or resource management. Unlike in the West, there was no incentive to invest time and money in improving production technologies or developing enhanced recovery techniques. These would only distract from the immediate goal of meeting the annual target. It was far easier to open up new fields than to maximise production from old ones. As Tchurilov explained:

> It was all too easy for the government to become complacent and believe that the oil Eldorado would go on forever. Back in the 1950s . . . we didn't have to worry about methods of enhanced recovery. The crude oil just came gushing out of the ground . . . As long as the oil gushers in the Volga/Urals region, and later in West Siberia and Kazakstan, were coming on line, no real investment was made in research and development. We just did not bother to develop advanced oil field equipment. All that our leaders wanted was quantity, quantity and more quantity.[36]

This pattern of Soviet oil development – exploiting one region at breakneck speed before moving on to the next – was reflected in Tchurilov's career. Born and raised in Grozny, the capital of Chechnya and heart of one of Russia's oldest oil-producing regions, Tchurilov's first job was in the Volga region, the core of the Soviet industry in the 1950s and 1960s. In 1964 he was moved to West Siberia, where Soviet geologists discovered the massive Samotlor field the following year. In 1973, he was sent to Russia's latest oil frontier: the vast Komi Republic, stretching from central European Russia to the oil-rich Timan-Pechora basin, straddling the Arctic Circle.

In Ukhta, Tchurilov found an Arctic oil industry which had changed little since the 1930s, when Stalin's northern drive opened up the Yareg and Voyvozh fields with gulag labour. The physical traces of the camps were still visible. The technologies applied were the same as they had been forty years ago. Typically for the Soviet oil industry, Tchurilov was forced to spend more of his time fighting a bureaucratic guerrilla war with Moscow, trying to persuade Gosplan that the grand production targets they had set for the Komi Republic could not be delivered.[37]

To Tchurilov and to others, it was clear that the effective development of this latest oil frontier – if possible at all – would require technologies available in the West, but unavailable in the Soviet Union. In some cases, an American embargo simply prevented the import of outside technologies and equipment. In others, the problem was a lack of hard currency. Even when Western equipment was purchased, Moscow was opposed to Western service companies being allowed to install it on Soviet soil. In the offshore Arctic, the problems were still more daunting. In 1977, a secret CIA memorandum argued that while the hydrocarbon potential of Russia's offshore Arctic was great, the 'technology for exploration and production of this region does not yet exist, even in the West'.[38]

Given the inability of the Soviet Union to develop Arctic resources by itself, it is not surprising that when the Russian oil and gas industry began to open up in the late 1980s and early 1990s, one of the first places foreign companies were allowed to invest was in the frontier regions of the Arctic and sub-Arctic. In 1987, Mikhail Gorbachev even raised the prospect of Arctic oil and gas development as a great cooperative endeavour for the Soviet Union and the West.

Later, in the early 1990s, the Timan-Pechora basin where Tchurilov had worked became the front line of Western involvement in Russian oil. In 1992 Conoco (later ConocoPhillips) made what was the largest ever foreign investment in the post-Soviet Russian oil industry, setting up a joint venture with Rosneft known as Polar Lights, with plans to develop the Ardalin field, above the Arctic Circle.[39] The field came onstream two years later.

A few weeks after the Ardalin field started producing oil, Russian President Yeltsin and US President Clinton signalled support for an even more ambitious scheme: the organisation of a Western consortium of oil companies – Texaco, Exxon, Amoco and NorskHydro – to develop Russia's Arctic Timan–Pechora reserves more widely. Meeting on the margins of the G7 summit in Naples in September 1994, the two presidents announced that the 'Timan–Pechora project [is the] top priority in the oil and gas sector'.[40]

But since the 1990s, most of the deals struck by Western oil companies for access to Russia's Arctic and sub-Arctic resources have unravelled. Facing apparently insuperable administrative obstacles, the Timan–Pechora consortium was effectively disbanded by the early 2000s. Almost uniquely, the French oil major Total has held onto its Production Sharing Agreement (PSA) for the Kharyaga field, but not without coming under tremendous pressure from Russian authorities and spending hundreds of millions of dollars on the field's development.[41] The various Sakhalin projects – sub-Arctic geographically, but Arctic in terms of conditions – have seen Western oil companies forced to relinquish some or part of their original positions.

This is the Putin system at work, reestablishing Russian companies – and the Kremlin itself – at the heart of the Russian oil and gas industry, while forcing foreign companies to accept secondary status. Moscow is determined that oil and gas development in the Arctic be defined as a national project and led by Russian companies.

Privately, however, many doubt Russia's ability to develop its Arctic resources on its own. One CEO of a leading Western oil major argues that, even today, the best Russian oil companies are still ten years behind their Western counterparts. The latest cycle in Russian resource nationalism, he argues, will be forced to an end when the Russian government realises that its resources cannot be fully developed without foreign technology, foreign money and foreign management.

In his view, Western companies will be involved in Russia's Arctic development because there is no alternative.

The best symbol of Russia's ambitions for its Arctic oil and gas future, and the best test of its ability to succeed, lies 550 kilometres north of the Kola peninsula, under 350 metres of the cold Barents Sea. Discovered by Soviet geologists in 1988, the Shtokman field is huge. Gazprom estimates that it contains 3.8 trillion cubic metres of natural gas – more than the entire proved reserves of the European Union, and more than half the proved reserves of the United States.[42] Gazprom strategists project Shtokman will come onstream in 2013, producing 23.7 billion cubic metres (bcm) of natural gas per year. Exports of Liquefied Natural Gas (LNG) to the United States are slated to begin the following year.

That scenario may prove too positive. With pack ice up to 3 metres thick and the risk of drifting icebergs weighing up to 10,000 tons, drilling Shtokman will be tough. The construction of a gas line to the shore – running for 550 kilometres along the bottom of a sea famed for its stormy weather – will be a major engineering feat. Once the gas has arrived onshore, Gazprom may face an even more daunting technical challenge. If it is to fulfil its stated ambition of becoming an exporter of LNG, it will have to construct an expensive and complex liquefaction plant, condensing the gas and cooling it to less than -150° Celsius. This is something that Gazprom has never done before.[43]

In October 2006 Gazprom surprised the world – and the five Western companies then negotiating to partner with Gazprom on the development of the Shtokman field – by publicly stating its intention to go it alone.[44] (The five suitors were officially informed of Gazprom's decision only after CEO Alexei Miller had announced it in a television broadcast.) It was a swaggering statement of confidence in Gazprom's ability to develop its most ambitious project.

Just nine months later, Gazprom reversed course, signing a framework agreement with Total for 25 per cent of the first phase of the Shtokman project. A few months after that, Norway's StatoilHydro was signed up with a further 24 per cent. As with the Nord-Stream and South-Stream pipeline projects, the deal took the legal form of a Special Purpose Vehicle set up to manage the project, registered in Zug, Switzerland. Although both Total and StatoilHydro will be keen to 'book' Shtokman's resources as part of their reserves, they actually only own a share of this SPV, rather than a portion of the field itself. Gazprom retains both ownership and control.[45]

Russian news reports of the Total–Gazprom deal presented it as the direct result of a conversation between then-President Putin and his French counterpart, President Sarkozy.[46] Coming at a time of rising tension between Europe and Russia, the deal certainly fitted a wider pattern of Russia's diplomatic strategy in Europe: favouring bilateral energy relationships with Europe's main players at the expense of both Europe's smaller countries and the European Union as a whole.

The deal with StatoilHydro involved a similar political element, with Norwegian diplomats and politicians actively lobbying for a Norwegian choice, and Putin apparently calling the Norwegian Prime Minister shortly before the deal was announced to indicate his personal support. Collaboration between Russia and Norway on Shtokman may act as a catalyst for closer cooperation in other Arctic areas, including in the resolution of an ongoing border dispute in the Barents Sea.[47]

In addition to any political pay-off involved, Gazprom had compelling technical and financial reasons for reconsidering partnership with foreign companies on Shtokman. It came to realise that, without Total and StatoilHydro's participation, the field might never be developed. An alliance with Total offered cash and, potentially, LNG technology.[48] StatoilHydro brought long familiarity with offshore drilling (both in the Barents Sea and in the North Sea) and the valuable experience of developing Norway's own offshore Arctic gas field, Snøhvit (Snow White).

Bringing foreign companies on board on the Shtokman project, even in such a secondary role, was not quite as prestigious as the Russian-only project some in the Kremlin might have wished for. But it may be a sign of things to come. As Russia turns to develop its Arctic north, the dilemma of whether or not to involve foreign partners will become sharper.[49]

MURMANSK, RUSSIA, 68° North: Nowhere stands to gain more from Russian oil and gas development in the Arctic than Murmansk. With around 350,000 inhabitants, Murmansk is the largest city above the Arctic Circle. It is also the capital of the Murmansk region, similar in size, population and population density to the US state of Montana.

You can already see the changes that the prospects of oil and gas development have brought. 'We used to have no hypermarkets at all,' a local tells me, pointing down to the new building on the edges of central Murmansk, 'but now, if trends continue, in just a few years from now, we will have more than one per head of population.' He laughs loudly. It's an old Russian joke, updated for modern Murmansk, the productivist fantasies of the Soviet Union replaced by the consumerist fantasies of modern Russia. But, like most Russian jokes, there's a serious point in the absurdity. As the promise of Arctic oil and gas development finally reaches the point of major investment, Murmansk is hoping to get rich, or at least less poor.

It is not a prepossessing place. In the Second World War, the city was obliterated by the German Luftwaffe, aiming to destroy the Soviet Union's northernmost year-round harbour, but not minding much if they missed. (One Russian tells me of a night during the war when his father, manning a fire brigade observation tower in the middle of the city, watched his own wooden house go up in flames.) Of all the Soviet regions, the Murmansk oblast was amongst the most heavily bombed. The city, built mostly of wood, was utterly destroyed.

After the war, downtown Murmansk was rebuilt as a monument of Stalinist neoclassical architecture, the main street lined with generously proportioned ten-storey apartment blocks. Later, as the city was developed into a major fishing port, a transport hub and home of the Soviet Northern Fleet, the hills around the city were covered with identical Brezhnev-era tower blocks. At its height in the 1980s, nearly 500,000 Soviet citizens lived here. But with the collapse of the Soviet system of Arctic benefits, and the decline of the city's military complex, many of Murmansk's tower blocks were abandoned. Now, with the city's population on the rise again, Murmansk faces an acute shortage of quality housing.

For those who know how, the city is a place of opportunity. Evgeny, a courteous and intelligent man with quick eyes and a quicker mind, runs his own translation business, shuttling between Russian oil executives and the Western oil executives – mostly Norwegian, British and French – who have come here to help develop Russia's Arctic oil and gas. When Evgeny bought his flat in 1998 it cost $8,000, but now it's worth perhaps $60,000. When his first child was born it was easy to get a place in kindergarten. He and his wife have had more difficulty getting a place for their second child, he says, but that, in a way, is a sign of progress.

For Victor Gorbunov, the government official responsible for economic development of the Murmansk oblast, things can only get better. His model for Murmansk's development is the Arctic city of Tromsø, on Norway's west coast. But there is a long way to go. 'The current salary in this oblast is 20,000 rubles [$700] per month, the tenth richest in the country,' he says, 'but in three to four years I hope it will be 40,000.' As Victor sees it, his main tasks are overcoming labour shortages and poor existing infrastructure.

Welders are in particular demand, Victor tells me. But most Russian welders trained in Murmansk want to work in Norway. Victor has been trying to get the federal government in Moscow to improve the visa regime for foreign workers so Murmansk can attract skilled foreign labour as well as exporting skilled domestic labour, but progress is 'very slow'. (At the other end of the scale, the city's hypermarket had to be built with Turkish and Chinese workers because there were not enough Russian workers to go round.)

Victor's primary infrastructure concerns for the future are electricity supply and the capacity of the port and railway network feeding Murmansk. Without upgrades in these areas, he worries that the substantial spending on the Shtokman field – and on other developments in the Barents and Kara Seas – will be redirected to Norway and Arkhangelsk. And, even with climate change, Victor thinks that not enough ice-breakers are being built to service Russia's oil and gas expansion in the more ice-affected areas north and east of Murmansk.

Oil and gas development is a spur to investment in other sectors, too. In 2008, the port of Murmansk announced a $6 billion programme of investments.[50] One element of the programme is the planned relocation of the current open-air coal terminal – a source of respiratory health problems across the city – to the other side of the inlet. But its centrepiece is the construction of an oil shipping terminal, expected to be a major terminal for Arctic oil. (Rosneft has taken a 15 per cent stake in the new port management company.)

Sitting in the back of a chauffeur-driven Volkswagen 4x4 driven at breakneck speed through Murmansk's potholed streets and up to the massive war memorial which dominates its skyline, Captain Vladimir N. Avduyukov speaks with pride about the the port. 'I have given my life to it,' he tells me. 'How can I fail to be proud of what we have achieved here?' Now an adviser to the Murmansk Maritime Port Administration, Captain Avduyukov first put to sea as a cadet in 1953, the year of Stalin's death, plying the Arctic cargo routes across to Siberia and up to the military installations on Novaya Zemlya. He claims to have gone as far north as 88° North, just shy of the North Pole.

These days, most of the old routes he used to run have been pared down. Decades ago he worked on a passenger ship which made thirty-two calls down the coast between Murmansk and Arkhangelsk – but 'now there's just one', he tells me. But, for Avduyukov, the future of Murmansk port is not in recreating that past. He expects oil shipments through the port to increase dramatically in the next decade. And, further into the future, he believes that the expansion of the Northern Sea Route under conditions of climate change will turn Murmansk into the dominant Arctic trading entrepot. 'I just hope for global warming,' Avduyukov tells me.

But this portrait of a booming Murmansk of the future, constrained only by the availability of infrastructure, is uncertain. Development of the Shtokman field could be delayed, and contracts to supply it could be taken up by Murmansk's rivals. Investment could be drawn off to the Yamal project. There are no guarantees.

Domestically, Murmansk faces considerable competition. Lukoil's plans to expand the Varandey oil terminal on the Pechora Sea may mean that there is not enough oil to fill the planned Murmansk terminal. Even though Rosneft has a financial interest in Murmansk port, some report that it places priority on shipments from Arkhangelsk and St Petersburg.[51] And while the Korean-built platforms for the Shtokman development will be assembled at the Sevmorput yard in Murmansk, much other work has already been taken up by the troubled Sevmash shipyard in Severodinsk, near Arkhangelsk. (Sevmash is attempting to transform itself from the home of the Soviet Union's nuclear submarine industry to the home of Russia's oil platform construction industry.)[52]

In Norway, businessmen and politicians believe that Kirkenes may play a strong supporting role in Russia's Arctic oil and gas development. Some technical equipment for the Shtokman field has already been shipped through Norway to avoid two weeks of customs' delays in Murmansk. When I visited Kirkenes, dozens of Russian fishing trawlers dotted the harbour. Despite Norwegian wages, the refitting service in Norway is quicker and cheaper.

Already, Murmansk is a divided city. Many stand to benefit from an uncertain – but mostly hopeful – future of oil and gas development, port construction and the expansion of shipping in Russia's Arctic. Some fear that they will be left behind, or that they already have been.

The fishing community of Murmansk is concerned its voice will be drowned out by companies with far greater investment and public relations budgets, and better connections in Moscow. Although approximately 10 per cent of the workforce of Murmansk is employed in the fishing industry – many more than will be directly employed on the Shtokman project in its operational phase – how can they compete politically with the likes of Rosneft or Gazprom? And what will happen to Murmansk's Soviet-era population: the thousands of citizens who came north when the subsidies were good in the 1960s and 1970s and who are now stranded here?

As the chief of the socio-economic department of Murmansk city, Victoria Shvets has to deal with the problems of both the Soviet legacy in the Arctic, and the more familiar growing pains accompanying the establishment of Russia's Arctic future. Her own family moved here in an earlier expansion of the Russian north, with the construction of the railways at the beginning of the twentieth century.

'Russia is not a country to prepare beforehand,' Victoria tells me. Although the Shtokman development has been talked about for nearly twenty years, the reality for most of that period has been economic decline. Now that things are beginning to turn, the city is struggling to respond to pressure on housing and public services. Local doctors already have one and half times the normal workload. And an explosion in the number of births – rising 5 per cent in every one of the last three years – is putting pressure on places in kindergartens in the city. 'We need more kindergartens, schools and hospitals,' Shvets says, 'but at the moment we have no solution to this problem.' Some oil companies have donated money directly to projects in the city – StatoilHydro has given some money towards a new sports facility – but it is a drop in the ocean.

There is frustration with Moscow as well. Although 2007 was the first year for a long time that the city was not in deficit, the growing centralisation of Russian government – referred to bureaucratically as 'vertical construction' – means that spending decisions are increasingly made in Moscow. Bluntly, 'we have to invest in what the federal or regional government tells us to'.

Murmansk's housing deficit, particularly for low-cost housing, is dire. Nothing was built in Murmansk in the 1990s. At that time, 'everyone was looking to their immediate bodily needs', as Shvets puts it. In 2008, the city entered a construction boom, with the construction of 28,000 square metres of housing, and with another 35,000 planned for 2009. But because of the earlier decade-long slump in Murmansk's housing market, there was no building infrastructure in place and everything had to be imported. As a result, most of the new housing is expensive, at 42,000 rubles per square metre. This is beyond the reach of those who Shvets euphemistically calls the 'stable income population': the many in Murmansk who get by on low pensions or government support. For them, Murmansk's housing boom is an irrelevance.

As I leave her office, Shvets points out one advantage of Murmansk which has lasted through the hard times and the good: 'Murmansk is the only place in the world you can see lilacs blossom in July.' Walking down Lenin Avenue, I see that she is right.

As the Russian Arctic comes increasingly into focus – and as plans for development of its resources take shape – protection of the environment will become a major battlefield.

Things have changed since the demise of the Soviet Union. Before the fall, environmentalism was widely mistrusted by government as a dangerous force with the potential to act as a focal point for public anger against the regime. Arctic economic development and military testing were largely conducted without full consideration of possible environmental or health consequences. In the mining town of Norilsk, in northern Siberia, huge amounts of copper and nickel oxides were released into the air. In Murmansk, there is still concern over the safety of the Kola Peninsula's Andreyeva B nuclear power station and the final decommissioning of nuclear submarines at the Anadyr shipyard on the coast of the Barents Sea. Further into Russia's Arctic north, the full consequences of the Soviet Union's extensive testing of nuclear weapons on Novaya Zemlya are not known. Of the ten most polluted places on earth identified by the American organisation the Blacksmith Institute in 2007, four are in the former Soviet Union.[53]

These days, environmental issues are much more widely discussed. Quality of life is a major public issue. (It is even incorporated into Russia's national security strategy.) Former Communist leader Mikhail Gorbachev heads an international non-governmental organisation devoted to the environment. Putin has criticised highly destructive practices in the oil and gas industry – such as the widespread flaring of gas – as 'barbaric'. But the environment remains a secondary issue when it comes to decisions on Arctic hydrocarbon development. When President Putin came to power in 2000 he placed environmental responsibilities within the Russian Ministry of Natural Resources which, conveniently, is also responsible for energy policy.

Chief amongst the political imperatives driving environmental policy is the drive for resource maximisation. Flaring gas, a frequent byproduct of oil development, is viewed as 'barbaric' primarily because it is a waste of a potentially valuable national resource, not because of the carbon dioxide emissions which it generates. Estimates of the amount of natural gas flared every year in Russia range from 20 billion cubic metres (bcm) to up to 50 bcm. (For comparison, Russia's overall gas production is around 600 bcm.) Estimates of the number of oil spills in Russia every year go as high as 20,000 – partly from the production sites, but also from shipping oil through a degraded pipeline network. The government's chief concern is over lost production more than clean-up.

In principle, environmental permits for oil and gas development are much tougher than in the past. But the system of acquiring permits is frequently ignored, and open to abuse. In 2006, in the midst of an investigation into alleged infringements of Russian environmental legislation at Sakhalin, Shell relinquished its controlling interest in the project to Gazprom for $7.5 billion. After Shell caved in, the investigation disappeared. Many concluded that the threat of ongoing environmental investigation had been a tool of Russian commercial interests, intended to pressure Shell into selling its stake in the project rather than improving its environmental management.[54]

At the Moscow headquarters of the World Wildlife Fund, in a dusty backstreet near Taganskaya Metro, Alexei Khizhnikov argues that real progress is being made in how Russian oil and gas companies incorporate environmental concerns into thinking about their strategy and operations. 'A year ago they [the oil companies] didn't want to talk to us,' says Alexei. Now, he is in the process of setting up long-term dialogues: with Rosneft on West Kamchatka in Russia's Far East, and with Lukoil on the Varandey oil terminal in the Russian Arctic.

The shift in environmental attitude comes partly as a result of partnerships of some Russian producers with Western companies – companies that are far more sensitive to allegations of environmental damage. For Alexei, the influence of Western oil companies on Russian environmental practices is unequivocally positive – whether it comes from general shareholder pressure or through the exchange of best practices on specific projects.

Khizhnikov cites Lukoil's relationship with ConocoPhillips as a positive example: 'A few years ago, they [Lukoil] used to try and hush up oil spills in the Komi republic, but they would think twice now.' Still, attentiveness to the environment is uneven. Lukoil's Moscow management team may be more accessible to environmental groups than in the past, but things are more varied in the field. In the Nenets Autonomous Okrug – an oil-producing region along Russia's Arctic coastline – there is a real distinction between offices operated in partnership with ConocoPhillips and those operated by Lukoil alone, some of which are still 'Soviet-style' in their environmental attitudes.

More starkly, Alexei reports a cultural divide within TNK-BP between British and Russian shareholders, some of whom apparently view environmental standards introduced by BP managers as 'expensive and unnecessary', tying the company in red-tape and stalling production.[55] The trend towards 'national champions' in the oil and gas industry, screened from the influence of Western corporations, might weaken the prospect of adoption of international environmental standards. It is no particular surprise that Gazprom, the oil and gas company least accessible to environmental groups, is also the least Westernised and the most closely linked to the state.

But Western companies are not the sole cause of a more subtle understanding of the environment in Russia. Much improvement comes from a wider cultural shift towards environmentalism in Russian society. The objectives set by government for improved resource management rather than Soviet-style, produce-and-be-damned, oil and gas development have also played a role.

Paradoxically, Russian companies' Arctic operations may be a catalyst for improvements in the quality of operations across the board. The levels of project management and technology required in the Arctic far exceed those required to develop the Soviet oilfields of the 1950s and 1960s. Forced to change the ways they operate by the difficulty of Arctic conditions, Russian companies are bound to adopt more up-to-date, and therefore environmentally friendly, methods.[56] And as Russian companies make substantial investments in places directly affected by climate change, such as on the low-lying Yamal peninsula, consciousness of the consequences of environmental mismanagement is likely to rise.

Russia's hydrocarbon development of the Arctic should, at the very least, be far cleaner than in the Soviet era. In the long term, there is no fundamental reason why Russian companies should not operate as cleanly and as efficiently as the best of the IOCs.

Nonetheless, there is still room for things to go badly wrong. Alexei puts it bluntly: 'If budgets need to be cut, environmental safety will be cut first, and production will be rushed . . . corporate standards and policy could easily be violated.' Government thinking on the appropriate balance between long-term resource conservation and short-term production could shift if the financial position of the Russian state deteriorates. And ultimately, whether or not the environmental standards of Russian companies improve, massive development of the Russian Arctic is bound to result in some environmental damage – whether from an oil spill on land, a tanker running aground at sea, or exploratory drilling offshore. The question is not if, but when.

The party of summer 2008 did not last. From a peak of over $140 in July 2008, oil prices fell to $40 by the end of the year, before recovering healthily in 2009. The economy shrank by 10 per cent in twelve months; Russia's financial reserves were slashed; the ruble was put under pressure. Companies which had been riding high – Gazprom, Rosneft and Lukoil – had to borrow money from the government to shore up their positions.

In Moscow, President Medvedev and Prime Minister Putin tried to pin the blame for recession on the United States, holding it responsible for exporting its financial crisis to the world, and accusing it of economic 'egotism' in the process.[57] But the truth is somewhat different. Russia's economic vulnerabilities are built in. No economy built on commodity prices can escape the risk of volatility to its global price. Russia cashed in during the good times; during the bad times things are less easy.

It may not last long. Historically, oil prices – and the gas prices which are linked to them – are still high. Many expect them to stay that way for years to come. With a historic lack of investment in productive capacity, and a steady upward push on demand from China and India, the International Agency predicts oil will average $100 per barrel to 2015, and $120 to 2030. If that is correct, the sharp contraction of the Russian economy in 2009 will be followed by continued expansion. The Arctic will be an increasingly important part of that, providing up to half the country's natural gas within the next ten to fifteen years.

In no other country, save Greenland, is the nation's future so bound up with the development of riches above the Arctic Circle. No other country's impact on the Arctic as a whole will be as great.

10
End of Empire in Alaska

'I do not believe that the apparent conflict between oil and the environment represents a permanent impasse. Instead it presents a challenge – a challenge to our engineering skills and a challenge to our environmental conscience . . . The development of the Prudhoe Bay reserves is of great importance both to the State of Alaska and to the oil reserve posture of this Nation'

President Richard Nixon, 26 September 1971

'I'll tell you this: that if Congress is truly interested in solving the problem [of high gasoline prices] they can send the right signal by saying we're going to explore for oil and gas in the US territories, starting with ANWR [Arctic National Wildlife Refuge]'

President George W. Bush, April 2008

'As an oilman he [George W. Bush] should know better'

Amory B. Lovins, Chair, Rocky Mountain Institute, July 2008

PRUDHOE BAY, ALASKA, 70° North: Apart from the distant silhouettes of the drill rigs, the low plain around Prudhoe Bay is featureless. The border between land and sea – where the North Slope turns into the Beaufort Sea – is rendered indistinguishable by the flat Arctic light, mixing everything into an ocean of green–grey.

Conditions here are more hostile to development than the traditional centres of the American oil industry, the balmy (though hurricane-prone) Gulf of Mexico or the flat dry plains of Texas. The lowest recorded temperature at Deadhorse, the nucleus of North Slope logistics, is -55° Celsius (-68° Fahrenheit). In winter, cold seeps into motor engines, causing them to seize up unless they are left running constantly; storms can shut down offshore fields and ice up those on the near inshore; shift-workers stay only a few weeks at a time, to avoid depression. Most of the year the light in Deadhorse is dull and the clouds are close. Then, for a few months of the year, there is virtually no light at all.

Despite the challenges of the North Slope's Arctic environment, the area developed into a bastion of US oil production in the 1980s, producing up to one quarter of domestic output. In some ways, the North Slope is still a citadel for corporate oil in a world where access to reserves is constrained by resource nationalism abroad and by geology at home. But these days, it is a citadel under siege.

America's love affair with oil has begun to sour. In the past, oil was an all-American product, associated with the American way of life at home, and with the extension of national power abroad. In the collective national consciousness, cheap gasoline underpinned the principle of individual freedom – symbolised by the automobile – which made America the envy of the world.[1] But these days, the country's dependence on it has come to be viewed as a national security vulnerability. An environmental movement has grown up which argues cogently that America must use less oil, not just for the sake of its security but for the sake of its environment. Oil companies which were once viewed as the bedrock of economic growth, are now more likely to be demonised. Arctic oil development has become a lightning rod for broader criticism for an American energy system which is broken. At Prudhoe Bay, the American oil industry is on the defensive.

'Prudhoe Bay is a place of rules,' says Joanna, handing me a protective plastic visor to put over my glasses. 'Wear this.' I soon find that it is difficult to go anywhere in Prudhoe without being shielded behind a barrier of fire-retardant, anti-shatter, flame-resistant material – whether it's the cracked windscreen of the Ford Super Duty that takes you from the terminal building in Deadhorse to my hosts' office building two minutes down a gravel track, or the double-glazing of the windows when you get there. It is unclear what I am being protected against, apart from myself, when ordered to hold the handrail at all times when walking up stairs in Prudhoe, even in the most basic two-storey office building or at the general store.

The intended message of this hyper-secure environment – despite the indistinct sense of paranoia it stirs in me – is a positive one: standards are non-negotiable, no one is above the rules, we care about our staff, safety comes first. It is not just about the Arctic. The basic rules of personal safety to which I am subjected by my hosts, Shell, are applied throughout the company, from London and the Hague to Prudhoe Bay and beyond. Every wall is plastered with a list of who to call in an emergency and statements of the core operating principles of the firm.

The rules are intended to minimise the risk of accidents, of course, but they also serve to protect against a more intangible and unseen threat to the modern oil industry: that careless operations will undermine a company's credibility and long-term reputation. No one wants to be forced to make a public apology to the nation, as BP did in 2006 when a string of environmental mishaps led to the temporary closure of the Trans-Alaska Pipeline.[2] And no one wants to lose a contract to develop scarce oil and gas resources because their company is viewed as unreliable. In a world where corporate oil is treated with mistrust, accidents can be expensive.

To counter suspicion of their practices and motives, Western oil companies now spend millions of dollars advertising themselves as ethically run businesses, servants of the public good rather than slaves to profit.[3] Companies such as BP, Chevron and Shell have taken the lead. Others, notably ExxonMobil, have adopted the new jargon of responsibility and sustainability only more recently. Former ExxonMobil CEO Lee Raymond was fond of arguing that oil companies are in the business of producing oil, and that that should be their focus. But corporate executives now spend much of their time speaking about their companies' public role in providing energy security, or in promoting economic development. Stakeholder management rather than shareholder management has become critical to corporate success. And a slip-up in one country can jeopardise prospects in another. 'I don't care whether you're in Brunei, Gabon or the North

Slope, whatever you do, people will hang that on you', says one oil company executive.

Shell is only one of the players on the North Slope, and far from the most important. The true masters of the North Slope are ExxonMobil, BP and ConocoPhillips, sharing ownership and control of the majority of the oil, as well as the Trans-Alaska Pipeline which carries it away.[4] (There are even dedicated BP/ConocoPhillips check-in desks at Anchorage airport.) The jewel in the crown is the Prudhoe Bay field itself, co-owned by the three main North Slope players, and operated by BP. But there are many other smaller oilfields scattered across the North Slope area, to which one or other company has the major (or even sole) claim. Overall, the area accounts for fully 42 per cent of ConocoPhillips' proven global reserves, and smaller shares of ExxonMobil and BP's reserves.[5]

Looked at from deep space, the core of corporate oil's North Slope citadel would be Deadhorse. All roads lead to it, and most goods and people on the North Slope pass through it. But what might seem broad transportation corridors from space, cutting confidently across the permafrost to link one oilfield to the next, are less impressive at ground zero. The transport corridors turn out to be gravel tracks. The buildings are remnants of an earlier Arctic boom. These days, there is more in Deadhorse to remind you of *Ozymandias* than Xanadu.

In the purest sense of the word, Deadhorse is a dump. Trucks, tyres, empty portakabins, rusting hulks of heavy equipment are all strung out for miles along the road. The cost of removing them is too high. A 2002 report on the current Dismantlement, Removal and Restoration regime on the North Slope – the responsibilities of oil companies to return land to its pre-development condition – concluded there was 'no specific guidance' on what infrastructure should be removed and that financial guarantees required from oil and gas companies were minimal.[6] Were future clean-up regulations to be tightened, the report warned, additional expenses would prohibit exploration and production. The industrial wasteland of Deadhorse is the result.

Perhaps the glory days of Deadhorse are behind it. The main Prudhoe field peaked more than twenty years ago. Since then, the sprawl of drilling across the North Slope has resulted in smaller and smaller fields being brought into production: Kuparuk and Alpine to the west; Lisburne and Point McIntyre around Prudhoe Bay; Endicott and Northstar out to sea.[7] Successive administrations – Democrat and Republican – have allowed areas open to oil and gas production to expand across the Colville River and into the previously closed area known as the National Petroleum Reserve-Alaska (NPR-A), but the production from these new areas has only partially been able to offset the decline in the main Prudhoe field.[8]

If nothing happens, North Slope output will continue to fall. Adding together fields currently in production, under development or with immediate development potential, a 2007 Department of Energy report concluded that North Slope oil production could enter terminal decline as soon as 2012.[9] Without exploration and development of new fields, output could fall below 300,000 barrels per day by 2025, the minimum required to keep the Trans-Alaska Pipeline operational.[10] In the Department of Energy's scenario, over 1 billion barrels of currently proven oil (not to mention further billions of undiscovered oil thought to lie onshore and offshore) would be stranded on the North Slope.

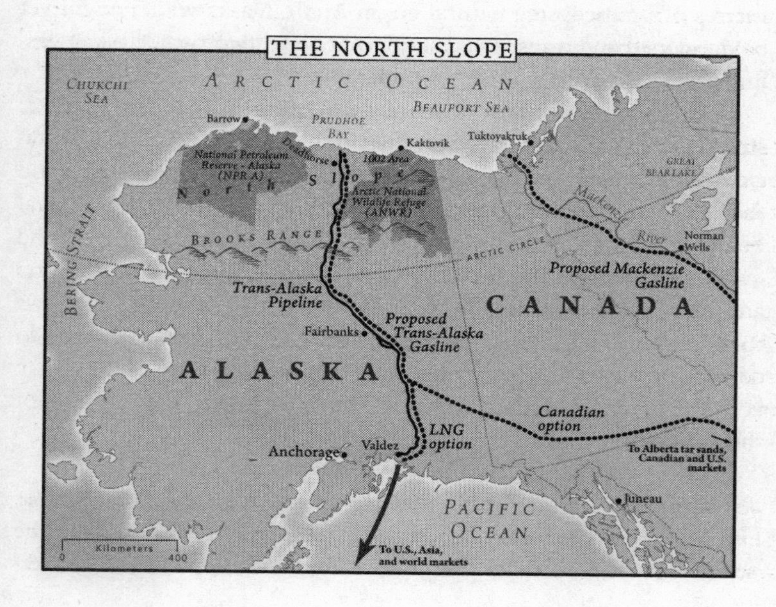

Yet many believe that the North Slope has a far brighter future as an oil-producing area, that the glory days of the 1980s can be revived and that the citadel of corporate oil on Alaska's North Slope can be reinvigorated by the application of new technologies and the opening of new fields. At a time of high prices and scarcity of easy oil, companies view the American Arctic as one of the most promising areas they can boost their production of oil and gas, from both offshore areas (principally the Beaufort and Chukchi Seas) or onshore areas historically off limits to exploration and production.

The United States Geological Survey (USGS) has estimated that, overall, Arctic Alaska may contain nearly 30 billion barrels of undiscovered oil, including perhaps 10 billion in the '1002 area' of the Arctic National Wildlife Refuge (though the accuracy of these figures is heavily disputed).[11] The Department of Energy believes that several billion more barrels of oil could be recovered from existing fields, or from heavy oilfields such as the Ugnu Formation, closer to the surface than the original Prudhoe field. Estimates for the potential for offshore oil run to over 10 billion barrels for the Chukchi and Beaufort Seas combined. USGS has further estimated around 6,200 billion cubic metres of undiscovered natural gas in Arctic Alaska which might yet be developed and transported south on one of two rival Alaskan gasline projects currently under consideration.

In Russia, the decision to extract oil and gas from the Arctic is straightforward, a national priority set in the Kremlin. The main obstacles are technical. In the United States, the democratic system (and the fact that Arctic resources are ultimately less crucial to the survival of the state) makes things more complicated. Expansion of oil and gas development in the Arctic is open to legal challenge in the courts and to political challenge in Washington. If the American Arctic is to be opened to further development, the industry will have to convince a sceptical America not only that development of the Arctic is in the interests of the companies, but that it is in the interests of the nation.

Locally, support is there. Alaska's political class – both Democratic and Republican – back further opening up of Arctic resources. Senator Lisa Murkowski believes that climate change may turn out to be the 'seminal event that allows the pace of Arctic development to accelerate'.[12] It is an acceleration that she welcomes, paraphrasing a popular 1950s song to gee up the vision of an Arctic boom:

> Way up North,
> Where the rivers are winding,
> Big nuggets they are finding . . .
> They're goin' North, North to Alaska,
> The rush is on.[13]

Most of Murkowski's fellow Alaskans share her hopes. They view further development of the North Slope as key to their future prosperity, counting on the revenues from oil and gas to build the state's infrastructure and fund generous state pay outs to its citizens.[14]

But nationally, where it really counts, things are far more complicated. The question of whether or not to open the American Arctic to greater oil and gas production has taken on the quality of a national psychodrama, both sides adopting the Arctic as a totemic cause, bound up with issues far beyond the Arctic itself.

In one camp, opponents of development – a broad alliance encompassing everyone from environmentalists and grass-roots campaigners to anti-capitalists and religious conservationists – present protection of the North Slope (and particularly the sensitive Arctic National Wildlife Refuge) as a litmus test of the Americas' commitment to the environment. They argue that the American Arctic should be put beyond the reach of large-scale economic development so that the 'wilderness which has exerted such a fundamental influence in molding American character' can be protected, and the values which it represents can be preserved.[15] For them, to develop the Arctic's oil and gas resources would be to cross the Rubicon. It must not be done.

In the other camp, proponents of development – a coalition of industry, national security hawks, and the vast majority of Americans who do not greatly care where their gasoline is from as long as it is cheap – view the development of Arctic oil and gas as a self-evident economic good, which will provide jobs for American workers and provide oil for American consumers. (For some on the political right, Arctic oil is a 'God-given endowment', as if this makes it God's work to exploit it.)

But proponents of development add a further crucial and highly contested argument, linking the North Slope with Iraq: that development of America's Arctic oil and gas would reduce America's dependence on imported oil and therefore the necessity for the 'foreign entanglements' which this entails.[16] In short, Arctic development would help secure the ultimate prize for America: 'energy independence'. The promise is not a new one.

On 7 November 1973, under pressure from the Watergate scandal and from the first oil crisis, President Nixon made a speech to the American people. In it, he promised them an end to energy insecurity: 'Let us set as our national goal, in the spirit of Apollo, with the determination of the Manhattan Project, that by the end of this decade, we will have developed the potential to meet our own energy needs without depending on any foreign energy source.' Nixon called his plan Project Independence. It was understood that development of Arctic resources would be a major part of that drive. The final authorisations for the construction of the Trans-Alaska Pipeline, held up for nearly three years in the courts, were passed a few days later.

Nixon's advisers had told the President that energy independence would be impossible to achieve in the time frame he had set, if ever. (By 1973, America's net petroleum imports stood at just over 6 million bpd, more than three times the figure in 1960, the year of Nixon's first and unsuccessful run for the presidency.[17]) But Nixon ignored them. His true purpose in raising what he knew to be a pipe dream was political: to demonstrate firm control of a difficult situation, a possible solution and, above all, a vision fit for a President. Not for the last time, opening up Arctic oil resources was presented as part of an energy-strategy-to-end-all-energy-strategies, rather than as the holding strategy that it really was.

After 1977, the year when the Trans-Alaska Pipeline opened for business, America's net imports of oil did begin to decline. But output from the Arctic North Slope was only part of the reason – a greater influence was falling domestic demand.[18] The high point of the North Slope's contribution to America's overall production of crude oil came in 1988, when output peaked at just over 2 million barrels of oil per day, one quarter of US production. But since then, it has declined rapidly. These days, the North Slope contributes around 14 per cent to America's domestic production of crude oil and somewhere between 3 and 4 per cent of America's total daily consumption of petroleum.

Yet in political debate – and in the public mind – the North Slope's central role in safeguarding American energy security has remained persistent. Indeed, it has been consistently inflated. During the 2008 Presidential campaign, figures of 17 to 20 per cent were commonly cited as Alaska's contribution to domestic oil production rather than the correct, lower figure of 14 per cent.[19] Worse, some politicians incorrectly implied that this 17 to 20 per cent share referred to total national *energy consumption*, clearly a much bigger number than domestic *oil production*. The reality is that the North Slope is far less important to overall American energy requirements than most people think.

Developing the resources of the American Arctic, meanwhile, is no quick fix for America's dependence on foreign oil. No one seriously suggests that Arctic production, by itself, could make up the shortfall between American imports of crude oil and domestic production. At over 11 million barrels of oil per day, the current shortfall is too large, equivalent to the entire daily production of Saudi Arabia.

And despite the political sloganising of some American politicians, the evidence suggests that opening up the Arctic would have a minimal effect on the price American consumers pay for gasoline.[20] In May 2008, the Energy Information Administration (EIA) calculated that granting leases in the '1002 Area' of the Arctic National Wildlife Refuge (ANWR) – one of the areas many consider to be most prospective for onshore oil in the US Arctic – would not affect the price of gasoline for ten years at least.[21] Twenty years after leasing, the EIA calculated the impact on the price of a barrel of oil would range from 41 cents, if ANWR's resource endowment turned out to be less than expected, to $1.44 if it turned out to be greater. On a gallon, that's chump change.

According to a Washington insider, the senator who commissioned the EIA analysis 'didn't like the report too much'. Understandably so. Far from supporting the argument made by Alaska's senators and by President George W. Bush that opening of the '1002 Area' would provide relief to hard-working American families, the report showed just how little it would do.

The conclusions of the EIA report do not mean that, beyond

Alaskans' interest in development, there is no *national* security or *national* economic argument for Arctic oil and gas development.[22] Clearly, Arctic oil and gas could be produced and, under certain circumstances, that would reduce dependence on foreign supplies. But what the report shows, in the end, is that the question as to whether to use the oil and gas resources of the Arctic is less important for America's overall energy future than decisions on, say, energy efficiency or alternative sources. Unlike in Russia, and despite the rhetoric, in America there is no overpowering national interest in the development of Arctic resources. At the corporate and the local level, however, interest is very real.

Back in Deadhorse, support for expansion for offshore oil and gas development runs strong. 'The oil industry changed my life,' says John Maketa. After graduating from high school in 1987 John went into the coastguard. But by the early 1990s he was working for Crowley Marine, shipping heating oil to Alaska's remote coastal communities. Since then he has worked for Exxon in Russia, Encana on the North Slope and now for Shell's logistics operations in Deadhorse, with responsibility for the company's offshore prospects in the Beaufort and Chukchi Seas. 'I was earning $20,000 a year,' he says. 'When I came to the industry in 1991 I started earning several times that – and it's only gone up from there. Enough scraping around. I can buy a house.'

John's job in Deadhorse sounds simple: 'Rope and soap,' he jokes. But providing supplies and general logistical support for Shell's ships in the Beaufort and Chukchi Seas requires the skills of a diplomat, an enforcer and businessman. Procurement is one problem. 'It's very competitive up here for goods and services,' he tells me. 'There was nothing up here in the 1980s and 1990s – so everything's run down.' Shell has to compete with the requirements of other companies' projects – John mentions BP and ENI, the Italian oil company – and the inherent complexity of a constantly moving set of issues to be resolved and needs to be met. On top of that, he has to manage the expectations of the head office back in Anchorage.

Nonetheless, John insists, there is no room for cutting corners on the ground: 'You have to have a plan for everything.' Careful planning – an attribute of Shell's corporate culture born of a company run by scenario-builders – can ultimately come at the expense of manoeuvrability. Says one employee, 'Twenty years ago you could get a decision taken real quickly. These days, everything is a hot potato.'

'Look at this,' says John. He shows me Blue Sky, an internet-based tracking system showing all of Shell's vessels off the coast of Alaska: five in the Chukchi Sea, off the north-west coast of Alaska, and a smaller number closer inshore, in the Beaufort Sea east of Prudhoe Bay. This is the new face of oil development in Alaska, on the federal Outer Continental Shelf.[23] Shell has already spent $3 billion in Arctic Alaska, with the possibility of 'billions more'. 'It's a huge gamble,' says Pete Slaiby, the general manager for Shell in Alaska, but, 'We think the numbers look good – we believe in the Arctic.'

The Chukchi boats are led by the *Gilavar*, a six-line seismic ship operated by Western Geco, a subsidiary of the international oilfield-services firm Schlumberger. Their job is to map the offshore leases that Shell bought for $2.1 billion in February 2008.[24]

With multiple seismic lines, each 6,000 metres long, the *Gilavar* provides Shell with a three-dimensional model of the subsurface geology. It is a major advance from the two-dimensional seismic of the 1980s, and it comes at a price. 'Right at the moment, it's a great time to be a seismic vendor,' says Barbara Bohn, the manager for Shell's seismic operations in Alaska. Operational seismic costs in Arctic Alaska are double the rates in the Gulf of Mexico, covering a similar surface area takes twice as long and the season is short. Boats from the Gulf of Mexico have to travel through the Panama Canal to reach Alaska. Shell spent tens of millions of dollars on seismic work alone in the Chukchi Sea in 2008.

In the Beaufort Sea, the main offshore focus of activities is on an oilfield known as Sivulliq (meaning 'first one' in Inupiaq) due north of the land border between the central Arctic region and the ANWR. Sivulliq was one of the fields discovered by Shell in an earlier phase of Arctic exploration in the 1980s.[25] Like the Northstar field, discovered by Shell in 1983 and then sold to BP in 1995, Sivulliq – formerly known as Hammerhead – was abandoned as commercially unsustainable in the early 1980s as the first Arctic oil boom came to an end and the oil price crashed. The reported costs of development then, at well over a billion dollars, were considered exorbitant.

These days, with long-term oil prices expected to range between

$80 and $100 a barrel, things have turned in Sivulliq's favour. For Shell, the question is not whether to develop the field, but how to get up to 200 million barrels of oil ashore. One boat is mapping ice scours on the sea floor to provide a basis for calculating how deep a pipeline would need to be built in order to prevent an errant iceberg from ripping it open. Another, the 115-foot *Norseman II*, is counting whales and other marine life in the area so as to provide an accurate baseline picture of activity.

But the activity John shows me on Blue Sky is only part of the story. Partly as a result of stakeholder concerns, but principally as a result of unresolved litigation, Shell's original plans for the Beaufort Sea for 2008 were scaled back from a two-ship to a one-ship programme at the beginning of the year, and the expected start date was pushed back from July to September 2008.[26] In June, plans were cancelled entirely.

In Anchorage, in a corner office overlooking the mid-town grid of asphalt, offices, multi-screen cinemas and shopping malls, Travis Purvis – Shell's well delivery manager in Alaska – explains what happened. 'I was disappointed, very disappointed,' he says. 'We were planning for success . . . Now some people think we're sharpening pencils up here.'

In 2007, Shell's plan had been to use two drill ships, the *Frontier Discoverer*, a 21,000-ton drill ship originally built as a log carrier by Japan's Mitsubishi shipyards, and the *Kulluk*, a drill ship specifically designed for the Arctic, built for Gulf Canada in the 1980s. The *Frontier Discoverer* was standing by in Dutch Harbour, at the southern end of the Aleutian Islands. The *Kulluk*, freed from its long-term overwintering site in McKinley Bay, just north-east of Tuktoyaktuk, was waiting across the sea border in the Canadian Beaufort Sea. But a legal petition from environmental and some Inupiat groups claimed that the Minerals Management Service (MMS), the federal agency of the Department of the Interior responsible for granting permissions to drill, had not taken sufficient account of possible damage to wildlife, including the bowhead whale. The 9th Circuit Court of Appeals confirmed the halt of drilling in August 2007, provoking a critical statement from Governor Sarah Palin.[27]

'We heard a lot from our stakeholders on pace, and on size and scope,' says Travis. So in 2008, with continuing legal uncertainty, Shell's plans were scaled back to one rig ship, the *Kulluk*, overwintering by Herschel Island, just inside Canadian waters. The rig was activated in May – with 106 people on board, and 'dozens' of contractors employed on shore bases. 'We had a detailed project plan,' Travis tells me – to start drilling operations in early September, drill 'top holes' preparing the ground for deeper drill holes in 2009, and then pull out in early November.

But in June the order came through to abort. Contractors associated with the project were laid off. Shell's plans had been derailed for the second year in a row. 'I find that very frustrating,' says Travis. 'Now, if we weren't ready, if we weren't convinced we had the right fleet, support and people I would welcome a delay, but we have a good programme, and we're excited to go.'

In December 2008, after a further ruling from the 9th Circuit Court of Appeals, Shell cancelled its 2009 programme for the Beaufort Sea – including the drilling of three exploratory wells. It intends to present a new plan for 2010 and 2011. Shell says the delay will cost several hundred jobs. Others take a different view. Chuck Clusen, from the National Resources Defense Council, told one newspaper, 'This is really a signal that Shell's plan was simply too much, too fast and too shoddy . . . the court has opened the door to a new administration to take a whole new approach, and hopefully a more precautionary approach, to America's Arctic.'[28]

An even more serious setback for oil companies like Shell would be the reinstatement of a presidential moratorium on oil and gas development on the Outer Continental Shelf – American waters beyond state jurisdiction. (One oil company executive described such a move as 'disastrous'.[29]) While legal rulings can be disputed and overturned, a political ruling against Arctic oil and gas development would close the door on further exploration and hopes for increased American hydrocarbon production in the Arctic.

FAIRBANKS, ALASKA, 65° North: Pam Miller is on campaign. When we meet, on the patio behind Fairbanks's best bookstore – Gulliver's Books – Miller is armed with maps, articles, data and an attitude, however charmingly put, that suggests that oil companies in Alaska cannot be trusted to develop Arctic Alaska. Their past record, she argues, is proof enough. From Miller's perspective, the history of oil development on the North Slope is one of broken promises on environmental protection, exaggerated claims for new technologies and a gradual creep of oil development.

Originally from Cleveland, Ohio, Pam has been in Alaska for twenty-five years. For eight years she worked for the US Fish and Wildlife Service – an agency of the Federal Department of the Interior. Now, Pam is serving as the director of the Arctic programme at the Northern Alaska Environmental Center, a grassroots organisation based in Fairbanks.

Oddly perhaps, on at least two important points, she is actually in agreement with some oil industry representatives. She sees offshore development as opening the door to ANWR. In private, this argument is echoed by advocates of the oil and gas industry – the reasoning being that ultimately, once oil and gas industries have taken root, barriers to new developments tend to fall.

And like many in the oil industry, Pam is deeply critical of the federal and state agencies involved in managing lease sales, assessing environmental impacts and monitoring compliance. An oil worker in Anchorage tells me, 'It's a mess, too bureaucratic.' Miller's criticism is that it is ineffective and biased. Some federal employees are uneasy about the speed and scale of leasing in recent years, Pam advises me, and the speed with which some environmental permits were granted. But they don't talk about it in public. (Separately a government official tells me, 'It's made pretty clear to people at my level that there's a company line and you'd better stick to it.'[30]) Pam warns that there is 'basically no oversight' from government once seismic or other exploration work has begun. The solution? 'We need a new federal agency,' she says.

Although the Minerals Management Service is conducting studies on everything from zooplankton and pinnipeds to coastal erosion and polar bear population, Pam argues such research needs to take place before any exploration or production activity is allowed, to provide an environmental baseline.[31] As development expands, government agencies need to focus not just on the environmental impacts of a single development, but on the cumulative impacts of many.

Miller is particularly critical of what she calls the '2,000 acres hoax', the idea that drilling on ANWR's '1002 Area' would only affect 2,000 out of 1.5 million acres on ANWR's coastal plain.[32] As she points out, even if construction sites themselves were limited to 2,000 acres the actual spread would be much larger because those sites would not be contiguous. From an ecological point of view, the '1002 area' is highly sensitive. It is so narrow – only 30 miles between the mountains of the Brooks Range and the Arctic shoreline – that one biologist tells me that 'if there's any disturbance, there's really nowhere to go [for caribou]'. The election of President Barack Obama has ended the immediate threat of the area being opened to development. Pam Miller is now focusing on attempting to achieve 'wilderness' designation for the coastal area, thus further strengthening its legal protection.[33]

But the issue may return. Those who campaigned for the refuge to be set up in the first place were well aware of its totemic status.[34] Since the refuge was renamed and expanded to its current extent, in 1980, there has been a guerrilla war in Washington between those who want the area opened to development, and those who want to keep it closed. For most of the 1980s and 1990s, the most vociferous proponent of opening the '1002 Area' was Republican Senator Ted Stevens. Stevens lost his seat in the 2008 elections, but his Democratic successor, Senator Mark Begich, has also long criticised the closure of the area to oil and gas development.

Proponents of opening the '1002 Area' have been able to rely on the political support of some of the coastal communities of the Inupiat who stand to gain from onshore oil and gas development. The community of Kaktovik, in the middle of the area, has been critical of outsiders without a sufficient understanding of local history or conditions. As one document, the so-called Kaktovik Papers, puts it:

> Our lands and waters have received much attention from those who want to search here for oil and those who want to make it into a wilderness. We see both of these outside interests to be potentially destructive. The most dangerous, in our view, to the things that matter to us, and surely the most insulting, is the notion that our homeland is empty or should be made so.[35]

Under certain conditions, the attitude of the local community is strongly favourable to oil and gas development. 'They are rough, some of these oil people, but so are we, and we think we can deal with them,' the report states. Ultimately, it argues, it should be the local community that determines what should and should not be a wilderness, not 'ten thousand government agencies with fifty thousand agendas'.

Opposition to offshore oil and gas development is more consistent. A specific concern must be addressed: that oil and gas development may harm the population of bowhead whales. For the Inupiat community, whales are sacrosanct. Hunting them is a core element of cultural identity. Companies that want to work offshore in the Arctic will have to demonstrate to the Inupiat that whales and development can co-exist.

PRUDHOE BAY, ALASKA, 70° North: A map of Alaska's North Slope, from the Canadian border in the east to Point Hope in the west, helps illustrate why knowledge about whale activity is imperative for oil companies. Much of the area offshore is divided into oblong blocks – red blocks for leases owned by Shell, grey blocks for leases owned by other companies and a far smaller number of green blocks run by native corporations, concentrated in the onshore area around Fort Wainwright.

But a closer look at the map reveals a second, more fluid, division, overlaid onto the first: large coloured areas indicating known whale migration and feeding areas, and smaller cross-hatched zones where seismic work – which is thought to disturb the whales – is banned for all or part of the year. Just north of Kaktovik, for example, no seismic work can take place between 25 August and 20 September. For oil companies wishing to conduct offshore work, any period of activity foregone is a major cost.

When a company enters discussions with Inuit communities about where and when drilling and seismic activity should be limited, the latest whale data is a powerful negotiating instrument. Bill Koski, a long-haired Canadian contractor who runs Shell's Marine Mammals Observation programme in Alaska is unapologetic: 'We need the best science; otherwise they will try and push the envelope.'

One part of Bill's job is to coordinate Marine Mammal Observers aboard each of the ships Shell currently has in the Beaufort and Chukchi Seas. Their task is to report sightings of whales and, if necessary, shut down operations, something which Bill says happens fifty times a year. But beyond the observers on each ship, Bill's team uses two Twin Otter aircraft for aerial surveillance of whales, and Directional Autonomous Seafloor Acoustic Recorders (DASARs) to monitor noise sources at sea. 'There's a lot of noise out there, and we want to know how much of it is ours,' Bill explains. Using technologies which recognise the individual acoustic signature of each boat, Bill hopes to better track how much whales are affected by Shell's activities.

This is not purely disinterested scientific research. The data Bill provides will be used to try and separate the general impacts of noise on whale migration from the impact attributable to specific oil and gas activities, so as to make the case that seismic or drilling work itself is not the problem. But a byproduct of corporate interest in the movements of bowhead whales in the Arctic is that, according to Bill, 'We now know more about this type of whale than any other.'

When Bill started working on the bowhead whale in the 1970s the scientific consensus – wrongly, as it turned out – was that the population was close to extinction. Now, Bill thinks that there may be as many as 13,000 bowhead whales in the Beaufort and Chukchi Seas – 'the largest concentration in the world' – compared to the 7,000 believed to be in the Canadian Arctic.

BARROW, ALASKA, 71° North: 'I've been in Barrow since 1969,' Taqulik Hepa tells me, with laughing eyes. 'I was born here.' Now, she is director of the North Slope Borough Wildlife Department, responsible for borough-level environmental management of an area covering nearly 95,000 square miles, from the Canadian border in the east to Alaska's coastline in the west, and south to the Brooks Range, separating central Alaska from the Arctic coastal plain. On the day I visit the Wildlife Department offices at Point Barrow, polar bears have been sighted near town, an increasingly common occurrence as diminishing sea ice reduces the bears' natural habitat. A morning jogger – Taqulik's eyes roll at the thought – was nearly mauled. Now, signs must be put up around the town to make residents aware of the threat.

For such a big job, Taqulik has a relatively small team, and a limited budget. She estimates it at $5.5 million annually, including all appropriations, money from the borough and some project-based grants. Federal government support comes in for stinging criticism. According to Taqulik, the Fish and Wildlife Service (FWS) staff down in Anchorage have 'no interest in coming up here' to help with Barrow's polar bear problem. Instead, she has asked her staff to keep a detailed record of the man hours they spend on the issue: 'I want to show them [the FWS] how much it costs to maintain the safety up here.'

There is a lot of relatively new work just monitoring the environmental assessments of proposed oil developments, particularly as interest in the Arctic development has returned. Cheryl Rosa, one of the team's biologists, complains that, 'I got a few months of my job before all of a sudden there were massive changes . . . All of a sudden we were inundated with Environmental Impact Statements.'

But at the core of the North Slope Borough Wildlife Department's work – and at the core of Inupiat culture – is the bowhead whale. Craig George has been working on whale populations since the late 1970s, when the International Whaling Moratorium, fearing that the population level was close to extinction, ordered a moratorium on the Inupiat hunt. 'Locally, the folks decided they wanted their own scientists,' he tells me, 'so we got heavily into whale counting.' But it took over a decade to figure out a trend in population levels, and to prove that the Inupiat villagers were right and the scientific consensus on whale numbers in the Beaufort Sea was flat-out wrong.[36]

Even with visual and acoustic surveillance, whales are hard to count. When the data was collated from annual counts over twelve years – from 1976 to 1988 – the results 'looked like the stock market'. But by then the trend was clear – there was a much larger population than previously thought, and, on average, increasing by 3.5 per cent every year. Given the high survival rate of bowhead whales and the relatively tiny size of the hunt (less than one half of one per cent are 'harvested' annually), the frequency of counts can now be reduced. The last was in 2001 and the next is planned for 2010. The numbers themselves are not disputed any more. This year, Craig tells me, 'There was one question: what kind of plane did you use? It's unbelievable.'

The fear is no longer that the bowhead whale will die out, but that it will be pushed further offshore by oil and gas development, while the wider marine environment on which they ultimately depend is degraded. Problems arise from the cumulative consequences of a broad range of factors: from invasive species and seismic work – which some claim deflects bowhead migration – to the risk of oil spills and disturbance of the sea floor from drilling. 'It's a matter of scale,' says Craig. 'Right now, there's arguably two offshore production islands, and life's going on, but what happens when you get to a Gulf of Mexico situation, with 4,000 oil-related structures?' Each exploratory well will leave what Cheryl Rosa calls a 'dead zone' around it, where the seabed has been churned up and the water column disturbed. Individually, these dead zones may be insignificant – but what happens when there are 'ten, twelve, forty, a hundred'? As Craig puts it, 'No one knows where the threshold is – I don't know, Cheryl doesn't know . . . But I bet you that the problem won't be linear.'

Unavoidably, drill rigs will cause some pollution (though companies insist this is contained) from their run-off.[37] The lubricating 'muds' used to facilitate drilling are, to some degree, toxic. In the Beaufort Sea, companies only use the least-polluting water-based muds. But, according to Cheryl, these are 'still not totally okay; they contain a whole bunch of things, often proprietary, which they [the oil companies] don't want to tell you about'. While companies claim the content of damaging chemicals is extremely low, Cheryl warns the muds may contain 'cadmium and mercury'. Drilling itself causes 'silt blooms', as the sea floor is churned into the water column. Some worry that these blooms may deflect whale migration.

Despite the focus on prevention, it is certain that increased oil and gas development will lead to oil spills in the Arctic, either from shipping or from production infrastructure. The break-up of Arctic ice in a warmer climate may increase the hazard. Taqulik warns that undersea pipelines may be at risk if icebergs increase in size and 'the historical depth of scouring changes'.

At Prudhoe Bay, the Alaska Clean Seas consortium runs a number of barges and skimmers that are ready to swing into action if there's an oil spill offshore, to contain it and then 'vacuum it up'. But in broken-ice conditions, mechanical responses cannot be used. Chemicals could be used to disperse the oil, but that merely shifts part of the pollution problem from the surface to the sea floor where part of the dispersed oil would come to rest. The most likely method of dealing with an oil spill in broken ice is simply to ignite it and burn it off.[38] (While America does have a National Oil Spill Response Test Tank Facility where new techniques for dealing with oil spills are supposed to be developed, Cheryl points out that 'it's not exactly Arctic conditions; it's New Jersey conditions').

Burning presents its own challenges. The pollution caused by burning oil means that it cannot take place less than three miles upwind of human settlement, and managing the process of burning oil is not necessarily easy. Craig remembers a test that he was invited to a few years ago:

> I witnessed one of them. They used a big flooded gravel pit and literally brought in ice with front-end loaders. They dumped the oil in this big elaborate thing, and then a helicopter came over and they threw in something to try and ignite it . . . and it didn't work, and it didn't work. And then, with this twenty-knot wind, the oil all pooled on one end of the pit, with the ice. And then they threw an even bigger grenade out, or whatever it was. Finally the oil lit, and they had this huge flame, which enveloped the crane which they'd used to make the oil spill, and it was like, 'Everybody, get back.'

The test was a few years ago, Craig admits, 'back in '85'. 'But I don't think things have changed much since,' he tells me. 'What if the weather's bad up here when they need to clean something up?'

The environmental consequences of climate change are another concern, not so much in terms of the impact of sea-ice retreat on the polar bear population but in terms of new species entering this part of the Arctic. 'There's been a smattering through time,' Robert Sudyam, a senior scientist on the NSB team tells me, 'but now red salmon and humpback whales are here with some regularity – it's a record for the Beaufort Sea.' The levels of invasive species may be sharply worsened by an increase in shipping from areas outside the Arctic, with the associated risk that microorganisms from other marine environments will be introduced to the Beaufort Sea when ballast water is expelled.

With a warmer Arctic, Robert worries that microorganisms from other environments may be able to survive and flourish. The trouble with such a scenario is that interspecies contact introduces a host of new variations: 'It's like, okay, this disease or this parasite is in area A, and then all of a sudden those animals are in area B.' The problem is acute with bowhead whales because they, charmingly, are 'naive' – they carry few parasites and are therefore particularly vulnerable to new ones.

From an environmental management perspective, there are questions as to who should be responsible for monitoring: companies or government agencies? Whatever the good work done by Bill Koski and Shell on monitoring whale movements from their base in Prudhoe Bay, some in the Wildlife Department in Barrow do not think enough is being done. 'Our impression is that they're not spending enough money,' Robert says. He estimates that Shell is spending 'a couple of million dollars' on environmental monitoring, yet their plans for exploration are 'gigantic'.

In some areas, such as the Chukchi Sea, the environmental baseline data is patchy or non-existent. Bowhead whale populations are relatively well understood. But for other animals – beluga, ice seal, walrus – information is scarce. 'They [the oil companies] will tell you, "Well, that's the job of the government, it's not our problem,"' says Taqulik, 'yet they're the ones pushing to do this development.' But there are misgivings about the dynamics of oil-hungry companies conducting environmental research themselves, where a particular result is so strongly in their interest – it's a case of the 'fox watching the chicken coop' as one scientist puts it. Local scientific voices can be outspent, outnumbered and outmanoeuvred by scientists contracted by the oil companies: 'Sometimes I feel like we go in a room and there's three of us on one side of the table, and thirty from the industry on the other side,' says Robert. (That said, Shell is thought to have done a 'reasonable job' in filling in some of the worst gaps in scientific knowledge, indeed 'probably better than would be done by the federal government'.)

In the 1980s, when whale numbers were in question, the Wildlife Department had the same problem which now faces the oil companies: its role in collecting the data was mistrusted. There was suspicion that higher-than-expected numbers of whales were the product of self-interested research aimed at justifying an end to hunt restrictions. Allaying concerns of bias required total transparency about the Wildlife Department's work, and calculations. Some feel the oil industry has not been so forthcoming. Robert remembers that 'the first two years industry was in the Chukchi doing seismic work, they were not telling us where they were', because they were worried that information on their whereabouts would reveal interest in particular tracts of the Chukchi Sea and alert rivals to areas they considered most prospective for oil and gas.

If development does take place, Craig urges, it must happen at the appropriate speed and with the appropriate environmental science to back it up. 'The bottom line for our concerns is the unknown,' he says. 'There's so many big question marks, whether it's about what's out there, in what densities, or whether it's about impact from sound.' From his experience, 'as we collected information, people's concerns diminished'. But that comfort level has not yet been reached.

Decisions made on North Slope development in the next few years will set the tone for decades to come. If government backs an expansion of oil and gas extraction, if prices continue to justify it, and if the right infrastructure is put in place, the American Arctic could enjoy an oil and gas renaissance. Natural gas from the North Slope could be pumped into American homes before 2020; oil from the North Slope could last several more decades, if investment decisions to maintain exploration and production are made now.

But the national debate – focused on the complex and emotive issues of 'energy independence' and the sanctity of the Arctic environment – is far from over. Oil companies will continue to argue that drilling in the Arctic is vital for the domestic industry, and that Arctic production would contribute materially to American energy security. Environmentalists will counter that the security argument is overblown while the risks of environmental damage from oil and gas development in the Arctic are consistently underplayed.

Locally, Taqulik is confident that the Inupiat community will stick together in insisting that development only take place if the environmental consequences are known and considered manageable. She insists that the values which bind the community – the protection of whaling in particular – are far more important than money. One example cited is that of Mayor Itta, who opposed leasing around Teshepuk Lake despite standing to gain from it financially.

It is not that Barrow does not need the jobs or money which oil and gas development might bring. Barrow's streets are littered with rusting vehicles. Many people get by on their annual cheque from the Alaska Permanent Fund and their dividend from the Arctic Slope Regional Corporation (one of several Inuit companies set up under the terms of the 1971 Native Claims Act). The social problems of Barrow – from drug abuse to violence in the home – are tragically persistent. But for the people of Barrow, the town is not just a place to live – it is their home. For locals, the important thing is to not lose control of development and to make sure that, if development does happen, it is carefully balanced with concern for the local environment. An example of the equilibrium which might be struck lies across the ocean in north-west Europe's Arctic petro-state: Norway.

11
Balance: Norway and the Arctic Model

'Norway was, and will always be, a polar nation'

Norwegian Foreign Minister Jonas Støre, June 2008

'We have never had a more aggressive exploration programme than we have today'

Helge Lund, CEO of StatoilHydro, September 2008

OSLO, NORWAY, 60° North: It's mid-April and there is still snow lying on the ground. The returning sun has melted most of it, but in the darker corners of the fields – as if swept up there to tidy the landscape – the last traces of winter remain, protected by the shadow of tall pines. Viewed from the air, there's a pleasing neatness to it all, as if the compact orderliness of the fields were an extension of Norway's well-organised, well-regulated society. This is a country where, weather permitting, the trains run on time. The country's oil industry, straddling an offshore zone from under 60° North to well above the Arctic Circle, is frequently considered a model of efficiency and sound public management.

Some Norwegians lament that the principles of social and political consensus are too restrictive. Others argue that stability has been bought at the price of excessive taxation. But from the outside there's something approaching Platonic perfection about it all. There's an elegant, almost classical, equilibrium between freedom and the state, between capitalism and social cohesion, and between national identity and a typically Nordic internationalist vocation. Local affinities – something Plato would have seen as the basis for any true democracy – are still strong, partly because the country's geography conspires against centralisation and partly because the country's history saw a constant shift of power between its cities. Bergen and Trondheim both served as capital city in Norway's medieval past; Stavanger has become the centre of Norway's oil industry. Society is made up of closely interlocking professional, social and regional networks. Shared values predominate. Even now, the population of the country is barely 5 million.

Above all, Norway is rich. Within Europe, it is second only to Luxembourg in income per head of population. In terms of quality of life, Norway ranks second in the world.[1] Despite the number of SUVs on Oslo's streets, and the country's high levels of consumption, the principles of good global citizenship, economic development and environmental sustainability are an integral part of national identity, at least when it is projected abroad.[2]

Every year, the Nobel Peace Prize is awarded by a committee made up of five members of the Norwegian parliament.[3] In 2007, the award was shared by Al Gore and the Intergovernmental Panel on Climate Change. The first secretary-general of the United Nations, Trygve Lie, was Norwegian. And although Norway has twice rejected membership of the European Union in national referenda, the country is a founder member of the North Atlantic Treaty Organisation (NATO) and a major contributor to peacekeeping operations.

Norway is one of the few countries in the world to exceed a long-standing global promise to devote 0.7 per cent of national income to overseas aid. The government's Oil for Development initiative aims to export Norway's own experience of managing resource wealth to countries in the developing world, including Angola, Iraq, Nigeria and Sudan.

The concept of sustainable development was the product of a United Nations commission led by a former Norwegian Prime Minister, Gro Harlem Brundtland.[4] Norway's sovereign wealth fund – into which a share of revenues from oil and gas development is paid – has specifically excluded investments in a number of companies on ethical and environmental grounds.[5] StatoilHydro, Norway's largest company and producer of much of its oil and gas, produces an annual report on its own environmental sustainability. (The Dow Jones Sustainability Index has consistently ranked it the most sustainable oil and gas company in the world.) StatoilHydro has become a leader in carbon capture and storage (CCS) technologies, advocating the use of CO_2 injection into oilfields, to enhance recovery rates and slow global warming at the same time.[6]

Norway's status as a model nation – independent, wealthy, well organised and responsible – is a recent phenomenon. Independence only dates back to 1905, when a constitutional crisis ended the personal union of Sweden and Norway under the Swedish crown.[7] In the early nineteenth century, Norway's ruling class was highly influenced by the tastes of Sweden and Denmark, themselves imported from Germany and France. Later, as nationalism swept through Europe's outer reaches, Norway's Viking past was popularised across the continent, providing a romanticised picture of the country's glorious and independent past. At the end of the century, Norway's national heroes were either the symbols of this distant and almost mythological past – the monarchs of medieval Norway – or the modern demigods of Arctic and Antarctic exploration: Roald Amundsen, Fridtjof Nansen and Otto Sverdrup.

Economically, Norway remained a backwater until the twentieth century. Occupying a thin strip of mountainous land stretching along Scandinavia's western and northern shore, the quickest means of travel was by sea. Internal communications were poor. Towns along the coast opened to the sea rather than towards the thick forests of the interior. For most of the nineteenth century, the country's economy was principally agricultural. Migration to Norway's cities was one way of escaping rural poverty. Mass emigration to the United States was another.

Particularly in Norway's Arctic north, the memory of poverty is strong. Far above the Arctic Circle, Tromsø's town museum is full of reminders of hardship. In one room, there's a mock-up of an Arctic cabin, completed by a recording of howling wind and the figure of a man, alone, bent over his desk writing by the light of a whale-oil lamp. Hanging outside is a Krag-Jørgensen bolt-action rifle, adapted for Arctic conditions with the addition of a harpoon. On the wall hangs the clock from a Russian tug which went down off Svalbard in 1933, stuck at the time of the disaster. To the side, there is an arrangement of the long pipes which men would use to try to smoke tobacco made wet by the permanently humid conditions of their lives. (One can imagine the frustration of multiple attempts to light the tobacco, the satisfyingly thick belches of smoke once lit, and anger at the accidental breaking of a pipe, one of the few possessions of desperately poor men.) These meagre artefacts are reminders of Tromsø's recent poverty, a yardstick against which the current comfort of contemporary Norwegian life can be measured, and a warning of the fickleness of fortune.

When it finally did come, industrialisation took several decades, overcoming obstacle after obstacle to the country's economic growth. In the mid-nineteenth century, cheaper grain imports freed up surplus labour, allowing much of the country to escape from agricultural subsistence. Much later, as transport improved, regional markets became national markets and Norway began a process of reintegration into the European economy. Demand for the country's natural resources – chiefly timber and minerals – allowed the development of basic processing industries. By the beginning of the twentieth century, hydroelectric power from Norway's abundant rivers and waterfalls gave some industries a major cost advantage. But foreign capital was crucial as well. In late 1905, just a few months after the separation of Norway from Sweden, it was the combined financial strength of French banks and the Swedish Wallenberg family which underwrote the foundation of NorskHydro, later the country's largest industrial conglomerate and, from 2007, one half of StatoilHydro. By the end of the first decade of the twentieth century, Norway had begun to catch up with its Scandinavian neighbours.

Real wealth arrived with oil. As late as 1958 the Norwegian Geological Survey told the government, 'The chances of finding coal, oil or sulphur on the continental shelf off the Norwegian coast can be discounted.' But the prospects for oil and gas – which had been discovered in large quantities off the coast of the Netherlands in the late 1950s – were never entirely eliminated. In 1962, the Oklahoma-based Phillips Petroleum tried to persuade the government to grant it an exclusive licence to explore for and produce oil on the Norwegian Continental Shelf (NCS). The licence wasn't granted but the government was spurred to establish a legal framework for oil and gas exploration. The focus was very much on the North Sea – the most southern part of the NCS. As one government official told me, in the 1960s, '62° latitude was considered the northernmost line for development.'

Ultimately, persistence paid off. On Christmas Eve 1969, Phillips Petroleum announced a major find at Ekofisk, in the middle of the North Sea. Later, the field would be estimated to contain some 3.4 billion barrels of oil. At the time, the message back to Stavanger was even more fantastic: 'I hereby declare that the North Sea, from here to the North Pole, is a huge oil basin.'[8]

Oil started flowing from Ekofisk in June 1971, just in time for Norway to catch the boom in exploration and production that accompanied the 1973 oil crisis.[9] By 1980 production amounted to over half a million barrels of oil per day. The Ekofisk field is still producing oil today.

Initially, foreign oil companies and NorskHydro seemed best placed to develop the resources of the Norwegian Continental Shelf (NCS). But in 1972, a recently elected Labour government established a new model for management of Norway's new-found oil wealth, aiming to use the creation of a home-grown industry as a catalyst for wider economic development. Instead of full nationalisation – which historical experience had demonstrated rarely yielded the best long-term results – Norway balanced its commercial and statist instincts in setting up a 'trinity' at the apex of the Norwegian oil industry.[10] The state would not just take the money from royalties and taxes, but develop its own company – Statoil – with regulations managed by the newly established Norwegian Petroleum Directorate and set by the oil ministry.[11] This approach would provide a system of checks and balances, preventing any one impulse – commercial, administrative, political – from becoming too powerful. The initial advantages afforded to Statoil would eventually be stripped away, and the development of the country's oil reserves would become an expression of a wider search for balance in Norwegian society.[12] These days, the Norwegian system is widely respected for its efficiency and transparency. 'We don't receive any complaints,' one official tells me, 'because everyone knows everything about everything.'[13]

At the company level, the search for balance has led Statoil (now StatoilHydro) to diversify away from Norway and has forced the

Above: In September 1909, explorer and American naval officer Robert Edwin Peary sent a telegraph to US President William H. Taft, announcing that he had raised the Stars and Stripes (pictured here) at the North Pole. He offered to put the place at the president's disposal. Taft responded, somewhat bewildered, 'Thanks for your interesting and generous offer. I do not know exactly what I could do with it'.

Left: In August 2007, a Russian-led mission reached the floor of the Arctic Ocean at the North Pole, and deposited there a small Russian flag. Here, Arthur Chilingarov, a leading advocate of his country's Arctic vocation, celebrates the feat back on shore. President Putin congratulated Chilingarov and his team; his foreign minister explained to concerned Arctic nations that this exploit did not constitute an official Russian claim to ownership of the North Pole.

Above: One year after the Soviet Union launched the *Sputnik* satellite, demonstrating its lead in ballistics and the space race, the USS *Nautilus* crossed the Arctic Ocean under the ice, demonstrating the United States' potentially counterbalancing lead in submarine technology, threatening the Soviet Union with the risk of surprise nuclear attack from the Arctic or an assured retaliatory capacity. In 1959, the USS *Skate* became the first submarine to surface at the North Pole, where the ashes of Sir George Hubert Wilkins, the Australian pioneer of polar submarine navigation, were scattered to the winds, in the ceremony pictured here.

Above: August 2009, Canadian Prime Minister Stephen Harper arriving on the HMCS *Toronto*, off th coast of Baffin Island, to observe military exercises in the Canadian Arctic. In recent years, Conservati governments in Ottawa have expounded a 'use it or lose it' philosophy for the country's vast Arctic hinterland, placing defence of the Arctic in a Conservative narrative of national identity

Above: The Soviet city of Murmansk, 68° North, during the Second World War. Occupying a strategic position as the only ice-free port in Europe to which Britain and America could ship supplies and matériel to the beleaguered Stalin, Murmansk was heavily bombed by Germany. Only founded in 1916, built mostly of wood, it was largely destroyed.

Above: The Russian city of Murmansk today. In the Soviet period Murmansk became the largest city in the Arctic, with some 500,000 inhabitants, at the centre of a dense military-industrial complex on the Kola Peninsula. After a long period of economic depression following the USSR's collapse, some hope Murmansk will now re-emerge as a major player in the Russian offshore oil and gas industry.

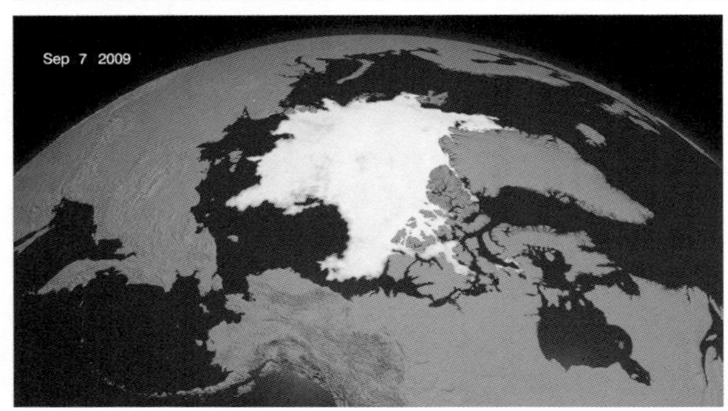

Left: Every summer, a large portion of the ice covering the Arctic Ocean melts. Much refreezes the following winter, but in recent years the overall trend has been one of decline. Some scientists worry that, beyond a certain point, this process will become irreversible, leading to a permanent reduction in Arctic ice, with profound consequences for both the local and global environment. Predictions of when the Arctic Ocean will be ice-free in summer have been coming gradually closer.

Above: Ny-Ålesund, the world's northernmost permanently inhabited settlement, 79° North, on the island of Spitsbergen. Originally built to support a nearby coalmine, the settlement now provides facilities for the scientific stations of ten countries: China, France, Germany, India, Italy, Japan, South Korea, the Netherlands, Norway and the United Kingdom. Research ranges from climate-related science, glaciology and permafrost, to prospecting for biological materials with potential commercial application.

bove: A satirical look at the Trans-Alaska Pipeline controversy in the early 1970s. Oil was struck at Prudhoe Bay in 1968, but only flowed south in 1977. Pipeline construction was held up by putes over environmental impacts and native land ownership. By 1973, the date of this cartoon, sts had spiralled. But, in an overworn argument, oil companies insisted development was a vital tional interest. In the end, President Nixon signed the final authorizations just five weeks after the outbreak of the 1973 Arab-Israeli war and the beginning of the first oil crisis.

Above: The Melkøya Liquefied Natural Gas plant, the first in the Arctic. Norway's StatoilHydro hopes that, despite difficulties in operating the plant, the experience will position it to win future business across the Arctic.

Inset: Alexei Miller, head of Gazprom. The lion's share of hydrocarbon development in e Arctic will take place in Russia, where gas exports by the state-controlled Gazprom are seen as key to the finances and geopolitical grand strategy of Russia, and where Arctic resources offer the best hope of maintaining or boosting production.

Left: In 1940, after more than two centuries as a Danish colony ruled from Copenhagen, the course of the Second World War in Europe led to Greenland coming under the effective control of the US. Denmark regained Greenland in 1945, but the US retained a major military presence in the north. Greenland would never be the same.

Right: Nuuk, Greenland's capital, today. In the foreground stands a statue of Hans Egede, the Norwegian missionary who refounded a European colony here in the 1720s, several centuries after a previous European population on the island had died out.

Right: On 21 June 2009, Denmark's Queen Margarethe II travelled to Greenland to transfer the acts confirming Greenlandic Self-Rule, enhancing the island's autonomy, and sharpening expectations of eventual independence. Given Greenland's small population – some 56,000 – even a small oil or other natural resource discovery might provide the financial wherewithal for Nuuk to become financially self-sufficient. Denmark currently pays an annual subsidy worth $11,000 for each Greenlander.

Above: A road across the highlands of central Iceland has been cut off by the waters rising behind the Karahnjukar dam, used to generate electricity for Alcoa's new aluminium smelter down at the coast at Reydarfjördur. The question of if and how to use Iceland's hydroelectric and geothermal potential has split the island. In part the question is environmental: what level of exploitation of Iceland's natural environment is acceptable to boost depressed parts of the country? But the question is also profoundly political: how can a small Arctic country maintain its identity and independence faced with the risks and opportunities of globalisation?

Below: Having allowed its banking sector to expand out of all proportion to the size of the economy, Iceland was hit hard during the financial crisis of 2008/9, which bankrupted the country's banks, forced the state to assume the banks' debts, and impoverished ordinary Icelanders. Here, Icelandic riot police protect the country's parliament from angry protestors.

Above: Where the land merges into the sea. Alaska's North Slope, August 2008.

company to play a complex political game to maintain its commercial freedom in Oslo. For the industry as a whole, the search for balance has meant finding a middle ground between the specific commercial interests of the oil industry, and the wider national interest in broad-based economic development. Partly in the interests of balance – and partly because Norwegian companies lacked the scale and expertise – foreign companies were allowed to retain their positions as field operators in Norway after Statoil's creation. And even though StatoilHydro is the lead operator in many of Norway's currently producing fields, lease blocks are almost always developed by a consortium of oil companies.

Norway's historic success in achieving an equilibrium between national, commercial, state and environmental interests may be harder to reproduce in the future. The decline of both the North and Norwegian Seas as oil-producing provinces is a problem. Having risen from nothing at all in 1970 to 528,000 barrels per day (bpd) in 1980, 1.7 million bpd in 1990 and 3.3 million bpd in 2000, oil production peaked in 2001. By 2007, production had fallen to 2.6 million.[14]

As a result of low domestic consumption, most of that oil is exported. (Norway exports as much as Kuwait though its proven reserves are one tenth as large.[15]) Any fall in production has a direct impact on exports. Technology is extending oilfields' lives and stemming declines in their flow. But without opening new areas to exploration and production – essentially the northern part of the Norwegian Continental Shelf, stretching far into the Arctic, potentially as far north as the Svalbard archipelago – production and exports will continue to fall. The oil and gas resources of the Norwegian part of the Barents Sea are expected to be nearly as great as those of the North Sea, but they could be greater.[16]

A fall in Norwegian production is not just a problem for Norway – which derives some 48 per cent of its exports by value from hydrocarbons – but also, potentially, a problem for the European Union.[17] Without a stable or increased supply of Norwegian gas, a growing dependence on Russian supplies will be accentuated. For Europe, a whole range of measures can be taken to reduce dependence: from helping LNG development, to diversifying its transport routes and freeing up its market internally. In the longer term, the best strategy will be to support energy efficiency at home and to develop alternative energy sources. But in the medium term – the next few decades – Norwegian gas will be important. As one European official I spoke to in Brussels puts it, 'Norwegian gas is expensive but secure.' For the future, 'Everyone wants a Norwegian production increase, but we can't just ask for it.'[18]

Matching the ambitions of StatoilHydro – in which the state still holds a 62.5 per cent interest – with opportunities in the North and Norwegian Seas alone is becoming impossible. In 2007, the merged StatoilHydro had revenues of 397 billion Norwegian Crowns ($57 billion). To grow beyond that, it will need to look outside Norway. Already, the company has pushed north into the Arctic and abroad into Russia, Canada, Alaska, Africa and elsewhere.

For Helge Lund, StatoilHydro's CEO, extending beyond the North Sea and the Norwegian Sea into the Arctic and beyond is a necessity of corporate strategy. Over the years the company has developed deep-water experience and subsea technology – 45 per cent of production on the NCS comes from subsea installations. Now, StatoilHydro is keen to apply that expertise elsewhere. The company trades internationally on its reputation as a technologically competent producer and as one which – despite the Mongstad refinery scandal of the late 1980s, and delays associated with the Snøhvit development in the Norwegian Barents – is capable of managing complex and environmentally challenging projects.

But the environmental and ethical problems that come with these developments are not easy to solve. And while the privatisation of Statoil in 2001 introduced the company to the private market – NorskHydro, in contrast, had always had private shareholders – the government's majority stake in the combined company means that all these issues are also acutely political.[19]

Opposition to Arctic oil and gas development has been persistent. Some environmental groups advocate extreme caution, warning that the Arctic is far more environmentally fragile than the Norwegian Sea, where most current activity takes place. Others argue that StatoilHydro's environmental record is not as good as it should be – in 2007 the company suffered a major spill in the North Sea.[20] Many oppose Arctic development completely, arguing that a country committed to climate change objectives cannot also be a major hydrocarbon exporter. Fishing groups, meanwhile, have urged strict controls on oil- and gas-related shipping and the scale of permitted developments. There is a gentleman's agreement with Russia not to drill for oil and gas in the disputed zone – 176,000 square kilometres of the Barents Sea, roughly equivalent in size to Norway's share of the North Sea.

Nonetheless, a series of major gas fields – collectively known by the name of the central field, Snøhvit – have already been developed by StatoilHydro, in the Barents Sea north-west of Hammerfest. A major Arctic oilfield known as Goliath is expected to be developed by ENI – the Italian oil company – before 2011.[21] Another oilfield, thought to be larger than Goliath, was discovered by NorskHydro in 2007, 65 kilometres due north of the Norwegian town of Honningsvåg.

Still, the oil industry argues that the opening up of Norway's Arctic to oil and gas exploration and development has already been too slow. In the 2008 licensing round, forty-six companies nominated 129 blocks to be opened for development in the Barents Sea. Just thirty were made available. Industry representatives warn that failing to produce oil in the Norwegian Arctic will not only reduce Norway's economic strength, but also displace production to areas with far lower environmental standards and weaken the country's capacity to invest in clean technologies for the long term.[22] As they see it, opening Nordland VI, Nordland VII and Troms II – areas where there is already geological data but where development is currently forbidden on environmental grounds – in the next few years would generate an additional 200 to 250 billion Norwegian crowns ($28 to $35 billion) in development investment between 2022 and 2040.[23] Despite the legal–diplomatic complications concerning the status of areas like the Svalbard archipelago, the industry's ambitions for the future extend still further north. In September 2009, the Norwegian oil minister visited the isolated Norwegian island of Jan Mayen, closer to Greenland and Iceland than the coastline of mainland Norway, advertising his country's ambition to search for oil far further into the Norwegian Sea.

StatoilHydro's investments in Russia, Canada and Africa come with their own challenges. Many ask whether the Norwegian government should hold a stake – albeit indirectly – in developments which may run counter to the political, environmental or ethical principles promoted by the Norwegian state. StatoilHydro's investment in

Gazprom's Arctic Shtokman field is politically controversial, with opinions divided on whether such an alliance potentially undercuts Norway's role in NATO or provides useful insight into Russian energy strategy and even a point of leverage over it. Equally controversial, though for entirely different reasons, StatoilHydro's $2.1 billion investment in Canada's Alberta tar sands is widely criticised for its potential environmental impacts. Over the border in Alaska, StatoilHydro purchased sixteen leases in the Chukchi Sea in February 2008, further defining the company as a leading Arctic investor. Finally, StatoilHydro's investments in Africa have led some to raise questions about the risks of the company – directly or indirectly – extending its deep-water exposure to the political and legal deep waters of offshore Angola.

Investing abroad is not a novelty for StatoilHydro. In the 1980s, as part of Arve Johnsen's strategy to develop a vertically integrated company protected against a fall in the price of crude oil, Statoil purchased downstream assets in Scandinavia.[24] In 1990, Statoil entered into a commercial relationship with BP. Ultimately this led to major investment in the Azeri–Chirag–Gunashli fields of the Caspian, as an independent Azerbaijan emerged from the wreckage of the Soviet Union and opened its doors to Western investors.

But StatoilHydro's investments in the Arctic, Russia and Canada are different. Not only are they of a different order of magnitude, but they also represent a fundamental shift in strategy: diversification away from the North Sea and Norwegian Sea and entry into an entirely new market, liquefied natural gas (LNG). To become something more like a typical Western IOC, StatoilHydro has rebalanced its strategy and objectives.

The Norwegian state is confronted with the same issue as StatoilHydro – declining production from mature fields encourages it to open Arctic areas. But squaring production in sensitive areas with Norway's reputation for ethical and environmental responsibility is a challenge. If it is met, the Norwegian model might provide a template for responsible oil and gas development elsewhere in the Arctic.

The government's answer to the dilemma of oil and gas development in the Arctic is contained in a 140-page document called the *Integrated Management of the Marine Environment of the Barents Sea and the Sea Areas off the Lofoten Islands*.[25] It amounts to a plan for Arctic development from Norway's northern coastline up to nearly 85° North, further north than any permanent human settlement on earth. The vast majority of the area covered by the plan will remain closed to oil and gas development for years to come. The plan's principal innovation is its search for long-term balance between environmental conservation, fisheries management, oil and gas.

Looking at the Barents Sea and Lofoten Islands as entire ecosystems, rather than as individual industries, individual species or the responsibilities of separate government agencies, the plan is intended to combine specific regulations with strong oversight, and to balance the objectives of economic development and environmental sustainability.

At the environment ministry in Oslo, officials are proud of their work, the result of cooperation between five different government ministries. Development of the North Sea in the 1970s and 1980s happened in 'bits and pieces' one tells me. But I am promised that for the Barents Sea, 'things will be different'. Already regulations on a range of issues – such as discharges – are stricter in the Barents Sea than in the North Sea.

One difference is the level of preparation before the publication of the management plan. It is, in the bureaucratic language of modern government, 'research-driven', with over thirty additional studies and reports drawn up between 2002 and 2005 to provide a scientific basis for regulation. (In some places, the sea floor was scanned with video cameras.) Compliance with rules on how close ships can come to the Norwegian coastline will be monitored from the radar station at Vardø. The Institute for Marine Research is already spending over 200 million Norwegian crowns a year ($28 million) on research and environmental monitoring in the Barents Sea.

Environmental oversight is only one half of the balance; the other is improving the geological understanding of the Arctic. In 2007–08, roughly the same amount – 210 million Norwegian crowns ($30 million) – was allocated to the Norwegian Petroleum Directorate to undertake seismic surveys in the Nordland VII and Troms II areas, off Norway's north-west coast.

The Norwegian government is trying to interest the Russian government in introducing a similar ecosystem approach for its own part of the Barents. But the chances of Russia adopting a plan as complex and rigorous as the Norwegian one are slim. 'It's difficult enough within our own borders,' says a Norwegian observer of discussions with Russia. 'It's going to add in another dimension with another country with maybe different views and agendas.' But it is not impossible over the longer term. The issue is regularly raised in the Norwegian–Russian Environmental Commission.[26]

More pressing is the review of the integrated management plan, slated for 2010. The arguments of a few years ago are likely to be rehashed: 'It's an opportunity for all sides to get more of what they didn't get in 2006.' It is also an opportunity for scientists to assess how resilient the natural environment of the Barents Sea is proving to be.

In some areas, the results are likely to disappoint – though the oil industry itself may not be to blame. In Tromsø, Birgit Njåstad and Per Arneberg – scientists at the Norwegian Polar Institute – explain that there is clear evidence of damage to sponge, coral and other communities on the sea floor in recent years, but the culprit is bottom trawling in the fishing industry, not exploration for oil and gas. Levels of persistent organic pollutants (POPs) in the Barents Sea are too high but the problem originates in the industrialised south, rather than activity in the Arctic. Overfishing has caused the decline of coastal cod, Greenland halibut and redfish in recent decades (though some may now be coming back). Global warming, meanwhile, is described as the 'driving force' affecting the Barents Sea ecosystem. 'In the media, the focus is on the petroleum issues,' Birgit says, but ultimately the development of oil and gas is one factor amongst many.

HAMMERFEST, NORWAY, 70° North: If you look out across Hammerfest harbour towards StatoilHydro's Melkøya natural gas plant, you can see two flames shooting into the air, just behind the hill which runs down to the sea. As night falls, the flames blaze brightly against the dark blue sky, lighting up the hillside with a sulphurous glow. There is something beautiful – and silently impressive – about it all. As if watching a volcano, I stand on the quayside in Hammerfest, transfixed, buffeted by the soft winds of an Arctic summer evening.

For StatoilHydro, these flames are an embarrassment. Initially, unable to fully process the natural gas coming from the three offshore fields collectively known as the Snøhvit (Snow White) development – Askeladd, Albatross and Snøhvit – the company has been forced to burn off, or 'flare', the excess. Particularly during the start-up period in 2007, carbon emissions resulting from this burning exceeded by several times the permitted rate, raising the possibility that Norway would exceed the international targets set for its carbon emissions under the Kyoto Treaty.

The short-term cost to StatoilHydro – in terms of lost revenue, reconfiguration of the plant and taxes on carbon dioxide emitted as a result – runs into billions of Norwegian crowns. The long-term cost to StatoilHydro – and to Norway's Arctic oil and gas developments in general – may be much higher. Snøhvit has demonstrated the challenges of Arctic development and led some to question the technical capabilities of StatoilHydro.

Sverre Kojedahl, spokesman for StatoilHydro in Hammerfest, is defensive. Snøhvit, after all, is a frontier project, he tells me. An LNG liquefaction plant – turning natural gas from the Barents Sea into more easily transportable liquid form – has never been built in Europe before, let alone in the Arctic. This is a first time for StatoilHydro, too. Once the plant is running more efficiently and at full capacity, large-scale 'flaring' will no longer be necessary. Beyond that, StatoilHydro is developing plans to reinject a large amount of carbon dioxide – released with the natural gas in production – back underground.

Kojedahl argues that Snøhvit should be viewed as a unique project, a template for the future. 'Most people living here [in Hammerfest] before had a very negative view of the future,' Kojedahl tells me. Now, with Snøhvit and the possibility of further developments – Goliath in the Norwegian Barents and Shtokman in the Russian Barents – Hammerfest is looking forward to taking some of the oil and gas business away from Stavanger and the south. Representatives from Gazprom have been to visit Melkøya, Kojedahl tells me, hinting that StatoilHydro's experience here will be used to parlay the company into more lucrative deals in Arctic Russia.

The first cargo of LNG from the Snøhvit development to Japan left Hammerfest in October 2007. Future shipments will head to plants in the United States, Spain and France, where the LNG will be turned back into gas for local consumption. Annual exports are expected to be around 5.7 billion cubic metres (bcm) of natural gas, about one quarter the size of the first production phase of Gazprom's Shtokman development.[27] (Some of those exports will be sold by StatoilHydro itself – others will be sold by other partners on the Snøhvit project, notably Total and Gaz de France.[28])

The Askeladd gas field was discovered in 1981, and the Snøhvit field a few years later. Plans to develop the field circulated from 1985 until the late 1990s. Then, in 1997, there was a concerted push: 'In the 1990s both the market and the technology were developing in the right way for us.' But it was five long years from initial planning to parliamentary approval. 'I met a lot of people,' Kojedahl tells me, working the communities of northern Norway in an attempt to persuade local inhabitants, particularly the fishermen, that the project was good news for them. Himself a southerner, he was never in any doubt of the value of the project to the north of Norway but, nonetheless, 'many times I thought that it might not happen. This wasn't a project until the parliament voted for it in [March] 2002.'

Before the future could be built on Melkøya Island, the past had to be uncovered. In the summers of 2001 and 2002 a team from the archaeological department of the University of Tromsø dug up remains stretching back 10,000 years, mostly stone tools and some later dwellings. In the final summer of the archaeologists' work, Statoil began construction.

The project itself is presented by Kojedahl as a technological marvel. Pipelines from the field run 143 kilometres under the Barents Sea to Melkøya. One of them, a 28-inch steel pipe protected by gravel and rocks, carries natural gas. Another carries the fibre-optic cables that allow the field to be exploited from a control base onshore. Others transport chemicals that prevent any ice plugs forming. Onshore, five gas turbines power a specially designed compressor that liquefies the gas.[29] About 6 to 7 per cent of the total energy potential of the natural gas extracted from the field will be used just to power these turbines.

The carbon dioxide from the gas turbine plant – estimated at 900,000 tons of CO_2 annually – is one of the major environmental worries for campaigners against Snøhvit. Carbon dioxide coming up from the field itself – fully 700,000 tons of it – will be separated from the natural gas and piped back underground. But until StatoilHydro builds a carbon capture plant for the emissions from the turbines, the Melkøya plant will be a major contributor to Norway's overall CO_2 emissions (representing around 2 per cent of the national total).

From the drawing board to production, Melkøya has been fraught with problems. Start-up production was delayed by nearly a year. Costs were much higher than anticipated, rising to nearly 60 billion Norwegian crowns ($8.4 billion). Deliveries to the plant were late – and sometimes incorrect.[30] And when the project finally came online there were problems with the seawater heat exchangers, a major and innovative part of the LNG technology used at Melkøya. The plant had to shut down for a while, and then operate below full capacity for months. In summer 2009, the plant was shut down yet again.

Arctic weather conditions have also disrupted production. In 2007, weather shut down an air inlet, forcing StatoilHydro to install additional shielding. Later, thick snow kept setting off an electronic sensor designed to detect leaking gas molecules, and part of the plant was automatically shut down. Worse, from Kojedahl's perspective, everything was reported in the media. The pressure took its toll: 'We are workers too. It was tough times, when you work through a problem and the door slams back in your face with another.'

Still, he thinks that there is room for expansion, maybe extending the life of the field from the current twenty-five or thirty years to nearly half a century. Physically, there's space on Melkøya Island. Standing underneath the gas flares, Kojedahl indicates the area where half a hillside was blown away with dynamite to make space for the development. 'The seagulls have made their home in the cliff now,' he tells me, the high-pitched screams of the seagulls the only thing that I can hear above the roar of burning natural gas, 50 feet away. In the future, the space could be used either to build a carbon dioxide capture plant, for sequestering some of the byproducts of the gas turbines, or else to build a second 'train' to produce LNG. 'No commitment has been made,' says Kojedahl.

But in Hammerfest many worry that the boom times associated with the construction of the Melkøya plant will be hard to sustain. Indeed, they may already be over. Jan-Egil Sørensen from PetroArctic, an organisation set up to represent local suppliers to the petroleum industry tells me that 'in dinner speeches they say that [Arctic development] will benefit local companies, but in reality . . .' Sørensen, smiling, shakes his head to indicate that dinner speeches may be one thing, but reality quite another.

It's not that Norwegian companies aren't benefiting from the development of oil and gas resources in the Norwegian Arctic. But many of those suppliers are the established Stavanger-based companies in the south, which grew up in the first flourish of the Norwegian oil and gas industry in the 1970s and 1980s, and they have no interest in establishing competition in the north.

Still, for Sørensen, the overall experience of hosting a major construction project has been a positive one. The increase in revenues for the town is for the long term. And given the decline of other industries (a frozen-fish factory which used to have 1,300 employees now has 200) the economic boost from StatoilHydro has been welcome. Before, 'We waited and we waited and we waited and we didn't do anything else.'

But it is an argument with an inbuilt ratchet. Major projects like Melkøya employ thousands of people in construction, but only a few hundred when they are up and running. When such projects are finished, suppliers and employees are looking for the next development, the next job, and the next boom. For Sørensen, development of further oil and gas fields can't come soon enough.

The model of balanced oil and gas development in Norway may not be perfect. But it is far better than the approaches taken in most countries. Some hope that the Norwegian example could provide a template for environmentally sustainable oil and gas development across the Arctic.

There are plenty of examples of collaboration on environmental issues in the Arctic. Norway has long-standing – though occasionally problematic – cooperation with Russia on fisheries' management. The environment is one of the main pillars of work of the Arctic Council, a largely (but not exclusively) intergovernmental body comprising Canada, Denmark, Finland, Iceland, Norway, Russia, Sweden and the United States. On some issues, notably fisheries, there is strong support for international regulation even in the United States. In 2008, Alaska's two then Republican Senators, Ted Stevens and Lisa Murkowski, coauthored a Senate resolution that called for the administration to negotiate an international framework for Arctic fisheries. Strengthened cooperation, including a regional agreement on Arctic environmental management, has been strongly advocated by several non-governmental organisations.[31]

But the adoption of an international system for the Arctic which mirrors the careful Norwegian balance between development and sustainability is unlikely. There may be room for an international agreement on strengthened environmental monitoring and on Arctic fisheries – after all, many other trans-boundary fisheries are subject to international agreements. But governments in the Arctic region will never allow supra-national bodies to have the final say on oil and gas development. Cooperation may improve, and the ecosystem-based approach should become standard practice, but it will be slow to come. Some doubt that a new legal regime for the Arctic would achieve much more than the existing international laws governing the sea.[32] In May 2008 the foreign ministers of five of the Arctic's coastal states – Canada, Denmark, Norway, Russia and the United States – specifically rejected a 'new comprehensive international legal regime to govern the Arctic Ocean'.[33]

Closer to home, there may be more immediate concerns for the Norwegian model. In 2010, the integrated management plan is up for review. Debate over the future extent of oil and gas development in the Norwegian Arctic will be fierce. Norway faces a major challenge over the next few decades in reconciling its ethical and environmental principles with the likelihood of growing Arctic development.

Ultimately, the success of the Norwegian model so far depends a lot on the bureaucratic efficiency and consensus-building capacity of the Norwegian political system. Transferring that balance and level of integration to other political systems will be tough; internationalising them will be next to impossible. But if the Norwegian model cannot simply be transferred, it can nevertheless act as an example of best practice. In countries where the administrative structures for managing development are already set – such as in Russia or in the United States – adopting foreign best practices may face bureaucratic resistance. In other countries such as Greenland, where the process of Arctic oil and gas development is only just beginning, there is a greater opportunity for the Norwegian model to become a model for the wider Arctic.

PART FIVE
Freedom

'The gateway through which we must pass is now wide open'
Greenlandic Premier Kuupik Kleist, 21 June 2009

12
Greenland's Search for Independence

'The people in that country are few, for only a small part is sufficiently free from ice to be habitable; but the people are all Christians and have churches and priests'

The King's Mirror, unknown author, circa 1250

'Is it to be expected that the Danish Government will keep this going for ever? Would it not be better and wiser for us first to recall our outposts and then gradually to withdraw the colonies and hand over the warehouses and buildings to the natives? In my own opinion, the very best thing we could do in the end would be to pack up all the stores, put them and the traders on board the Company's nine ships, and set sail with the whole back to Denmark. This will have to be done sooner or later, but perhaps not until there are no natives left behind to inhabit the land'

Fridtjof Nansen, *Eskimo Life*, 1891

'Today is a day of freedom'

Danish Prime Minister Lars Løkke Rasmussen, 21 June 2009

On 21 June 2009, Greenland and Denmark celebrated the beginning of a new constitutional relationship – extended Self-Government for Greenland, falling short of full independence – with ceremonies and speeches in Nuuk, Greenland's capital. 'It is as if we have woken up to newly fallen snow in which we are to set the first footprints for those who come after us to follow,' intoned Greenlandic Premier Kuupik Kleist, perceiving the opening of a gateway to eventual independence from Copenhagen after centuries of colonisation and three decades of limited Home Rule. Danish Prime Minister Lars Løkke Rasmussen, in reply, spoke of freedom, responsibility and solidarity, welcoming a day 'when we loosen the ties that might be too tight, but keep hold of those that strengthen'. He celebrated the 'great solidarity and mutual respect' which had led to a negotiated constitutional settlement, confirmed in a referendum of Greenland's voters. In front of an audience including the President of Iceland – a former Danish colony turned independent state – the Queen of Denmark presented the Self-Government Act to the Greenlandic parliament, and Copenhagen's grasp on the Arctic's largest island became a little fainter.[1]

Self-Government is not the same as independence. Ties between Nuuk and Copenhagen have not been cut entirely. Greenland is perhaps a nation – but it is not yet a nation state. Copenhagen will continue to pay an annual subsidy to Greenland of over 3.4 billion Danish crowns ($660 million), over $11,500 for every one of Greenland's 56,000 residents. In return, Denmark retains control of foreign and defence policy, while its Queen remains Greenland's head of state.[2] The Self-Government Act raises the possibility of eventual independence for Greenland, but makes it clear that this could only come about as a matter of negotiation agreed by both sides, rather than secession pronounced unilaterally.[3]

But if Self-Government is not a blueprint for Greenlandic independence, it points to the means by which independence may eventually be achieved: exploitation of the island's mineral wealth. The country covers over 2 million square kilometres, nearly the size of Western Europe. Though most of the land is covered by ice, the mineral potential of Greenland's on- and offshore areas is considerable. Exploitation of this bounty might just provide the economic basis for independence and the creation of the first truly Arctic state, with a majority Inuit-origin population. (In Canada, the federal territory of Nunavut was formed in 1999, but there is little prospect of independence in the short term.)

At the heart of the Self-Government Act there is a deal. If Greenland's income from mineral extraction rises, the subsidy from Copenhagen will fall by half that amount.[4] But what happens if and when Denmark's subsidy is reduced to zero?

According to the Act itself new negotiations would then be opened, with the 'distribution of revenue from mineral resource activities in Greenland' as one of the points on the agenda, leaving open the possibility that Denmark might insist it continue to receive some income from mineral exploitation on Greenland.[5] But for many Greenlanders, the answer is both obvious and simple: full independence. For them, independence depends almost mechanically on rising resource income cancelling out financial dependence on Copenhagen. Harnessing the mineral potential of Greenland – from the land at the fringes of the ice, and from under the seas beyond its shores – is the country's ticket to freedom.

NUUK, GREENLAND, 64° North: It is just after nine in the morning when the utility truck screeches into the parking area of the Nuuk Seamen's Home. It's my ride. 'Come,' shouts the driver across the gravel, revving the engine for emphasis. 'We're late.'

At the harbour, we are met by a deckhand, half-raising his hand in greeting and guiding the truck to the edge of the quay – and to our boat, one of dozens arranged along the marina. We unload 10-litre jerrycans onto the bow, hands smearing with rust, grit and thick engine oil. It's silent, efficient work. There is an immediate and unspoken agreement on who should do what. Formed up into a rudimentary production line, we wordlessly pass the jerrycans between us – from truck to quayside and from quayside to boat. Ten minutes later we have eased out of the bay and are gunning north-east towards Qussuk Fjord, a fine morning mist suspended above the sea.

The boat is smaller than I had expected – not much more than 15 feet long. It looks more suited to the Florida Keys than to plying the working route between Nuuk and west-coast mine camps. But it is fast, easily picking up to 26 knots once we hit the open water, the Mercedes-Benz engine whining above the slap of the boat against the water.

Occasionally, as I scan the mist, more solid patches of white catch my eye, momentarily acquiring form and depth before sinking back into the mist: icebergs, calved from the glaciers that feed into the fjord.

We skirt them effortlessly, changing direction a couple of degrees, barely dropping speed as we pass by. Before long, the mist begins to dissipate, vaporised by the warming summer sun. And then, quite suddenly, it is gone.

Blue skies, bright sunlight glancing off the surface of windless water – it takes the sight of an iceberg to jolt me out of consideration of what it would be like to jump in. But this is Greenland. Even in September, when the water is at its warmest, the temperature never gets above a couple of degrees Celsius. I turn my back against the sky and go inside.

'You have to kiss a lot of frogs before you find the prince,' says Jock, a former academic geologist from South Africa with the kind of straightforward manner and laid-back ebullience which turns every problem into a challenge, and every challenge into a solution. 'What we're looking for is a seam of rock about two to three metres thick, with a concentration of about five parts per million – a needle in the proverbial haystack.' He doesn't seem fazed.

Jock's current work is as a consultant to Impala Platinum Holdings Limited (Implats), a Johannesburg-based mining company that supplies one quarter of the world's platinum. Historically, Implats has focused on production in South Africa and Zimbabwe. These days, they are pursuing a more aggressive strategy abroad. Alongside Greenland, Implats is involved in exploration in Botswana, Canada, Madagascar and Mozambique.

On this trip, Jock is hosted by Ole Christiansen – a barrel-chested Greenlander who runs Nunaminerals, a mineral-prospecting company formerly owned by the government of Greenland, but now nearly two thirds in private hands.[6] Nunaminerals discovered the prospect in 2005 – one of several being explored for platinum, gold, copper, iron and nickel across Greenland.

With Ole is his colleague Paul, an English geologist with seven years' prospecting experience in Norway and a PhD in the geology of the 66,000-square-kilometre (25,000-square-mile) Bushveld Complex, site of Implats's South African platinum mines.

Paul has been in the area of the prospect for two four-week stints already, walking the land, taking samples, trying to understand the history of the rock to evaluate where there is the greatest chance of sufficient mineral-bearing formations. It is slow, patient work – the gradual accumulation of information, the testing of geological hypotheses, the constant refining of one's understanding of what lies beneath.

At this stage, Nunaminerals is selling a prospect, with no guarantee that the prospect will turn out to be good. Essentially, all there is is enough geological information to suggest the possibility of recoverable minerals, and enough confidence to warrant investment to find out exactly what is there.

Paul is the investment; Jock is the customer. He needs to be convinced about the prospect's possibilities in order for any additional investment to be forthcoming. An initial exploration programme lasting eighteen months should cost $1 million and will buy Implats a 20 per cent interest in the prospect. Beyond that, the company has an option to increase its stake to 60 per cent and enter into a joint venture with Nunaminerals – but only with the commitment of several million more dollars to exploration and a $1 million payment to Nunaminerals for previous exploration costs.

At each stage, rational assessment of risk and reward has to be set against experience and intuition (what Pauls calls 'a feeling for where things are'). At each stage, new investors may have to be found, more money borrowed, new shares issued. At each stage, geological information can be creatively repackaged to bring in new investment. 'There's a lot of marketing in the mining industry,' one geologist tells me. 'When companies need to raise new cash they drill a hole three metres away from the best hole they already have and then cry, "New shaft confirms northward extension of the prospect" – yeah, three metres northwards.'

The mining giants of the modern world – BHP Billiton, Anglo-American, Rio Tinto, Xstrata, companies that calculate their annual revenues in tens of billions of dollars and employ hundreds of thousands of people – are often not involved until the quality of information about a prospect is sufficient for them to pile in with the kind of investment which only the behemoths of the industry can provide.[7] 'But if companies like us find stuff here in Greenland, they'll make their move,' says Jock. And when they do? The rest is foreplay. (Nunaminerals already has an alliance agreement with Rio Tinto Mining and Exploration Limited for the identification and assessment of potential projects in Greenland.[8])

Several mining companies are already here. Flying down the west coast of Greenland the previous day I had caught sight of an industrial warehouse and a spit of land north of Nuuk – the Seqi olivine mine, owned by Swedish mining company LKAB. Started in August 2005, it is a strategic investment for LKAB which should provide olivine, used in steel-making, for several decades.[9]

Canadian-based Quadra Mining expects to open a 200-million-ton molybdenum mine in Eastern Greenland in the next few years (for an overall investment of nearly $1 billion). Ironbark, an Australian company, is working on a lead–zinc deposit in northern Greenland, increasingly easy to access as a result of retreating ice. In February 2008, a subsidiary of the British Angus & Ross mining company applied to reopen the lead–zinc mine at Black Angel, shut since 1990. South of Kangerlussaq, Hudson Resources, from Vancouver, is exploring for diamonds. And while mining for uranium is currently forbidden and many Greenlandic politicians are opposed to it, several companies have expressed an interest in the known potential of Greenland's deposits, particularly in the Kvanefjeld complex in the south.

Overall, the Greenlandic Bureau of Minerals and Petroleum expects seven mines to open over the next five years, creating some 1,500 jobs in a country of just 56,000 permanent residents. And this does not include the additional jobs – potentially several thousand in the construction phase – should Greenland ultimately go ahead with Alcoa's plans to build a 340,000-ton-a-ear aluminium smelting facility at Maniitsoq, on Greenland's west coast.[10]

Just how many jobs go to Greenlanders will depend on their ability to compete against international mining engineers. In order to improve their chances, a mining school has just opened in Sisimiut. A new graduate can expect to earn 400,000 to 500,000 Danish crowns ($80,000 to $100,000) a year, double what the average fisherman earns.

On the boat, Jock clears the cabin table of the half-eaten cinnamon buns and Styrofoam coffee cups from our aborted breakfast. He rolls out a geological map of the area, placing a book on one edge and a camera on the other to prevent it from snapping shut like a blind. And he points.

'We wanted to call it "giraffe" – but the word for giraffe in Greenlandic is unpronounceable, so we called it "squid" instead.' With a bit of imagination, I can see his point. A distended block of green at one end of the formation, with tentacles streaming out behind it to the south-west. The official name for the platinum prospect, in Greenlandic, is Amikoq. 'Last year we ran traverses every five hundred to two thousand metres, but the only positive results turned up at the edges, so now we're focusing on those,' says Jock, pointing at the three narrow sections of purple and blue, circled on the map. 'That's the real prospect now,' he tells me, a fraction of the 124-square-kilometre (48-square-mile) property.

The season available for assessing mining prospects in Greenland is limited, and we're already in July. In a couple of months there will be snow on the ground. 'On about twenty to thirty per cent of days, fog means that there is no helicopter access,' says Paul. And that is during the Greenlandic summer. The rest of the year, no prospecting takes place at all.

Seasonality is one reason why Greenland is presented as a frontier for mining companies. But Jock is philosophical. 'South Africa was the mining frontier once,' he muses. Obstacles that appear significant when a few hundred thousand dollars have been invested to stake a prospect will seem trivial if the prospect turns into a sure thing, and investment budgets increase from a few hundred thousand dollars to a few hundred million.

The lack of onshore mining infrastruture is dismissed as not a problem. 'The fjord is our highway to the mines,' Ole says, gesturing out towards the water, a mile or more from one side to the other. The depth readings from the boat's sonar have rarely indicated a water depth of less than 60 or so metres – frequently, the bottom of the fjord is out of range. When this area is commercially mined the necessary export infrastructure can be built close to the mine itself.

'We're in a much better position than in northern Canada,' explains Jock. In much of the continental Canadian Arctic access is limited to air – year-round access, but prohibitively expensive for anything other than vital supplies and people – and ice roads which have to be rebuilt every year. 'Global warming really screws up their logistics,' says Jock, as ice roads melt earlier in the year and can only be rebuilt later in the following winter. 'In Greenland, it's expensive, sure – but we've done the calculations.'

In many respects, Greenland is considered highly predictable. In many frontier operating environments the chief risks for mining companies are political – a change in regulation of foreign mining interests in Russia or Latin America, for instance, or the overthrow of a friendly government in Asia, or the outbreak of war in central Africa. In Greenland, these risks are viewed as minimal. The legal framework for mineral development set up by the Danes will remain under Self-Government. Licences are issued for up to fifty years. The importance of mineral development to the Greenlanders' long-term objective of independence from Denmark gives mining companies confidence that positive attitudes towards their presence are unlikely to change.

'Companies like Nunaminerals make a big difference, too,' says Jock, nodding at Ole. With a 2-million-crown ($400,000) loan from the Bank of Greenland, Nunaminerals has set up a Canadian-run laboratory in the basement of its office building, allowing it to analyse rock samples for clients in Greenland. 'It's a huge advantage. In South Africa it would take two to three months to turn samples around – from the lab analysis to the user. Here it's a week. I should have the results by email in a few days, and we can discuss it before I leave.' It almost sounds easy.

The base camp for the Amikoq prospect consists of a central cluster of rounded structures which look as if they have been airlifted directly from the Korean war and dropped around two sides of a grassed-over, not-quite-flat parade ground. Further off, there is a diesel generator, a washing area and a number of smaller two-man tents. Behind the camp, a mountain rises gently up to 2,000 feet above sea level, cheating perspective in its massiveness. To one side, just out of view, is a helipad.

Besides the platinum prospect on which Paul is working, the camp supports geologists working on iron ore and gold prospects. Forced to operate with a drill bit not designed for the drill shaft, work by the Australian team looking for iron ore has been slow and progress uneven. One day they had to drill through a layer of ice, the next the ice had melted and refrozen in the drill shaft. A day's drilling had been wasted, an enormous expense given the cost of basic logistics and the short length of the season.

The team looking for gold is more upbeat. 'We've progressed very quickly,' says Rasmus, a sun-burned Dane working for Nunaminerals, his face topped with tightly curled blonde hair. 'In 2005 there was nothing, now there's thirteen holes bored over fifteen kilometres – one of them is two hundred metres deep.' It is expensive to get the drill gear into place, requiring the help of a helicopter to lift the gear several kilometres into the interior, where it can be reassembled and primed. 'You have to be pretty confident you're going to find something,' says Rasmus. But once in place, the drill will work two twelve-hour shifts every day.

In the evening, we take the boat back to Nuuk. Ole falls asleep in the forward bulkhead, lulled by the motion of the boat and the steady whine of the engine. All of a sudden, we slow, and then – stop. We've lost one of the boat's two propellers. 'It's just been serviced,' the captain tells me. 'They must have forgotten to reattach it properly.' It is nothing serious, the boat can still get back to port – in mid-July there will be light around the clock and we are only fifteen miles from Nuuk. But our speed is cut to ten knots, barely faster than a slow jog. We can see the lights of the old town winking at us across the fjord, stubbornly maintaining their distance even as we phut-phut towards them.

The air is completely still as we get close enough to make out the individual buildings of Nuuk. It is like archaeology in reverse. First, the wooden houses of old Nuuk, laid out around a statue of Hans Egede, the Norwegian missionary who landed in Greenland in July 1721, re-establishing a Scandinavian presence after a gap of two centuries. Then, a slab of European 1960s prefabricated social housing, an oversized breeze-block stuck onto the headland, its sharp vertical and horizontal lines in stark defiance of the roughness of the coastline. 'Nuuk's answer to Beirut,' says one of the passengers. 'Not bad inside, though,' he adds, 'and hard to get – either you wait a long time or you have a job that entitles you to it.'

Almost next door, at Nuuk's southernmost point, surrounded by water on three sides, there's a large green-painted house. No one knows who lives there, but, according to the claims of one of the passengers, it's worth 7 million Danish crowns ($1.4 million).

Finally, on the shore opposite the harbour, the unfinished new district of Qinngorput is creeping along the edge of the fjord, connected to the rest of Nuuk by a jet-black ribbon of freshly laid tarmac. It is the future: transparent green-glass buildings with hints of brushed steel and wood. But it's not for everyone. One of the women on the boat – smartly dressed, not more than thirty-five years old – grew up in Nuuk. But now she is considering moving away. 'Nuuk's too big these days,' she says, 'too big.'

COPENHAGEN, DENMARK, 55° North: You can look at a geological map of something as familiar to you as the outline of your own country, and yet fail to recognise it. There are no towns or roads and, unless you look very carefully, there are no coastlines to tell you whether you are looking at land or sea. Unless you pick out the scale you could be looking at the precise geology of a single square mile of the earth, or the impressionistic geology of half a continent. In place of the familiar lines by which the uninitiated can read most maps, there is a kaleidoscope of colours – orange, pink, purple, blue, green – chosen for their brightness and contrast, each to mark out the extent of a different rock type.

To a non-expert the overall effect is one of chaos: more an abstract arrangement of colour than anything else. But to a geologist, each map conveys a multiple layer of meanings. First and most simply, a map tells a geologist the location of a particular type of feature in a particular geographical area. But the arrangement and extent of these features – their shape and their relationship to one another – can tell another story, or several. I look at the maps in the Copenhagen office of Lotte Melchior Larsen at the Geological Survey of Denmark and Greenland – piled up on every flat surface in the room, rolled up in every corner and hung on every wall – and I realise that I am looking at much more than the present. With experience and intuition, a geologist may be able to see several million years of the earth's history in a single map. To be sure, each map is an image of the earth frozen in time – but each map is also loaded with a knowledge of the past events which made it so.

Larsen herself is a diminutive woman in her fifties, but with the energy and inquisitiveness of someone much younger. When I arrive, she comes to greet me at the door of her office, elegantly navigating the narrow fjord of floor space that the bookshelves and filing cabinets have not yet occupied.

'We are not the ones who find the minerals,' she explains. 'That's what they have mining companies for. But they always start with a map. And that's where we come in.' She has been making geological maps of Greenland for over thirty years.

At first, Larsen's work on the Disko Bay area in western Greenland was informal, driven more by personal interest than by the directives of the Geological Survey itself, still less by the commercial designs of resource companies. Her thesis was on the geology of a different part of Greenland.

But she became interested in western Greenland through a colleague, and she was able to use her background in geochemistry to help explain apparent similarities between rocks from west and east Greenland. When a colleague suggested mapping work, she agreed. The colleague – who had been collecting data in western Greenland since 1969 – later became her husband. 'I went into this area because I fell in love with a man. I'm still married to him,' she explains. And although her work has since taken her to the prospects of the north-east of Greenland, and the east of the island, 'I'm still working on western Greenland. When you've been there it's a sort of sickness,' Larsen explains. 'You want to go back.'

Initially the maps were an informal project: 'We thought from the start that these maps would be absolutely uneconomic and unuseful – but scientifically interesting.' But when she started publishing the results of her work, interest in the Disko Island area and the Nuusuaq peninsula began to increase. 'These uneconomic rocks turned out to host possible deposits of platinum and nickel,' she says, 'and now they are hosting oil – the oil companies are paying us to put them into three-dimensional models.'

To explain what makes Greenland geologically special, Larsen draws out a large map of the North Atlantic. She points at similarly coloured zones on different islands between the European continent and Greenland. 'See these rocks here,' she says. 'They are the same as on Skye [in Scotland] or in the Faeroe Islands, because it's all part of the same thing, something that happened 62 million years ago, something that made the North Atlantic break up. Imagine,' she says, looking at me earnestly. 'You had a very great continent and then,' she pauses for effect, 'suddenly it broke to pieces.' And Greenland was at the centre of it all, a dagger of land pointing south. 'Here you would have found molten lava flows that were fifty, eighty, up to three hundred cubic kilometres. You wouldn't find anything on that scale now. We had volcanic eruptions here where the ash ended up in Denmark.'

Geological timescales are several orders of magnitude greater than the familiar timescales of history: a generation, a century, a millennium. Things that would be impossible in the span of human history – the creation of a new ocean, for example – become possible in geological time. And the Atlantic ocean, as Larsen reminds me, is geologically young. If Larsen describes the birth of the Atlantic – a process which took many thousands of years – so dramatically, it is because, for her, it is one of the most recent instalments in the long drama of the earth's evolution. And it isn't finished – the continents are still drifting away from one another.

But the reason for Larsen's apparent diversion into the geological birth of the Atlantic isn't simply for the purposes of enlightening a non-geologist. It is also to explain the geology of Disko Bay. 'See this rock,' says Larsen, picking up and handing to me a piece of rock studded with beads of shimmering material. 'This is a very special rock,' she tells me. 'The tiny crystals there are metallic iron. It's one of the very few places where metallic iron occurs in nature. It was in a lava flow, see.'

'The Inuit knew about this place,' she tells me. 'Their name for it begins with a word which means "knife" in Greenlandic – and this iron would be good iron for making tools. It isn't cast iron – that would break – it's malleable iron.'[11] Twentieth-century geologists knew that the presence of iron was likely to be an indicator of the presence of other, more valuable, metals. 'Iron acts to concentrate other metals,' Larsen explains. 'Nickel wants to sit with a metal a thousand times more than in a silicate [lava flow], and platinum ten thousand times more. So where you find iron like this, you are likely to find platinum and nickel.'

For other resources Disko Bay may require a rethink of the most basic tenets of modern geology. Oil is generally found in sedimentary basins where a barrier of rock, a salt dome or some impermeable formation has caused the oil to rise and collect in a reservoir. But in Disko Bay there is oil in the lava flows – far from acting as a seal, as is normal, the volcanic rocks are acting as a reservoir. It is almost unheard of: 'If you told me that oil would get into a lava flow I would say, "No, it can't." But it did, you can see it. I have no arguments for refusing it. We don't know how it gets there – perhaps microcracks in the rock – or how long it took to get there. There's a lot of things we don't know.' The only other place that Larsen can think of where there are similar occurrences is offshore Australia.

One result of the unusual combinations of the geology of Disko Bay is that the offshore areas are often difficult to understand properly. 'The oil is there,' says Larsen, 'no doubt about it. The sedimentary basins, too – like the Nuusuaq basin. But the oil companies' theories might be wrong.' And finding out where the companies are right and where they are wrong is made more complicated by the presence of volcanic rock. 'It's hard to see through,' Larsen explains. 'The seismic waves disappear, they are deflected. Except when they use very low frequency seismic.'

With all the years of mapping she has conducted on Greenland, Larsen sees herself at the beginning of the process, rather than at the end. 'Look at the map and all the ice,' she tells me, pointing at the white ice cap which dominates the island. 'All this is areas that you have been unable to access,' she says. 'What we've mapped is less than 5 per cent and it's taken years. Now we're carefully considering where we should do the new mapping – in cooperation with Greenlanders, according to their wishes, and their wishes are to map those areas which have the largest potential for economic deposits.'

Larsen hopes they find oil, 'even if that probably means that Greenlanders will say "bye-bye" to Denmark.' This is the central irony of the geological survey. Though the survey is financed principally by Denmark, its results may ultimately provide the basis for Greenland's independence. One of the geological survey's current projects is to map the seabed around Greenland with a view to claiming an extended continental shelf under the terms of UNCLOS. Within a matter of years, however, ownership of the seabed could pass to a newly independent Greenland, free of all ties to Copenhagen.

Historically, mining has been a major part of Greenland's economy only episodically. For the most part, it has been based around the products of the sea – first sealing and whaling, and then fishing.

Up until the twentieth century, the centre of Greenlandic society and economy was the Danish-controlled Royal Greenland Trading Company, established as a monopoly in 1774 (a string of earlier companies with concessions on Greenland had gone bust).[12] The trading posts of Royal Greenland were the symbols of Danish sovereignty, the tools of Danish control and – along with the missions of the Greenlandic church – the cement for colonial society.

By the middle of the nineteenth century, Royal Greenland had established thirteen colonies. In principle, these were outposts of European empire, ruled from Copenhagen. In reality, the distance between them and Denmark itself meant they were managed as individual enterprises at local level.

This was colonial rule with a light touch. Although the managers of the trading posts were almost always Danes, the total number in Greenland was never high. The legal system distinguished between Europeans – subject to Danish laws and proceedings – and Greenlanders – subject to a more limited set of laws based on local traditions rather than European imports. The Royal Greenland company was not simply a rapacious commercial organisation designed to extract maximum profit from Greenlandic trade, indeed it rarely made a profit at all. Royal Greenland was as much an instrument of government as anything else.

Even as criticism of the company's policies grew at the beginning of the twentieth century, the notion of a trade monopoly was unquestioned. It was accepted as a means of guaranteeing economic stability and protecting Greenlanders from the unfettered ambitions of commercial enterprises. Even C. W. Schultz-Lorentzen, a fierce critic of Royal Greenland and a staunch advocate of greater economic and political opportunities for Greenlanders, defended the principle of a state-run monopoly, warning that 'the sole aim of private capital is to enrich itself at the expense of the Greenlandic population'.[13]

Until the twentieth century what little mining that did exist on Greenland tended to be small scale and artisanal. The most significant Greenlandic mine in the nineteenth century was the cryolite mine at Ivittuut – opened in 1854 – which provided a steady stream of royalties to the Danish state, partially offsetting the deficits of trade in other goods.[14] Initially, cryolite was used to make enamels. Later, its most important use was as a key flux in the electrolytic processing of bauxite to make aluminium. In the late ninteenth century, aluminium was more of a precious metal than an industrial one. When the Washington monument was finished with a small aluminium pyramid in 1884, aluminium was as expensive as silver and that pyramid alone represented one twentieth of American production.[15] Demand for Greenland's cryolite was correspondingly limited.

At the beginning of the twentieth century, interest in exploiting Greenland's mineral resources re-emerged. But with fears that an influx of capital and foreign workers might undermine the traditional structure of Greenlandic society, mining was only permitted under strict conditions. A concession granted in 1902 – which covered everything other than cryolite – stipulated that no local labour was to be used in the mines and there was to be no contact with the local population. A few mines opened over the next couple of decades – the Josva copper mine in 1905, the Amitsoq graphite mine in 1915 and a coal mine at Qulissat in 1924 – but these were risky, short-term enterprises. Only the cryolite mine at Ivittuut operated continuously throughout the pre-war period.

By the 1920s, the traditional foundations of Greenland's economy were beginning to disintegrate. The number of seals caught declined from eight per head of population in 1910 to below five per head in 1935. Backed by government development policy and helped along by a warm period for Arctic climate, fishing became increasingly important. Greenlanders had owned 157 wooden boats in 1900; a few decades later they owned nearly ten times that number. By the late 1930s, Greenland was exporting salt cod to, of all places, Portugal.

Just as Iceland had been transformed from an agrarian economy to a fishing economy over the course of the late nineteenth and early twentieth centuries, so Greenland was transformed from a hunting-based economy to a fisheries-based one. In the process, the relationship between Greenlanders and their environment was fundamentally altered. Hunters were being turned into industrial employees. The construction of new fish-processing plants required a concentration of the island's population in fewer and fewer places.

At the same time, production at the cryolite mine at Ivittuut took off. Throughout the 1920s the demand for cryolite boomed on the back of technological improvements in aluminium smelting, a drop in the price of aluminium and an increase in its use, including in the construction of aircraft.[16] Now, what had once been considered a marginal addition to Greenland's economy had become its main source of revenue.

By the late 1930s, Greenland was exporting 50,000 tons of cryolite, mostly to Denmark.[17] With the occupation of Denmark by German forces on 9 April 1940, exports shifted. Henrik Kauffmann, the Danish ambassador to Washington, signed an agreement with the United States that put Greenland at the service of the Allies. Exports of cryolite to North America grew from 42,000 tons in 1940 to 62,000 tons in 1941 and 85,000 tons in 1942, much of it to be used in the construction of warplanes.

For the first time since the Dutch trading expeditions of the seventeenth century, Greenland's trade had been redirected outside Denmark.[18] The war had demonstrated that Greenland could survive without direct Danish support. With the stationing of American troops on Greenlandic soil, the island's cultural isolation was shattered.

Denmark emerged from the Second World War with its north Atlantic realm under threat. Iceland, independent since 1918 in all areas but foreign policy, and connected to Denmark through the monarch, held a referendum on its political destiny in May 1944. Reykjavik's ties with Copenhagen were formally severed.[19] The Faeroe Islands – occupied by the Allies during the Second World War – were moving towards greater self-government, a goal finally achieved in 1948 under Denmark's overall sovereignty. With Iceland lost and the Faeroe Islands on the point of virtual secession, stalling any moves towards Greenlandic separatism was a point of Danish national pride.

Increasingly, however, it was also a matter of Denmark's national security. With an American military presence on Greenland from the Second World War Denmark's diplomatic position was bolstered.[20] Denmark's hold over Greenland was a diplomatic asset. In 1946, United States Secretary of State James F. Byrnes reportedly suggested the US purchase Greenland outright, reviving schemes for American ownership of Greenland which dated back to his nineteenth-century predecessor William H. Seward.[21] When Denmark joined NATO in 1948, some argued that an American guarantee for Denmark's territorial integrity had been purchased with Greenlandic soil, though Foreign Minister Gustav Rasmussen assured parliament that Greenland's facilities would serve a purely defensive purpose.[22]

It was in this context that Denmark parried the Greenlanders' post-war request for the transfer of the central administration of the island from Copenhagen to Nuuk with a substantial increase in government investment, and a doubling of salaries for government employees. Denmark was acting no differently from other colonial powers.

(Despite hardship, rationing and financial weakness at home, French and British administrators similarly boosted spending on their colonies in the 1940s and 1950s, combining a self-interested policy of delaying independence with a gesture of post-war munificence.)

But unlike other colonies, Greenland had never been a net economic benefit to its colonial master (apart from a few years during the First World War and a few years after 1935 when the cryolite mine was expanding). Copenhagen had frequently been forced to balance the deficits of the Royal Greenland Trading Company with state funds. Now, Greenland's dependence on those funds became greater than ever before.

Officially, Greenland's colonial status ended in 1953 with a constitutional referendum confirming its full integration into Denmark. In 1954, the United Nations General Assembly voted forty-five to one to accept Denmark's decision to de-list Greenland as a non-self-governing territory.[23] However, economics and politics pulled in different directions. Greenland was no longer officially a Danish colony – but its economic dependency was near total. Government investment in Greenland rose five-fold between 1948 and 1958. More and more Danes lived in Greenland, and more and more Greenlanders were on the payroll of the Danish state.[24]

A fundamental tension between two conflicting objectives – social justice and economic self-sufficiency – lay at the heart of state-backed modernisation policy. The only way that Greenlanders could achieve anything approaching equality of incomes with mainland Danes in the short term was with substantial government support. But the scale of government assistance tended to crowd out private enterprise and to reinforce dependency. Increasingly, Greenland was caught in a vice: growing political aspirations on the one hand, and growing economic dependence on the other.

Demands for autonomy increased through the 1960s and 1970s. While Danish-educated Greenlanders brought the sixties campus politics of anti-colonialism back to Nuuk, the growing political mobilisation of Inuit groups across the Arctic in the seventies gave a sense that history was moving their way.[25] But few wanted a total break from Denmark. Even the more radical voices proposed greater self-reliance rather than independence.

In the end, it was the very practical and traditional question of fishing rights which pushed Greenland on the first step towards Self-Government. Just as anger at fishing rights granted to the Faeroese had led to concerted Greenlandic opposition to Copenhagen in the 1920s and 1930s, so fish served as the catalyst for Greenlandic hostility to Danish entry into the European Economic Community (EEC) in the early 1970s.[26] Greenlanders believed that Danish membership of the EEC would force Copenhagen to grant wider European access to Greenlandic fishing grounds, just as the Danish government had given in to Faroese pressure fifty years previously. While the rest of Denmark voted in favour of joining the EEC in a referendum in October 1972, Greenland voted against.

Seven years later, in 1979, Greenland was granted limited Self-Government in the form of Home Rule. (The suggestion that Greenlanders might be granted full control of the island's mineral resources was categorically rejected. Danish Prime Minister Anker Jørgensen, frankly told an interviewer, 'No dice.'[27]) In 1984, Greenland formally left the EEC, reducing the Community's total territory by half.[28]

COPENHAGEN, DENMARK, 55° North: The house of Denmark's former Foreign Minister Uffe Elleman-Jensen is, from the outside, surprisingly modest. But like so many things (and so many people) in Denmark, that modesty is deceptive. Inside, the house is elegant. One room flows into another through wide-open double doors in a succession of neutral tones, each corner a showroom of Danish design. It is a house built for entertaining – which suits a former foreign minister, and suits the informality and natural affability of the man himself. A controversial figure in Danish politics, and foreign minister for eleven years over the 1980s and 1990s, Uffe Elleman-Jensen has managed the trick of being both hard-nosed and soft-hearted. It's a combination which made him one of the most successful Danish politicians of the twentieth century, and which helped the West win the Cold War.

Elleman-Jensen's attraction to the Arctic is long-running. 'See that man,' he says, pointing to a heavy black sculpture of a man's head, beak-nosed, a heavy hood pulled halfway back across his skull. 'Do you recognise him? Knud Rasmussen, the Arctic explorer of the 1920s. I read all about him as a child.' A large part of north-west Greenland was subsequently named after Rasmussen, at the north of Baffin Bay.

For Elleman-Jensen, as for many Danes, there is an emotional link to Greenland – a sense that Greenland is part of Danish identity and that it has been well served by Denmark. (As one Dane told me, 'Every Dane dreams of going to Greenland . . . They don't say the same thing about the Faeroes.') Hedged in by political correctness, many Danes will, if pushed, admit that they think there's something churlish about Greenlanders' desire for independence.

Elleman-Jensen still professes a love of Greenland, but he is no longer quite in thrall to it. The relationship has become touchy, not least because Elleman-Jensen was edged out of the job as chairman of Royal Greenland in 2002 after going public with an allegation that a prominent Greenlandic politician had received a substantial pay-off from the company without board approval. The sense of hurt is all the greater because Elleman-Jensen believes he did a good job of defending Greenland's interests while foreign minister in the 1980s and 1990s. (Recalling the negotiations to allow the exit of Greenland from the European Economic Community, Elleman-Jensen remembers bringing a group of Greenlanders to Brussels and telling them, at the appropriate moment, to loudly berate the European Council, in unintelligible Greenlandic. 'We scared the German state secretary,' he laughs, 'and we got a very good deal.')

In the 1980s and 1990s, one of Denmark's main bargaining chips in discussion with allies was Greenland's strategic position. Elleman-Jensen brings out a map, centred on the North Pole. 'I used to use that map all the time,' he tells me. 'I pointed at the map and said, "Well, where do you see the shortest distance from the nuclear installations on the Kola peninsula to the United States? Take a look at the map and you can see – it goes over this region,' he says, drawing a finger slowly and carefully from Murmansk to the central United States, stabbing decisively in the vicinity of Chicago.

'In the eighties there was always nagging in NATO, we are not spending enough and so forth. But the substantial expenditures that we Danes have to keep stability – that should also be counted as part of our defence expenditure.' Ultimately, if the price of supporting Greenland could be counted against defence expenditures, then block grant payments to Greenland were not just a downpayment against the loss of pride which would be entailed by moves towards independence in Greenland, but a shoring up of security guarantees. 'Some Greenlanders now use that against us, saying that they want the money that we saved back,' says Elleman-Jensen. 'My answer to that is, "It was your bloody security as well as ours."'

Even at the height of the Cold War, Elleman-Jensen suggests that Greenlanders were naive about the threats they faced and the wider strategic game played by Moscow. The Soviet Union sought consular representation in Greenland and on the Faeroe Islands, which Denmark denied. 'Some of the Greenlanders even accepted invitations to Moscow and so on,' he says. Elleman-Jensen recalls a confidential briefing he gave to the foreign affairs committee of the Home Rule parliament. 'I told them that we had reason to believe that there were nuclear weapons in Greenlandic waters,' he says. 'Soviet nuclear weapons on board Soviet submarines.' But the Greenlanders did not want to hear it: 'A light went out.'

Later, when new surveillance technologies made Greenland less critical to NATO security, the United States began to close down its bases at Kulussuk and at Narsasuaq. 'And then suddenly it went from "Yankee, go home!" to "Yankee, please stay,"' says Elleman-Jensen. 'Greenlandic politicians came up with plans to allow certain kinds of waste to be stored at Mestersvig, and for British Nimrod planes to have a back-up base to Keflavik [in Iceland],' he remembers. 'Things changed very quickly.'

And now, as Greenland seriously considers the possibility of full independence? The world looked a safer place in the warm afterglow of the West's Cold War victory, but now? With Russia resurgent, the United States in decline, and an international consensus on global institutional reform seemingly out of reach, can Greenland really give up Denmark's cover, with the friendship of NATO, the EU and all that this implies? Elleman-Jensen is sceptical. 'The idea that they could move alone in the world is absurd,' he tells me. 'They would probably make a deal with the Americans.'

Perhaps. But Greenlandic independence is no longer as far-fetched as it was in the 1980s. Then, the economics of Greenland's position were stacked against it. Now, the prospect of income from natural resources exploitation offers the possibility that the island could become economically self-sufficient within a generation. Onshore mining may soon be eclipsed by offshore oil.

NUUK, GREENLAND, 64° North: Hans-Kristian Olsen is no average national oil company CEO. There is no sign of a dark-grey suit or a pair of hand-crafted $2,000 brogues. Instead, Hans-Kristian is sporting a clean blue-striped shirt over a white T-shirt and jeans, and his haircut probably cost less than mine. There is no Harvard-educated adviser, high-powered executive assistant or portrait of a ruling family in sight. Nothing, in fact, to distract from Hans-Kristian himself, an open book of a man – boyish, earnest, serious – with one of the biggest jobs in his country. If Greenland constructs its independence on the foundations of oil, Hans-Kristian will be able to lay claim to being one of independent Greenland's founding fathers.

Nunaoil is not an average national oil company. While Saudi Aramco of Saudi Arabia, PDVSA of Venezuela, Sonatrach of Algeria or SOCAR of Azerbaijan all have a workforce of tens of thousands and billions of dollars in capital investment, Nunaoil had only five staff members until a few months previously and runs on a budget of 10 million Danish crowns ($2 million). The company does not even have a full-time legal adviser. In the vernacular of the oil industry, companies range from the super-majors – ExxonMobil, Shell, BP – to the minnows. Nunaoil might qualify as a super-minnow.

As of yet, Nunaoil does not produce any oil – indeed, it is unlikely that any oil will be produced on Greenland for at least ten years. What the company does have, however, is an automatic 12.5 per cent participation in the exploration phase of any oil and gas development in Greenland. Calouste Gulbenkian – the famous Armenian 'Mr Five Per Cent' of early twentieth-century oil – made his fortune and endowed his museums with less.[29] Why should Nunaoil be any less successful?

'Nuna' means 'land' in Greenlandic. But most of Greenland's oil prospects are offshore, either off the west coast, from south of Ilulissat to the southern tip of the island, or, further into the future, off north-east Greenland, in an apparent geological continuation of the prolific Norwegian basin. 100,000 square kilometres of offshore areas are ready for exploration and an additional 120,000 square kilometres will be opened in coming years. Existing licences are clustered in three areas around Greenland's western and southern shores. A much bigger area, around Baffin Bay, is due to be opened for bidding in 2010, while a licence sale for areas off north-east Greenland is slated for 2012.

'Licence periods are divided into three sub-periods with different obligations,' explains Hans-Kristian as we sit in his office in a modest green-painted wooden house in a Nuuk backstreet. First, a three-year period in which the company which has won the licence commits to acquire seismic data. Second, a three-year period during which at least one well must be drilled. Finally, a four-year period of more active appraisal and exploration.

The company can opt to pull out at any point, lose the licence and put all the information it has gathered into the public domain. After each sub-period of the licence, the company is obliged to give up a fixed percentage of the area and release collected data on the relinquished acreage. As far as efficient oil and gas development is concerned this stipulation has two useful consequences: it forces licence-holders to concentrate their efforts on areas most likely to be productive, while simultaneously providing opportunities for other companies to stake claims.

Nunaoil's particular advantage, besides its automatic 12.5 per cent stake in the initial exploration phase, and an option to buy into subsequent development stages, is the on-the-job training in the international oil industry that Nunaoil receives in the process. 'Part of the vision is that Nunaoil will eventually own and run its own licence blocks,' says Hans-Kristian. The model for Nunaoil is DONG (the majority state-owned Danish energy company)[30] and Norway's StatoilHydro[31], both of which operate outside their home waters.

But there is a long way to go. Labour and capital are constrained. 'Young Greenlandic people are being educated, but it takes time and experience to develop a workforce,' explains Hans-Kristian. 'So, for the moment, we're very dependent on foreign workers.' And while billions of dollars may ultimately accrue to Greenland's government in taxes, in the short term they must rely on foreign capital.

Lacking money and staff, Nunaoil's best strategy is to sponsor exploration work where it can improve its visibility relatively cheaply and acquire some of its own data. Nunaoil has co-sponsored some offshore ship-based seismic work, but at $1,000 to $1,500 for each kilometre-long seismic line, covering even just the most promising areas of a single 10,000-square kilometre block is prohibitive. Aero-magnetic 'overview mapping' conducted from aircraft is cheaper but provides less detailed information. For the last couple of years, Nunaoil has sponsored aero-magnetic surveys of the coastal areas of north-east Greenland.

Exploration for oil on and off Greenland has been sporadic. In the late 1970s Greenland felt some of the rush for Arctic resources which hit Canada and Alaska. Five offshore holes were drilled in 1976–77, but all came up dry. A further offshore hole was drilled in 2000, but Statoil and its partners abandoned the Qulleq prospect after spending 300 million Danish crowns ($60 million) with no result. Onshore, Arco and Agip held a lease in Jameson Land in eastern Greenland which they relinquished in 1990. A few years later, in 1996, GrønArctic Energy, a Canadian company, spent C$12 million ($11 million) on a 3,000-metre-deep hole near Nuussuaq. It, too, was dry.

At the same time, both private companies and the state have sponsored seismic projects to better understand the geology on and around Greenland. In the mid-1990s, a consortium of major oil companies – BP, Exxon, the Japanese National Oil Company, Shell and Texaco – spent millions of dollars on the Kanumas project, with Nunaoil managing multi-million-dollar seismic prospecting east and west of Greenland.[32]

Hans-Kristian estimates that any new offshore drill hole now would cost upwards of 500 million Danish crowns ($100 million). But he argues that it is worth the investment. Even if past holes have proved dry, there simply have not been enough holes drilled to draw any definitive conclusions about the oil and gas potential in general. 'In the North Sea,' he points out, 'it took fifty wells at least before the first commercial oil was found'.

The oil companies seem to agree. 'In the 2007 licensing round, we expected two or three of the West Disko licences to go,' says Jorn-Skov Nielsen, Head of Department at the government Bureau of Minerals and Petroleum, the body responsible for managing all licensing rounds in Greenland, 'but in the end we got seven out of eight,' he tells me. There has been a significant increase in seismic work offshore, amounting to 22,000 kilometres in 2008.[33]

Besides ExxonMobil and Chevron, the companies investing in exploration in Greenland are mostly small. Husky Energy, a Canadian company with leases in south-east Asia, bought into three of the Disko Bay leases. But it is Cairn Energy, a British company with an eclectic portfolio covering Albania, India, Papua New Guinea, Peru and Tunisia, that has placed the greatest stake in Greenland. Cairn bought into two of the West Disko leases, as well as the Open Door area in the south-west and a further minority stake in two west

Greenland leases. In total, the company holds part of leases covering over 52,000 square kilometres of sea off the coast of Greenland.[34] The terms of the licences commit oil companies to an investment of 10 to 12 billion Danish crowns ($2 to $2.4 billion) over the next decade. Drilling for oil may begin in the Atammik and Lady Franklin prospects as soon as 2010.

With disarming frankness, Jorn-Skov shows me why climate change is likely to help oil and gas exploration off Greenland. The number of days when offshore areas are ice obstructed has been falling in recent years. 'Of course,' he says, 'you have to be careful to look at the long term before deciding on the trend, but . . .' Data from the Danish Space Centre confirms it. Icebergs are already less of a problem on western Greenland than in eastern Canada. Jorn-Skov argues that, in the future, weakening transpolar currents will mean 'much less ice in north-east Greenland – the Kanumas area will be strongly affected'.

Just how much oil and gas is out there is unknown. The United States Geological Survey (USGS) has estimated 31.4 billion barrels of oil equivalent (including 8.9 billion barrels of oil) off north-east Greenland, mostly in the North Danmarkshavn and South Danmarkshavn basins.[35] In the West Greenland & East Canada basin, USGS estimates run to around 7.3 billion barrels of oil, and a further 1,500 cubic metres of natural gas.[36] But uncertainty around these figures – as with with every other frontier region for oil and gas in the Arctic – is high. For the South Danmarkshavn basin alone, USGS estimates ranged from a fifty-fifty probability of 3 billion barrels of oil, to a 5 per cent probability of 14 billion barrels of oil and an equal probability of nothing at all.

Still, for oil companies faced with rapidly declining reserves, even such speculative potential is attractive. A single find of a few hundred million barrels of recoverable oil would transform the fortunes of a small oil company, and make a major contribution to the proven reserves of a super-major. For Greenlanders, the threshold above which oil and gas discoveries would buy independence from Denmark is equally low. The USGS mean estimate of 31.4 billion barrels of oil equivalent in north-east Greenland alone would translate into nearly 500,000 barrels of oil equivalent per head, several times the per capita resource endowment of (much more populous) Saudi Arabia.

The shift in Greenland's economic focus towards the exploitation of natural resources from under the land and sea floor, is a challenge to the more traditional sea-based Greenlandic economy. Axel, a Greenlandic fisherman who works for the 2,000-strong national association of small fishermen, is upfront about the risks he sees for the future of fishing in the new Greenland: 'We just don't know if we will survive when the other investments come in,' he tells me when I visit the association's headquarters, down by the port in Nuuk. Like many Greenlanders, Axel was drawn to the capital city by the prospects it offered to him and his family rather than by any particular desire to live there. Most Greenlandic fishermen and hunters remain in smaller communities up and down the coast. Their way of life may be dying.

The government has tried to help the local fishing industry by forcing Royal Arctic, Greenland's domestic shipping monopoly, to keep freight rates artificially low.[37] But the challenges for Greenland's fishing industry run deep. A new 27-foot fishing boat costs up to 3 million Danish crowns ($600,000); a steel-hulled ship twice as long will cost 10 million crowns ($2 million). It's the kind of investment which most Greenlandic fishermen will never be able to afford. In the north of Greenland, hunters already survive on the wages of their wives – 'Her salary buys the bullets,' Axel laughs.

He is sceptical of the idea that fishermen and hunters will seamlessly become construction workers, engineers and oil-platform workers once investment in natural resources comes flowing in. A recent survey sent out to members asked them whether they would consider working in another profession. The response was unequivocal: 'Most couldn't care less, they like the freedom of hunting and fishing.'

Axel's attitude towards climate change, typically for Greenland, is highly ambiguous. For hunters in the north, it is a problem: 'The ice isn't thick enough for hunting with dog sledges any more; you used to be able to hunt for eight months of the year on the ice around Qanaaq – only three months now.' Moreover, weather has become more changeable in recent years and more difficult to predict, undermining the traditional knowledge of Greenlandic hunters: 'Normally, hunters can look at the sky and see the weather for a day or two ahead, but now the weather changes fast.'

But for fishermen, a warming Arctic climate could bring benefits

– fish are heading north. After disappearing in the 1990s, cod are returning to the fishing grounds off western Greenland. Shrimp are now caught off Upernavik, a thousand miles north of where they were caught five years ago. Just as a previous period of natural global warming led to a boom in Greenlandic fishing in the 1920s and 1930s, perhaps anthropogenic climate change will benefit west coast fishing communities by increasing the size of its more valuable fisheries. 'Maybe there's nothing to worry about after all,' says Axel, hacking off two pieces of smoked trout, one for himself and one for me.

The challenge to the most traditional part of the Greenlandic export economy – sealing – is cultural rather than environmental. 'Most Europeans will never understand hunting for seals,' Axel explains. 'It is difficult for a hunter and a farmer to meet.' A few years ago a full-time hunter could get 450 to 500 Danish crowns ($95 to $100) for a single seal pelt; today, the price has fallen to between 175 and 200 Danish crowns ($35 to $45). The reason? Alex blames 'that stupid French actress, whatever her name is – Brigitte Bardot – she wants people in hunting areas to be vegetarian.' Although new markets may open in Russia or in China, currently the bulk of seal products go to Western Europe. If those markets don't buy, the economics of sealing will tilt further against the traditional sealing economy, and the seal hunt could become a purely local practice, reverting back to a time before European traders set foot on Greenland's shores.

Aleqa Hammond, Greenland's finance and foreign minister, leaves no room for doubt about Greenland's destiny: 'My children will have their own country.'[38] Her vision of what an independent Greenland will be like – a savvy and practical nation at ease with globalisation and as close to China and the United States as to Denmark – is a million miles from the Greenland in which she grew up. But it is a vision which she now believes is within grasp.

Hammond argues that while independence from Denmark should not be rushed, the economic basis for it is increasingly firm. She ticks off the various mining and industrial projects which she believes will offer Greenland economic self-sufficiency from Copenhagen in the future. Arctic climate change may help: 'Lots of things are popping up as the ice retreats – Greenland's rivers and lakes have never been as large as they are now, boosting hydro power.'[39]

The outlook of Greenlanders has been transformed over the last thirty years. When Nuuk won limited Home Rule from Copenhagen in 1979, Hammond was just thirteen years old. 'It was the year of my first communion,' she tells me. Life choices were restricted: 'If you were a Greenlander you could only become a teacher or a priest.' But things have changed dramatically since then: 'We take it for granted that Greenland has its own government, that my lawyer will be a Greenlander, or that if I fly to Copenhagen I will fly Air Greenland with a Greenlandic pilot.' Deference to Denmark has declined, and a sense of Greenlandic pride and national self-confidence has begun to replace it.

The way Hammond sees it, Danish administrators have never understood Greenland, and Greenlanders have never been Danes in any real sense: 'We are not Nordic, our language and culture are different, we look different, our race is different; geographically we are part of North America.' It is a point of view which angers many Danes who have made their lives in Greenland. For one, full independence would be 'a denial of history – an illusion'.

Hammond's creed of self-help, national self-determination and financial independence – a kind of Greenlandic Thatcherism – is appealing on one level. But are Greenlanders ready for what Hammond's vision of the future might mean? Full independence from Denmark would not only imply the removal of Copenhagen's financial support (which will in any case run down as income from resource extraction rises). It would also mean the removal of a psychological safety net for Greenland's citizens. 'It's like Stockholm syndrome,' says one local journalist. 'The Danes have held us hostage, and now we've become psychologically dependent on them.' Are Greenlanders ready for the responsibility and hard choices which independence would entail?

Hammond is confident that Greenland can rise to the challenges of independence and resource development. Others are doubtful. Some Danes question Greenlanders' ability to make tough decisions. One businessman tells me, 'people don't face problems up here, they look away – the civil service, for example, is falling apart.' In Greenland, he argues, the flow of money from Copenhagen has taken away the habit of setting priorities and sticking to them: 'So you say, "Seal-hunting – that's important, let's spend some money on that," and "Fishing – that's important, let's spend some money on that," but where the money isn't spent is on social security and education.'

One of the most awkward questions for Greenland's future – whether the Self-Rule government will continue to financially support remote settlements which would not otherwise be economically viable – is neatly evaded by Hammond. She simply says that 'Greenlanders have never, *by their own wish*, closed down a settlement – ever.' Perhaps. But Greenlanders will face the same policy dilemmas as the former Danish administrators. Will their answers really be any different? Should Greenland accept the 'concentration' of its population in fewer and fewer settlements, thus changing patterns of settlement which have lasted for over a thousand years?

And what of the impact of the major capital investments – mines, oil and gas development, a new aluminium smelter – on which the economic self-sufficiency of Greenland is supposed to be built? 'Sometimes I think our economic plans sound more like the five-year plans of the Soviet Union or the DDR,' says Henrik Leth, the outspoken head of the Greenland employers' federation, warning of the risk that wage inflation caused by the scale of inward investment into the minerals sector could simply crowd out other sectors of the economy, damaging them irretrievably. 'People seem to think that we can just take people in early retirement and re-employ them, or that fishermen will go straight into the factory,' Leth tells me, 'but people are not machines.'

Some argue that Greenland's government does not know what it is getting itself into. They fear that Greenland's political system will be swamped by the sheer scale of investment, that its government will be taken over by special interests, that the country will replace dependence on Denmark with dependence on multinational corporations, and that Greenland's cultural traditions will be lost in the headlong rush for economic self-sufficiency. Above all is the fear that an influx of large numbers of foreign workers will irrevocably change Greenland's society. (Several thousand may be required for the construction of Alcoa's aluminium smelter alone.) Many Greenlanders are keen to change voting rights to ensure that workers brought in for just a few years will not be able to vote in Greenland's elections.

Ultimately, the crux of the matter is simple: winning the relative freedom of Self-Rule is one thing, but turning autonomy into independence will be problematic. The constitutional bargain which came into effect on 21 June 2009 is an experiment. The emergence of the world's first truly Arctic state is not a foregone conclusion.

If successful, the transformation would be extraordinary – turning Greenland from a colony with a single lifeline to the outside world, dependent on financial support from Denmark, into an open and wealthy internationalised economy within a generation. But the promise of freedom, however alluring, is difficult to realise for a small nation in an era of globalisation. Across the Denmark Strait lies Iceland, an inspiration to Greenland in seeking independence in the past. These days, it is also a reminder of the difficulties of making freedom work.

13
Selling Iceland: The Grand Illusion

'No power could have forced my father to believe . . . that there existed in Iceland people who wanted to hand over her sovereignty to foreigners the year after the establishing of the Republic or, as it is called in modern terms, "sell the country"'

Halldór Laxness, *Atómstöðin*, 1948

'Icelanders may be well-off nowadays, but it is as if a particular myth is fixed in the mind of the nation, the myth that we are a poor little country that needs something to save us. The war saved us from the Depression, and then came the Marshall Plan once the war was over; then it was the herring that saved us, and when that disappeared we had the aluminium plant and the Búrfell hydro scheme to save us . . . This is a pattern, a myth, a metaphor our thinking has come hooked on'

Andri Snær Magnason, *Dreamland*, 2008

'Iceland's current financial and economic crisis is unprecedented in scope for any advanced industrial country relative to the size of the economy'

Moody's *Global Sovereign Report: Iceland*, January 2009

Iceland's history is punctuated by catastrophe, cataclysm and calamity. Some were unavoidable: the natural disasters which have periodically swept through Iceland's close-knit communities, decimating the population and devastating the hopes of national development. In the early eighteenth century a smallpox outbreak wiped out one third of Iceland's population, as high a proportion as were killed in Europe by the Black Death in the fourteenth century. A few decades later, in 1783, Iceland's geographic situation abreast the European and North American tectonic plates brought disaster when the Laki volcano erupted, affecting weather patterns across Europe and inviting famine and disease into Icelanders' homes. In 1780 the population of Iceland had been just above 50,000; it would not reach that figure again until 1825.[1]

Other catastrophes have been self-inflicted: the dark civil war which ravaged the countryside in the early thirteenth century and which ended with what would later be viewed as the national betrayal by chieftain Gissur Thorvaldsson, when he 'sold' the country to the King of Norway, in 1262. Still other catastrophes, on a radical nationalist reading of Icelandic history, have been simply imposed from outside, though often with the help of quisling Icelanders: the British 'invasion' of 1940, or the stationing of American military personnel at Keflavik up until 2006. Now, Iceland faces the consequences of another catastrophe, the meltdown of the country's financial system and an explosion of its national debt.

In Iceland, history is an inescapable reference for contemporary politics and culture. In 1996, at the tail end of a washed-out summer, I travelled around the island alone, visiting villages of no more than a house or two with a petrol station twenty miles down the road. I remember the small island of Flatey (literally, 'flat island') and my host there, who spoke about a passage in a thirteenth-century saga – the burning of Njáll Thorgeirsson's homestead, with him and his family inside – as if it had been yesterday, and as if it had been his own family who had died.[2]

The history of the country's brushes with foreigners has taught Icelanders to be ambivalent about the outside world. On the one hand, Icelanders like to think of themselves as global citizens. Most speak perfect English, sometimes with the accent of their favourite American or British television stations. Yet, for all this apparent worldliness, there is a powerful strain of isolationism in Iceland too. Many express pride in the linguistic, cultural and genetic continuity their isolation has promoted. Independence is considered the prime national virtue.

Sometimes, this love of Icelandic uniqueness and the desire to protect it has slid into outright racism. In the 1950s, for example, African-American soldiers were secretly banned from Iceland, despite Icelandic protests that no such policy existed and that, indeed, it would be 'antithetical to the Icelandic mind'.[3]

Outsiders have traditionally been seen as conquerors as much as saviours, as threats to the country's cherished independence as much as guarantors of it. In foreign affairs, Iceland has sought neutrality, a principle enshrined in the country's constitution since 1918.[4] Although Iceland was one of the early members of the United Nations it has, until now, always avoided the embrace of the European Union.[5]

Many Icelanders instinctively mistrust defence organisations such as NATO (despite Iceland's membership) calling for the country to become more like the Åaland Islands, a demilitarised archipelago in the Baltic Sea.[6] In 1945, plans for an American military presence were scaled down in response to Icelandic concerns, and the country avoided becoming the North Atlantic 'battleship' it might have become.[7] During the Cold War, the country was viewed in Washington as an alarmingly unreliable ally.[8]

Nonetheless, more recently, successive right-wing governments tried to stall the departure of American forces from Keflavik airbase. Some argue it was only the personal pleading of Prime Minister David Oddsson (1991–2004) with President Bush – and the fact that Iceland was nominally an American ally in Iraq – which kept US forces there as long as they were. When America did finally pull out in 2006, Icelandic diplomats scrambled to build new relationships with NATO allies, and to get recognition of the island's strategic importance. When I visited in May 2008 there was a Danish frigate lying at anchor in Reykjavik harbour and the American F-15s had been replaced by French Mirage 2000 jets, a token force on a six-week tour of duty.[9]

Economically, Icelanders seem to have a knack of riding the good times, surviving the bad times, and failing to prepare for either. 'We are not a nation of chess players. We are a nation of ludo players,' writes one.[10] Perhaps because the national economy relied for so long on the changeable and unpredictable fortunes of fishing, cunning and quick thinking are more highly valued than forward planning and strategy. The regularity of disaster throughout Iceland's history, and the historical volatility of its economy, have bred a certain fatalism into Icelandic culture. These days, it is fatalism's flip side which the country will require: equanimity in the face of catastrophe.

In the first years of the twenty-first century the Icelandic economy boomed, expanding by more than one third between 2000 and 2007. Much of the growth was in the financial sector, pumped up on the economic equivalent of steroids: cheap money. Icelandic banks, unleashed by domestic liberalisation, took full advantage of the global liquidity glut, growing their balance sheets to over nine times gross domestic product (GDP). At the time, Iceland was lauded as a 'Nordic tiger', more akin to the high-growth economies of East Asia than its more staid European cousins. But when the global financial system began to freeze up, so did international confidence in the ability of such a small country to support such an outsized financial system.

For a while, at the beginning of the financial crisis, high interest rates – intended to stave off inflationary pressures from Iceland's rapid economic growth – managed to keep the Icelandic currency afloat. But within a matter of months it was all over. In essence, the Icelandic banking system went bust. The Icelandic crown went into free fall, thus increasing the cost of servicing foreign debts. When the government was forced to step in, national debt exploded. Under a worst-case scenario drawn up by a major financial ratings agency in 2009, the country's gross national debt will grow from 28.1 per cent of GDP in 2007 to 176.8 per cent in 2010.[11] Even under a more moderate scenario, debt would reach 146.4 per cent of GDP. In 2008–09 the Icelandic economy contracted more sharply than almost any other developed country.

By any standards, that is an economic calamity to rank with the earthquakes, volcanic eruptions, and famines which have marked Icelandic history. Lives may not have been lost, but life-horizons have certainly been lowered. The financial crisis has had a broader impact too, emphasising the risks to small economies within the global system. When good times turn bad, freedom may prove to be costly. It is a message which may have resonance across the Arctic, in Nuuk and in Iqaluit as much as in Reykjavik.

Iceland used to be the epitome of small-state independence, and a model for aspiring small countries such as Greenland. The country's view of globalisation was one of opportunities. When I visited Iceland in May 2008, it seemed as if the population of Reykjavik, some 200,000 strong, further north than any other capital city in the world, truly believed that their city could become the summit of globalisation. In 2007, it was a confidently independent Iceland (population 313,000). which had the gall to begin negotiations for a bilateral Free Trade Agreement with China (population 1.3 billion). In 2008, Iceland even ran (unsuccessfully) to become a non-permanent member of the United Nations Security Council.[12] From the fringes of the Arctic, it seemed, one could look down on the rest of the world and yet remain somehow separate from it. And this was precisely how Icelandic businesses and banks saw the world: as a game-board full of opportunities, just waiting for a cunning Icelander to sweep down from the Arctic and snap them up.

The country's banks were in the vanguard of Iceland's economic internationalisation. Financial institutions which had, up until the early twenty-first century, been traditional retail banks, taking deposits and lending money, began exploring and exploiting the more exotic corners of global finance. Retail banking was supplemented by investment banking, often in support of Icelandic entrepreneurs. The language spoken in Icelandic banks became that of seizing opportunities, taking risks and exploiting prospects. Having once been highly conservative, Icelandic banking became exciting. I remember speaking to one banker who reeled off the commercial opportunities that his bank was looking at: Latin America, the United States, Russia, China.

Still, in amongst all the confidence and bravado, the warning signs were there. 'You could sell almost anything a year ago,' one fund manager told me in mid-2008, 'not now.' I remember a conversation with the owner of my guesthouse, as he told me of his plans to redenominate his loans into Japanese yen, so as to avoid the high and rising interest rates on the Icelandic crown. 'They're killing me,' Jon told me, emphasising his individual financial plight with a universal physical gesture, drawing his hand slowly across his neck. I remember asking him whether he blamed the government for the economic problems; his worrying response was that, in Iceland, 'People forget after fourteen days.'

Now, I can only hope that Jon never got round to changing his debts into Japanese yen. If he had, those debts would now be next to impossible to repay, having doubled in value as the Icelandic crown has crashed against the yen. What seemed an apparently clever financial wheeze in 2008 would have shown itself to be too clever by half, pushing Jon – like Iceland itself – to the brink of bankruptcy. Some of the banks I spoke to that May are now either bust or state controlled.

Trust, one of the essential ingredients of any healthy relationship between government and the governed, has evaporated. Although Iceland's president has survived the political cull, the make-up of the country's parliament has shifted radically – politicians who used to be in opposition are now in government, bringing with them their misgivings over allowing international business too much of a role in the Icelandic economy. And a sense of Iceland's proud, world-beating independence has given way to an uneasy feeling that the nation has been cut adrift. To some, the departure of American personnel from Keflavik airbase in 2006 now looks less like a reassertion of Iceland's sovereignty and more like an abandonment.

As the financial crisis unfolded in the second half of 2008, many Icelanders felt that their natural allies were unsupportive or, in the case of Britain, downright hostile.[13] As things turned from bad to worse, it briefly looked as if Iceland might seriously entertain a diplomatic volte-face, and accept a $5.4 billion loan from Russia.[14] Having historically avoided membership of the European Union (EU), which the local fishing industry feared would allow in hungry foreign trawlers, in July 2009 the country applied to join, hoping to exchange freedom outside a broader alliance of European states for the security of being inside (though even an accelerated path to full membership will take several years).

Some economists dispute the advantages to Iceland of membership of the European Union and the euro. The equation of economic size with economic strength is flawed, they argue. While economic volatility is in the nature of small economies – borne out by Iceland's tumultuous economic history – their average growth rates may actually be higher than those of larger economies, over time. 'Small gymnasts always win the Olympics,' Dr Tryggvi Thor Herbertsson tells me when I meet him in May 2008 at the minimalist offices of Askar Capital, the finance company of which he is CEO.[15] As for the euro, while membership removes an element of currency risk for business investment decisions, it also removes a tool for managing the economy. Without the possibility of fluctuations in the value of a country's currency, any changes in its economic circumstances must be reflected in wages and prices. That can be painful.

But what is a valid economic argument against membership of the euro in normal times – including a normal recession – may not apply in a crisis. When the question is not managing an economic downturn, but avoiding economic annihilation, the euro offers clear benefits. The fact that the euro is so heavily used in the global economy means that a run on the currency is unthinkable. The small Icelandic crown had no such in-built protection. And while the collective economic heft of the twenty-four euro-states makes them credible lenders of last resort, things are different for the government of Iceland. When confidence in the integrity of the country's banking system began to fail there was nothing that Reykjavik could do to prevent a run from turning into a rout. Iceland became toxic. In 2009, a senior Icelandic official announced quite simply that, 'The krona is dead. We need a new currency. The only serious option is the euro.'[16] While the origins of the crisis were in the financial sphere, the consequences of the crunch are in the real economy. Life savings have been wiped out and pensions have been destroyed.

Even with its banking system wrecked and its currency on life support, however, Iceland still has tremendous natural resources. First amongst these, though often ignored, is the entrepreneurial talent of its highly educated people. A university education, and particularly a postgraduate degree from a foreign university, is commonplace in Iceland. 'Even the girl on the phone has a university degree,' boasts one businessman. As the country seeks to reinvent itself in the wake of the financial disaster of 2008–09, the contacts, skills and international outlook of Iceland's population will be invaluable. These, more than anything else, are likely to keep Iceland a wealthy country.

But Iceland has other natural resources. The economic potential of geothermal energy from deep under Iceland's hills and hydroelectric energy from the country's rivers and glaciers has long been recognised.[17] A third natural resource – the oil and gas which may be found off the north-east coast of Iceland – has only just begun to attract attention, with the licensing of the Northern Dreki area between Iceland and the Norwegian island of Jan Mayen, above the Arctic Circle.[18]

Regardless of their country's current problems, many Icelanders believe that the development of natural resources will ensure the country's future wealth. But, as in Greenland, is the environmental and political price of developing these resources worth paying? If development goes ahead, will Iceland enhance its freedom and security or, as critics suggest, simply give up its identity?

REYKJAVIK, ICELAND, 64° North: It's May 2008 and I'm walking to the headquarters of Glitnir, then one of Iceland's main banks, housed in a long red and grey building on the outskirts of the city. I'm late. When I tell my host that I decided to walk rather than take a taxi he shoots me a forgiving smile. 'No Icelander would have done that,' he laughs.

Arni Magnusson is managing director of the bank's Sustainable Energy Team. His job is to drum up sustainable energy business opportunities for the bank, at home and abroad. By all accounts there is no shortage of them. Magnusson tells me about Glitnir's equity participation in geothermal projects in Iceland, plans to build up activities in North America, and a joint venture, known as Shaanxi Green Energy, being developed with China's largest oil company, Sinopec. He draws a picture of a booming geothermal market across the globe. 'The US and Mexico, of course; Latin America's very interesting, all up the east coast; New Zealand, obviously; the African Rift area, though it's a difficult market to play with; in Europe, it's mainly Italy; a bit in Kamchatka in the east of Russia.'

Iceland's political class has been hugely helpful to Iceland's energy industry. President Olafur Grimsson (re-elected unopposed in 2008) has travelled the world spreading the gospel of green energy – Iceland is the only country in the world to get all its electricity from renewable geothermal and hydroelectric – and raising the country's international profile in the process. At home in Reykjavik, Grimsson has played host to a number of advocates of global green energy, including former US Vice President and Nobel Prize-winner Al Gore. Grimsson's globe-hopping and conference-jumping activities have boosted Iceland's political profile and created business opportunities for his country's energy companies. 'He's been very active, he's been very hands on,' Magnusson tells me.

'We [Icelanders] have a unique platform to stand on,' says Magnusson. 'I've seen figures which estimate the country's potential at 5,000 megawatts from geothermal . . . That's probably not realistic,' he confides, 'because some of the reservoirs are under glaciers or in environmentally sensitive areas.' Still, even if Iceland only develops a fraction of that potential, say 2,000 megawatts, the business opportunities that would create are tremendous. Magnusson does a back-of-the-envelope calculation. 'In the US, development costs are about $4 million per megawatt,' he tells me, 'and they're a little lower in Iceland, say $3 million per megawatt – so that's a $6 billion investment required to develop 2,000 megawatts.' In Icelandic terms, that is a huge amount of money, about $20,000 for every man, woman and child in the country.

Some of that investment may be in a project known as the Iceland Deep Drilling Project (IDDP). All Iceland's main energy companies – and a number of foreign companies, including Norway's StatoilHydro and the US aluminium company Alcoa, are involved in a consortium known as Deep Vision. The idea is, in itself, quite simple: to drill several kilometres into the earth's crust and to harness the energy of geothermal systems at up to 600° Celsius (1112° Farenheit). But implementing it is complicated. At the moment the project is focusing on three geothermal systems in Iceland: the Reykjanes peninsula, Krafla and Hengill. 'The scientists believe that we could get ten times more power at five kilometres than two kilometres, increasing from 5 megawatts to 50 megawatts [per geothermal borehole],' says Magnusson, 'and you'd see the profits increase, too.'

Internationally, Magnusson estimates that investments in geothermal may top $40 billion over the next twenty years.[19] The United States Geological Survey, better known for its estimates of the world's oil and gas reserves, has estimated the potential geothermal resources of the US at nearly 40,000 megawatts.[20] One way or another, should that be developed, Iceland will be keen to capture a large share of support business associated with it. 'In a world where everyone is extremely keen to have energy, we should be in a good position,' says Magnusson.

Across town, in a high-rise office block in one of Reykjavik's sprawling suburbs, I speak to Thorstein Hilmarsson, an executive with Landsvirkjun, Iceland's national power company. Landsvirkjun operates thirteen hydroelectric plants and two geothermal power stations across Iceland. In the last few years, the company's production of electricity has increased rapidly. 'Since 1997 we have probably grown electricity production in Iceland by around 200 per cent,' Hilmarsson tells me. But he estimates that Iceland is still harnessing less than a third of the country's assessed electricity production potential from hydro or geothermal. That is low compared with countries such as Norway or Switzerland. Hilmarsson thinks that Iceland could double its electricity production, 'if we want to go all the way, that is'.

Maximising Iceland's electricity production, whether from hydro-electric power or from geothermal sources, is a long-standing goal. 'Framework plans' have been developed. Maps of the country's hydrological resources have been drawn up. The National Energy Authority has produced assessments of Iceland's energy potential.[21] More recently, Hilmarsson tells me, a 'co-ordinating committee' has been established to produce a classification of potential energy resources in the country, based on economics, technical feasibility and, of course, environmental and alternative-use considerations.

On one level, there is nothing sinister about this. It's just the National Energy Authority doing its job. But many Icelanders are suspicious. At what precise point does an assessment of energy potential become a plan to develop that potential? At what point do discussions with end-users of that energy become contractual obligations? Who is really in charge of making decisions on development: the people, or the bureaucrats, or the companies who might benefit? Who, ultimately, decides what is to be protected and what can be used?

Opponents of 'maximisation' of Iceland's energy resources argue that it is economically unnecessary, environmentally harmful, politically dangerous and, above all, unimaginative. 'This was an example of technocrats taking group decisions in their own office room and creating a national plan according to their own livelihood – state companies serving their own interests,' says Andri Snær Magnason, the most articulate critic of Landsvirkjun's plans, as we talk in a bar in downtown Reykjavik. 'The economists are just stuck in their own lingo,' he tells me. 'They couldn't think of people philosophically.' According to Magnason, the only way that Landsvirkjun could fully develop Iceland's stated electricity potential would be by damaging Iceland's precious natural environment:

They [the electricity companies] said they had thirty terawatt-hours they could produce – so I did my 'two-plus-twos' and added up all the unused rivers, and got to sixteen terrawatt-hours. We've already harnessed seven, so together that made twenty-three. I phoned them [the electricity companies] up and asked them how they got their final seven and they said, 'Oh, yes, you have to look at plan B.' And this was the salmon rivers and the creeks, where you could get an extra five. But the only way to get to thirty was to dam all the rivers and waterfalls in Iceland, including Gulfoss [Iceland's most famous waterfall]. I mean it's almost funny.

Some suggest that Iceland's identity as a nation is at stake. Others complement this traditionalist line with an attack on a broader front. Saving Iceland, an activist network devoted to stopping development in Iceland, draws together a mixed bag of grass-roots environmental organisations, anarchists and anti-capitalists from around the world.[22] They have attempted to articulate opposition to the wholesale use of Iceland's energy resources in terms of a global struggle against big business. In their eyes, Iceland is one front in a much wider war. In recent years, their focus has been the controversial billion-dollar Karahnjukar Dam, built to provide electricity for Alcoa's aluminium smelting on Iceland's east coast.

As a spokesman for Landsvirkjun, Hilmarsson is at the sharp end of much of the opponents' criticism. 'You've got to have a thick skin to work here,' he tells me. Hilmarsson rejects the notion that his company plans to harness all the energy potential of Iceland. Some areas will remain off-limits to development. As for the Karahnjukar project, he describes it as 'the hardest political issue in Iceland since the Cold War'.

But his confidence in the development of more of his country's energy resources is undiminished. 'When the economy goes down, concern for the environment goes down, people care more about jobs,' he tells me. Another businessman is more direct: 'We can always debate everything, but when it comes down to it, I don't see how we're going to develop as a nation without developing our resources.'

The exact scale of those resources is hotly debated. Some, like Magnason, argue that the official figure is unrealistically high, based on harnessing every single stream, river and glacier in the country. Others argue it may be an underestimate. Most of Iceland's hydrological power comes not from rain-fed rivers but from glacier-fed rivers, which are sometimes 'black like coffee' because of the large quantities of silt they carry. As climate change accelerates the melt rate of

Iceland's glaciers, those rivers are actually swelling. It is as if the country's vast frozen reservoirs of water are being gradually unlocked. 'We're getting five to seven per cent more out of those glaciers than anticipated,' Hilmarsson tells me.

He is not particularly worried as to what happens when the glaciers have melted entirely – that is so far into the future as to be almost irrelevant. 'We will have paid off the Karahnjukar Dam within thirty years, we've got a contract [to sell the dam's electricity] for forty years, and the dam will last for maybe a hundred and twenty years,' he tells me. The time frame for the disappearance of Iceland's glaciers is several times longer than that, even in a worst-case scenario. In the meantime, global warming may bring another benefit to Iceland's hydro economy, Hilmarsson says: more rain.

Some of the benefits of global warming may already be here. In 2008, a Milan-based company, Becromal, decided to set up a new manufacturing plant in Akureyri, Iceland's second city. Becromal makes high-specification aluminium foil for use in electrical goods. The manufacturing process, which is power intensive, uses more electricity than the city of Akureyri itself. But when Landsvirkjun signed an agreement to provide electricity to the plant, the power company decided that there was no need to build a new power station, because electricity production capacity was already higher than expected as a result of increased available hydro energy. In these cases, Hilmarsson tells me, it is more about optimising the utilisation of existing capacity than anything else.

Becromal is a showcase deal for Landsvirkjun, a small high-tech company with no need for expensive new infrastructure. But Becromal is the exception, not the rule. The majority of Landsvirkjun's electricity production goes to large-scale energy-intensive industries – Iceland's three main aluminium smelters account for nearly 70 per cent of Landsvirkjun's electricity sales.[23] That proportion may increase in the future. Alcoa has been working with the Icelandic government and the municipality of Husavik, in northern Iceland, to scope plans for another smelter, to be powered by geothermal energy. Construction could begin there as early as 2012.

Hilmarsson sees the symbiosis between big energy and big industry as a simple result of history and economics. Back in the interwar period, he explains, when Norway, Switzerland and other European countries were expanding their hydroelectric electricity production and even selling it abroad, Iceland's isolated household market was simply too small and too fragmented to support the necessary investment. 'The biggest leap forward was just before the Second World War,' says Hilmarsson, 'when the first part of a fairly sizeable river [the River Sog] was harnessed and then Iceland had enough electricity not just for lighting, but even for electric stoves.' But there was a problem: 'No one had electric stoves.' In a marketing technique not unlike that used by mobile phone and cable television operators, the company which produced the electricity gave away electric stoves at a hefty discount in order to create a market for their electricity.

The disconnect between power production and consumption – with production rising in large steps when each new power plant was built, while consumption grew more steadily – resulted in swings between huge production surpluses in some years, and power shortages in others. It was a textbook economics problem. 'Electrification was taking place, people were buying things that needed power,' he explains, 'and every time you got up to a maximum people said, "Okay. Let's build a new power station." But every time it was a new problem. Who was going to put up the money? The town council of Reykjavik? The state? International aid?'

Although the World Bank put up one third of the investment capital required to set up Landsvirkjun, the real answer to Iceland's electricity problems turned out to be aluminium. The aim, Hilmarsson explains, was 'to get a big user of electricity, which would provide the basis for investments in a power plant'. The capacity needed to ensure a secure supply of electricity to domestic consumers would piggyback on the much larger industrial demand from foreign investments. The obvious investment to look for was an aluminium smelter, requiring huge amounts of electricity to turn alumina (aluminium oxide) into useable aluminium metal. In 1969, Alusuisse built Iceland's first aluminium smelter, at Hafnarfjördur near Reykjavik. It's on your left as you come into Reykjavik from the airport.

In the 1960s it was household demand for electricity which drove the need for aluminium smelters. The smelters themselves were merely a means to an end. These days, however, the relationship is different. Instead of pent-up domestic demand forcing the electricity company to find ways to boost supply, the electricity company is looking at its unused power potential and trying to think up ways to develop it and sell it. Instead of the aluminium companies courting the electricity companies, it is the electricity companies (and the government) which are courting the aluminium companies. (Some view this as a kind of moral corruption. Andri Snær Magnason quotes one American's picturesque description of three Icelandic diplomats trying to attract the attention of a power-intensive industry boss at an international conference: 'It was like seeing thirteen-year-old girls trying to get off with Mike Tyson.'[24])

Over time, the logic of meeting demand has been replaced by the productivist logic of maximising supply. The Karahnjukar dam project, including a 690-megawatt power station, is the perfect example. It is barely connected to the national grid. Pretty much the sole consumer of electricity from Karahnjukar is the Alcoa smelting plant at Fjardaal. For Landsvirkjun, this makes sense. The company gets a steady stream of income from Alcoa, guaranteed by the terms of a forty-year contract.[25] But the question remains: Does it make sense for Iceland?

Hilmarsson argues that it does. Above all, it creates jobs in the east of the country. Proponents of the Fjardaal smelter claim it has created several hundred jobs in the town of Reydarfjördur and surrounding areas, saving the region from inevitable economic decline. Some suggest that the number of jobs created in Iceland as a whole may be as many as one thousand.

These numbers are open to debate. Opponents of the smelter argue that the jobs created by the smelter might have been created by another form of employment. Is it not a failure of imagination to believe that only aluminium can provide jobs? Andri Snær Magnason, for one, argues that the view that only Alcoa could save East Iceland is psychologically attractive but deeply flawed:

> Where before there was pessimism and torpor in Iceland's rural areas, now there is boom and a sense of expectancy. A lot of people got together and saw one type of engineering, one metal on the world markets, and one contract with one company as the basis for the future. But is the future genuinely bound up with this single possibility? Or is it rather that a mixed and varied view, a broad faith in the future, might constitute a threat to how these people view the world?[26]

Magnason's sceptical views have found a powerful echo amongst many Icelanders. *Dreamland*, his book on the project, has sold some 20,000 copies in Iceland (equivalent to 20 million in a country as populous as the United States).

Besides increased employment, Hilmarsson has another argument in favour of the Karahnjukar project and the Fjardaal smelter. As he sees it, if aluminium is not smelted using hydroelectric power in Iceland it will simply be done elsewhere, with far worse environmental consequences. If the world is going to consume more aluminium, then surely it is preferable for this aluminium to be produced in the cleanest way possible.

Amongst globalists, there is some sympathy with this view. Year after year Alcoa wins sustainability awards in recognition of its commitment to reducing greenhouse gas emissions, made possible by increasing operational efficiency and by the switch to hydroelectric power sources.[27] But many see Hilmarsson's argument as pure sophistry. If the world recycled more aluminium or simply used less of it – and much is now wasted – then there would be no need for a new smelter anywhere. Sure, it is better to make aluminium using hydroelectric power than coal-fired power, but converting alumina to aluminium produces large amounts of carbon dioxide in any case.[28] The Fjardaal smelter emits over 500,000 tons of carbon dioxide per year – tiny in global terms, but nearly a fifth of Iceland's total emissions.

To some, the notion that Alcoa is genuinely committed to sustainability and the environment is absurd. It is, they suggest, to misunderstand the way in which businesses work, where the only thing that counts is the bottom line. Energy efficiency cuts the company's energy bill directly, now. Use of hydroelectric power (rather than power from gas-fired or coal-fired power stations) reduces the company's exposure to the potential risk of a high price of carbon in the future. Meanwhile, all other things being equal, manufacturers of branded goods – aircraft or cars, say – would prefer to source their aluminium from relatively 'clean' producers, rather than their dirtier rivals, giving Alcoa a competitive advantage.

That is too harsh. The messy reality of Alcoa's investment in Iceland is that many people on both sides of the fence truly believe that they are acting in the world's best interest, in Iceland's best interest or in their community's best interest.[29] Both sides in the debate reject the perceived absolutism of the other. Opponents of the aluminium industry often suggest that Alcoa and others are seeking to develop all the country's energy potential at once, and to turn the country into a virtual colony of heavy industry. Supporters of the aluminium industry sometimes accuse environmentalists of wanting no further electricity production in Iceland at all or of being against any number of abstract nouns: growth, progress, the future.

Instead, for the most part, both sides like to present themselves as the reasonable middle ground, the authentic voice of Iceland. The most thoughtful voices in the debate – both in favour and against – are, for the most part, scrupulously polite about one another personally, though tempers have occasionally frayed.

When I tell Andri Snær Magnason that I'm going down to Reydarfjördur to see the smelter and the dam for myself, he asks me to pass on his best wishes to Tomas Sigurdsson, the man who runs Alcoa's Fjardaal plant. Perhaps it's tongue in cheek. After all, the two have fallen out enough in public – the Reykjavik intellectual against the American-educated engineer – spilling ink across the pages of Iceland's newspapers in the battle of numbers, statistics and ideas.[30] Or perhaps Magnason's request is a reflection of Iceland's tiny size, where everyone shares a relation, or several, so making enemies does not make sense. As I ask Icelanders about the smelter debate, time and time again I feel as if I am intruding on a family quarrel, to which outsiders are not invited.

REYDARFJÖRDUR, ICELAND, 65° North: The road from Egilsstadir to Reydarfjördur passes through a landscape too broad to be captured in a glance. The arc of the valley is so subtle that at the bottom, where the road runs by a fast-running stream, it seems almost flat. But the edges of the valley are steep – black rocks thrust up into the low clouds. One side of the valley is higher than the other and it feels as if the whole world has been tilted slightly, everything just a few degrees out of place. The effect is something like vertigo. As I barrel along the road in the cheapest rental car I could find, I feel dizzy, as if I might lose my bearings at any moment, slide off the edge of the world and fall into the sky.

Reydarfjördur is a one-horse kind of town. Today, its focal point is the petrol station, the successor to the staging-posts and inns which used to dot the Icelandic countryside. There's not much else: a couple of shops, one restaurant and an uninviting hotel. On the steep hill behind the town, up and away from the harbour, there are a couple of new tower blocks, each eight storeys high. At the top of the town there is a cluster of tidy blue bungalows, each with a small front garden. A couple of them have paddling pools. These are smart new houses, comfortable and well kept, with a brand new car in every driveway. But there is something almost comical about them, set against the massive backdrop of the fjord itself.

A few miles on from Reydarfjördur, up a gravel track to the left, is a ghost-town of brightly coloured portakabins, the housing for the construction workers – up to 1,600 of them – who built the smelter. Most of the Poles who came over to Iceland during Reydarfjördur's construction boom have now gone home. The portakabins are for sale, waiting to be shipped out to some other building site in some other corner of the world: Greenland, Dubai, China. A few Poles remain, however: a couple of monks who used to say mass for their compatriots, representatives of the Roman Catholic church in a country of Lutherans.

The smelter itself is out of sight from Reydarfjördur, out towards the open water of the Norwegian Sea beyond. Like all aluminium smelters, the main physical features of the plant are the 'pot rooms' which stand like two giant cathedral naves – each several hundred metres long, and fifty metres wide – side by side on the flat ground. They house the 'pots', 336 in all, where alumina is converted into aluminium. Next to the 'pot rooms' there's a storage silo – with capacity for up to 85,000 tons of alumina – and the casting house, where the molten aluminium is graded and cast for the world market. The most expensive products made at Fjardaal are 'rods' of aluminium, made using an elaborate machine imported from Italy. Cheaper than the 'rods' are the T-bars; cheapest of all are the 'sows', graded by the purity of the metal which they contain. Most of these products are shipped straight out to Europe. 'They wanted the whole process to be as lean as possible,' an Alcoa employee tells me, 'so there's virtually no inventory.'

Though built on Icelandic soil, Fjardaal is an international operation: on the day that I visit there's a Japanese ship, with a Chinese crew, unloading its cargo of alumina from Australia. Because the process of making aluminium is far more expensive than the raw material itself, it makes economic sense to import alumina from the other side of the world, and smelt it in Iceland, where electricity is cheap. At Fjardaal, everything is about optimising the process, to get the most out of three raw inputs: alumina (from Australia), electricity (courtesy of Landsvirkjun's Karahnjukar project) and carbon anodes (imported from Norway). Given that the basic technology of smelting aluminium is well known, the only way for Alcoa to increase profit margins is to concentrate relentlessly on cutting waste and streamlining production.

When I visit, everything is just coming out of the start-up phase – 'A very stressful time,' one employee tells me. The walls of the corridors are covered with multicoloured graphs showing various indices of the plant's efficiency: the number of anodes replaced per day, the ferrous content of the aluminium produced, the total production of the plant (around 1,000 tons of aluminium a day). Where the numbers are written in green, the plant is fulfilling its targets; where the numbers are written in red, the plant is not yet operating at its maximum rate.

Fjardaal is the first smelter that Alcoa has built in twenty years – 'State of the art,' says one of the plant's employees – and clearly, there is a certain amount of genuine pride in what has been done here. The general atmosphere at the plant is open, friendly and relaxed. Everyone I speak to at Fjardaal – from the head of public affairs and the man who shows me around the plant, to the manager of the smelter and the employees who work in the warehouse, the 'pot room' and the casting house – seems willing to give me his or her time.

But Alcoa has clearly also spent a lot of money on public relations. One result is a glossy handbook entitled *Fjardaal Takes Off*, singing the praises of the company's various sustainability initiatives and ticking off the ways in which the plant is changing the life of Reydarfjördur.[31] The book covers everything from the parking habits of the Alcoa staff – they are requested to back into the parking spaces at the smelter, so that they can leave more quickly should there be an emergency – to the random alcohol tests to which workers may be submitted and a tree-planting project which Alcoa says will result in 10 million trees being planted, worldwide, by 2020, absorbing some 250,000 tons of carbon dioxide (just under half Fjardaal's annual emissions). A fund has been set up for local projects, including money for a local woman who wanted to preserve the Icelandic tradition of storytelling. 'She's doing an excellent job,' I'm told.

Investment in public outreach makes sense for Alcoa. Nationally, the plant is controversial and there has been significant and vocal opposition to overcome. Alcoa plans to be in Reydarfjördur for at least forty years – possibly as long as eighty years – so investment in the community makes sense. But Fjardaal is also Alcoa's showcase plant. If it is seen as a success, life will become much easier for Alcoa in getting the final go-ahead to build a second smelter in Iceland, near the northern town of Husavik, or to build the smelter planned for Sisimiut in Greenland. Getting strong political support in East Iceland may end up paying dividends elsewhere. Magnason disparagingly calls this the 'lemming effect'.[32]

Locally, despite a few isolated opponents, support for the smelter has been strong.[33] Well over two thirds of the population of north-east Iceland are in favour of Fjardaal, according to opinion polls commissioned by Alcoa.[34] The main reason is the jobs brought in by the investment. Although it is difficult to measure the number of indirect jobs created – whether in the fjords region or in the country as a whole – the plant itself employs around 450 people, one fifth of whom are East Icelanders who have moved back since the smelter was built. These are relatively well-paid, stable jobs. Some employees drive seventy minutes into Reydarfjördur every day.

According to Gudmundur Gislason, the thirty-eight-year-old chairman of the local council, Alcoa has changed the region for the better. 'I always say that the major change from this project was that people got faith again, you know, in the future,' he tells me. Gislason isn't from Reydarfjördur itself, but from Neskaupstadur, two villages north of here. Things weren't so bad up there, he tells me, with a good fishing industry providing local jobs. But in Reydarfjördur, 'there wasn't much going on'.

Alongside his job as an elected councillor, representing the left-leaning Party of the Fjords, Gislason used to run a restaurant up in Neskaupstadur. 'It was a typical Icelandic restaurant,' he tells me, 'serving everything from pizza to steaks, and with a dance house at the back' – but now he works as a subcontractor to the plant, for a recycling company. He remembers when the go-ahead for the smelter was given. 'It was so clear when the decision was made,' he remembers. 'People started to paint their houses again.' For the future, his hopes are mostly that the population of the fjords will expand – 'from 5,000 to 7,000 in the next ten years would be good. It's possible.'

As far as Gislason is concerned, Iceland's natural resources are there to be used. 'You know, if there's oil under the ground, you should go for it,' he argues, 'and if you have a lot of water running down to the sea, why not change it into energy?' And he is sceptical of arguments made against the project:

> A lot of people in 101 [downtown Reykjavik] confused the smelter and the dam – they confused everything. They have this view of life, but they have never lived outside of Reykjavik and they don't give a shit if people in areas like this have to move from their homes to Reykjavik. *They* don't give a damn about it, but *we* give a damn about it.

Gislason's views make him a strong advocate for further developments in other parts of Iceland. 'I've spoken to the politicians from Husavik [where Alcoa is considering a second smelter] a couple of times,' he tells me.

It is far from clear whether electricity generation for heavy industry will play the same role in Iceland's future that it has played in the country's immediate past. Will the financial crisis lead to a rethinking of the whole idea of internationalisation of the country's economy or will it scare Iceland into seeking even greater foreign investments in the energy sector? Will those investments still be viable? Plans for a new smelter at Helguvik in south-west Iceland – slated to produce as much as 250,000 tons of aluminium per year, three quarters as much as Fjardaal – are hanging in the balance as Century Aluminium and its Icelandic subsidiary Nordural consider whether they can finance the necessary investment.[35] Despite rumours that Alcoa was ready to pull out of plans to build a smelter at Husavik in northern Iceland – the so-called Bakki project – the environmental impact assessment process for the plant was officially launched in early 2009.

Some contend that the economic crisis has made it more necessary than ever for Iceland to exploit its natural resources to the full. But others argue that Iceland's age of heavy industry is over. The 'focus on aluminium is outdated', says Steingrimur Sigfusson, leader of the

Left-Green party and Minister of Finance since February 2009. 'We think that that kind of big solution is not a very good one,' he tells me. 'It comes with a high price, you have a huge boom, and then there's a hangover.'[36] Instead, Sigfusson's vision of the future is 'small and incremental'. 'We want to do the whole thing in a much more self-sustainable way,' he tells me. Instead of large hydroelectric plants requiring the construction of dams, reservoirs and high-tension power cables, he sees a future of mini hydroelectric stations – built without dams – and geothermal power plants – which tend to be small and more readily scaleable. As to the industrial consumers of Iceland's electricity, Sigfusson expresses a preference for smaller, high-tech companies: 'Let's invite some clean future-oriented businesses, small and medium-size enterprises, the ones that fit.'

Iceland's electricity companies have had to follow suit. According to Thorstein Hilmarsson, Landsvirkjun will not sell electricity to any aluminium smelters for which it is not already under contract.[37] But the company's hand was forced. In 2007, a local referendum in Hafnarfjördur rejected plans for an expansion of the existing smelter there, owned by Alcan. 'After that we put everything on hold,' Hilmarsson tells me. It became clear that the company would have to follow a more diversified strategy in the future. Hilmarsson talks brightly about 'a new step for Icelandic energy'.

Particular attention has been paid to attracting server farms, which require huge amounts of electricity to run their computers. Iceland has been touted as 'the ultimate location for data centres'.[38] New fibre-optic cables have been laid in the Norwegian Sea.[39] A report commissioned by the Invest in Iceland agency boasts the country provides 'a new location concept, attracting new business and enhancing their image by building on cost attractive, green supplies [of electricity]'.[40]

The Karahnjukar Dam, focus of so much of Iceland's domestic politics in recent years, is a few hours' drive away from Reydarfjördur, back towards Egilsstadir, down the edge of shimmering Lagarfljót Lake and then up into the Icelandic highlands. It is a barren-looking landscape, covered in rubble and rust-coloured grass. In a few places, patches of snow break up the monotony of the terrain.

The initial impression of the landscape here is one of lifelessness – the area looks more like the moon than the earth. But that impression is misleading. On the way to and from the dam I come across a herd of reindeer and a flock of geese, brilliant white against the dull background. When the dam was originally assessed by Iceland's national planning agency, back in 2001, it was initially rejected on the basis of potential damage to this fragile highland environment. (The minister of the environment overturned that assessment, allowing the dam to go ahead, while attaching a string of environmental conditions to the project.[41])

At the height of the dam's construction in 2007, there were over 1,000 workers up here – half of them from China. Now, the place is empty. There's a visitor centre at the site of the dam, but it's closed when I visit – the neatly stacked chairs are visible through the window. The reservoir stretches into the distance, but seems only half full, covered in patches of ice. What I cannot see from this side is the full height and extent of the main dam – 730 metres long and 200 metres high. Over the next months and years the reservoir will be topped up by meltwater from glacial rivers feeding into it. Eventually the reservoir will cover 57 square kilometres of the highlands, flooding the feeding areas and breeding grounds of local fauna. Already, a road which used to cut across the landscape has been severed by the rising water. Road signs frantically warn drivers that what used to be a road is now a dead end.

Next to the visitor centre a small white boat has been hauled up onto dry land, presumably to be used in an emergency. Someone with a dark sense of humour has baptised it in thickly painted red lettering: *Örkin – The Ark*. Perhaps it's just a light-hearted joke, a parting shot from one of the staff at the dam as the reservoir began to flood. But it is also a troubling metaphor for modern Iceland – a small boat struggling to stay afloat on a sea of Arctic troubles, searching for a saviour.

Before the deluge, Icelanders felt invincible. But that feeling has now been replaced by fears for the country's viability. Thousands of Icelanders have deserted their country, taking their skills to Europe, Asia and America. The illusion that tiny Iceland could exploit the opportunities of globalisation without any risk to its independence or statehood has been cruelly shattered. In the past, Iceland's physical location secured it from the violence in the world around it. But the country's Arctic isolation could do nothing to protect it from the cold winds of global recession.

As a result, just as Greenland takes a step towards the dream of freedom, Iceland is rediscovering the attractions of security in numbers. Reykjavik has sought assistance from the International Monetary Fund – more used to dealing with the risk of Latin American defaults than Nordic ones – to secure the country's financial future. Iceland's application to join the European Union, formalised in July 2009, is a signal for help in rough seas.

But in considering his country's future, Steingrimur Sigfusson, as a true Icelander, invites me to recall its history. 'Back in 1918 there were a lot of doubts that this small group of fishermen could survive as an independent country,' he reminds me. '"How on earth are you going to manage?" people asked – but we did.'[42] It is a message with resonance across the Arctic – the future is a choice.

Epilogue

From 20,000 feet nothing could look more eternal and more vast than the Greenland ice cap. Look to where the horizon should be, and the ice cap and the sky fuse into a single expanse of blinding white. Look down at the ice cap, and only the neat shadow of our plane – a child's drawing on a blank page – provides a sense of scale. Stare too long at the shadow of the plane against the ice and the image burns itself onto your retina. Look away, and the shadow of the plane stays, fading with each blink.

Further on, closer to the coast, the flat surface of the ice begins to undulate in response to the unseen land beneath. In one place, a single razor-thin line cuts deep across the white, like the fluid first line of a draughtsman's sketch. In another, a zigzag line fractures the unity of the icescape. And then, as the lines proliferate, the ice cap shatters. Order gives way to chaos, brilliant white gives way to matt grey. The texture of the ice cap becomes as rough as elephant hide.

Pools of cyan meltwater have collected in a few places, impossibly blue against the dullness of the elephant-hide grey. Feeling their way towards one another a few have burst their icy enclosures to form distended figures of eight. And then, strung out across the ice, are the beginnings of streams of meltwater, coursing knowingly down towards the far-off sea.

The glacier is not an an ice field now, but a flow, relentless and insistent. The immobility of mass has given way to the constant pull of gravity. As the glacier approaches the sea, the flow lines that indicate its course are bunched closer and closer together, like the abrasions of a steel comb. Where land underneath has offered resistance, the glacier twists and turns to avoid it, clinging to the obstacle, curling tightly round its edges, enveloping it and, eventually, wearing it away.

In some places the glacier ends in a final expression of energy, projecting itself heroically across the flat land. In others, the glacier's end is less dramatic, petering out into grey-green mud, cut through with winding channels of faster water, bearing only the lightest and the finest silt down towards the coast. And finally, at the very edge of Greenland and the beginning of Davis Strait, the delta of glacial meltwater joins the sea. A cycle is complete. The water of the ice cap – evaporated from the sea countless years ago, deposited as snow, crushed into ice – returns to its source.

For hundreds of thousands of years, the history of the Arctic has been determined by these natural cycles, beating a slow rhythm through geological time. Now, things have speeded up. Nature's Arctic empire, untouched by human influences, has been disturbed. The balance of natural and human influences has shifted, decisively. Human choices – to protect or to develop the Arctic – are key to its future history.

Our ideas of the Arctic – permanent, pristine, unchanging – will persist long after they have been overtaken by Arctic change. But slowly, bit by bit, our ideas of the Arctic will have to adapt. As they do, a little bit of our sense of earthly eternity will be lost, for ever.

What will replace it? Perhaps the image of the Arctic as a zone of global cooperation, a focus for scientific research and global environmental stewardship. More likely, however, is the image of the Arctic as a battleground, fought over not just by states but by the different economic and political interests which are jostling for their part of the Arctic future, trying either to develop its economic potential or to protect its environment. A battleground does not mean war, but it does mean conflict and competition: political, economic, cultural and diplomatic.

There is no fatality to such a conclusion – though history and current events support it. The institutions of global governance will likely avert some potential for conflict, or channel it into bureaucratic resolution. Economic and environmental groups may yet achieve the kind of cooperative approach which would put their objectives in balance, rather than in more confrontational opposition to one another. Arctic countries could find a way of articulating a common vision of the Arctic's future, and cooperating to achieve it. Perhaps. But as the Arctic enters the course of global history and as its uniqueness is taken from it, the likelihood of the Arctic escaping the realpolitik of the rest of the world seems low. We can no longer deal with the Arctic as we would wish it to be – in the future, we will have to deal with the Arctic as it is.

Acknowledgements

Writing a book, I have come to realise, is a collective effort. This is particularly the case for a book which draws so heavily on interviews with experts, and chance conversations in various corners of the Arctic with taxi drivers, restaurateurs, hotel owners and those who just like to talk and who recognised in me someone who was likely to give them a good hearing.

I was set on the path towards *The Future History of the Arctic* by friends who had written books themselves – Henry Hitchings, Ben Skinner and Parag Khanna – and who were consequently able to give me realistic encouragement on the question of whether I should write the book, and sound advice on how to go about it. In New York, my agent Jennifer Joel of ICM took my first proposal and made it better, before finding it a home at Bodley Head (Random House) in the United Kingdom and PublicAffairs in the United States. When I met the perennially elegant Will Sulkin for the first time, and spoke to Lisa Kaufman on the phone, it became clear that Jennifer had found not just two publishers, but the right two publishers: serious, experienced, interested. So it has proved. For that, above all, many thanks are required.

I wrote the last word of the first draft of the book in Melbourne, Australia, having just spent two relaxing days down in Point Lonsdale, Victoria, at the house of a family friend, Leanna Darvall – just one of many wonderful spots where I have been able to catch a few hours, writing, and felt myself intellectually refreshed by the company of others. I would like to thank friends who, by providing board and lodging in some cases, the occasional nudge in others, or simply good company and distraction in places both north and south, have made the whole process more fulfilling and fun than I ever expected it could be: Teresa Drace-Francis and Ronald Grover (and Teresa's never-ending family), Victoria de Menil, Alex Burghart, Nicola Donnelly, Cassandra Florian, Maria Sanchez-Marin Melero, Johannah Christensen, Ditte Steen, Alissa de Carbonnel, Dario Thuburn, Nina Eldor, Alice Boyle, Leo Tomlin, Naureen Khan, Josie Rourke, Jacky Klein, James and Camilla Smith (and Hubert), Tannaz Banisadre, Aurelie Vandeputte, Edouard Sebline and Cicely Fell, Emma Castagno, Jasper Goldman, Valerie Nussenblatt, Keith Campbell (who provided me with a lesson in Canadian politics on the A303 to London) and my parents' friends in Australia who I really feel are more family than friends.

My former colleagues at the World Economic Forum provided me with opportunities for enjoyment and edification in equal measure as I considered how and when I should start writing. My particular thanks go to Jesse Fahnestock, Randall Krantz, Shruti Mehrotra, Delphine Angelloz-Nicoud (who put up with the first few months of a long process), Christoph Frei, Gareth Shepherd, Thierry Malleret, Nicholas Davis, Viktoria Ivarsson, Sylvia Lee, Johanna Lanitis, Benita Sirone, Anastassia Aubarikova, Michele Petochi, and Micky Obermayer's '05 cohort. Successive Davos meetings of the World Economic Forum provided an extraordinary opportunity to hear politicians, academics and business leaders discuss their views of the world. From the many people I met in the course of my work at the forum who, in one way or another, have informed this book, two individuals stand out: David Victor, who is more clear-headed about the interlocking issues of energy, security and environment than anyone I know, and Alyson Bailes, whose understanding of the complexities of Nordic and wider security questions is without par. When it came to the hard job of leaving the forum and starting field research, Borge Brende – Norway's former environment minister and forum colleague – sat down in a café in Switzerland, opened his Norwegian address book, and generously provided me with a racing start.

Research for this book has brought me into contact with an astonishing range of people. I have accrued debts in many cities and places, in conferences in Sweden, Norway and the United States as well as in interviews and in informal discussions not listed here. I hope this book may to some small extent repay those debts.

My debts are heaviest in Norway, where I spent time in Oslo, Svalbard and the cities of the north, and on many bumpy plane rides in between. In Oslo, I was fortunate to talk to, amongst others, Turid Sand and her colleagues at the Norwegian Ministry of the Environment, Bard Bredrup Knudsen and his colleagues at the Norwegian Defence Ministry, Karsten Klepsvik and Geir Westgaard at the Norwegian Foreign Ministry, Mette Gravdahl Agerup at the Norwegian Petroleum Ministry, Arild Moe and his colleagues at the Fridtjof Nansen Institute, John Skogan and Johnny Skorve at the Norwegian Institute of International Affairs, Pal Prestrud and Grete Hovelsrud at the Centre for International Climate and Environmental Research (CICERO), Rolf Tamnes at the Norwegian Institute for Defence Studies, Bjorn Brunstad at Econ and Olav Orheim, former director of the Norwegian Polar Institute and all-round Arctic expert.

In the north, in Hammerfest, StatoilHydro's Sverre Kojedal proudly showed me around the world's first Arctic LNG plant, while Jan-Egil Sorensen provided me with an understanding of how a small town had been affected by such a substantial investment. To the east, in Kirkenes, I was fortunate enough to talk to Rune Rafaelsen, Atle Staalesen and Thomas Nilsen at the Norwegian Barents Secretariat as well as Norwegian Border Commissioner Leif-Arne Ljokjell, Christoffer Alviniussen and the soldiers at the Elvenes border station who guard their country's frontier with Russia. A few degrees further south, in Tromso, Kim Holmen, research director at the Norwegian Polar Institute, put Arctic climate change into global context. His colleagues Per Arneberg and Birgit Njastad provided me with an insight into the local impacts of development and climate change in the Arctic, and the politics of the environment in Norway. On the spur of the moment, Jesper Hansen at the Arctic Council secretariat took me down to the site of the sinking of the German battleship *Tirpitz*, underlining the long-running strategic importance of the Norwegian coast.

Further north, on Svalbard, I spoke to to Elisabeth Bjorge Lovold and Tor Punsvik at the governor's office, bevore travelling to Ny-Ålesund, the world's most northerly permanently inhabited settlement and a hive of international scientific research activity. I extend my particular thanks to Rasik Ravindra, Director of the National Centre for Antarctic and Ocean Research in Goa, India, who made it possible for me to visit India's Himdari research station. There, I was generoulsy hosted by C. G. Deshpande, and was able to meet B. C. Arya, Rakesh Misra and Druv Sen Singh. Beyond the Indian research station my thanks go to the other scientists and administrators who gave me their time and insights, patiently explaining their work and, just as interestingly, the role of the scientist in informing public policy on climate: Mike Kendall, Dorothea Schulze, Marten Loonen, Sebastian Westermann, Trond Svenoe, Christiane Hubner and Elin Austerheim, Wei Luo and the scientists at the Yellow River Chinese research station.

In Helsinki, I spoke to Goran Wilkman of Aker Arctic, and the experts of the Finnish Institute of International Affairs – Tapani Vaahtoranta, Hanna Ojana and Hiski Haukalla – who enlightened me on Finland's security dilemmas. In Denmark, I was fortunate enough to speak to Mikaela Engell of the Danish foreign ministry, the genial and insightful Uffe Elleman-Jensen and his successor as foreign minister, Niels Helveg Petersen. Soren Thuesen, Frank Sejersen and Bent Nielsen gave me a thorough background on the politics and history of Greenland. At the Geological Survey of Denmark and Greenland, Christian Marcussen provided me with an insight into Denmark's Arctic mapping work, while Lotte Melchior-Larsen gave me a captivating lesson in Arctic geology.

On Greenland itself, the then foreign and finance minister Aleqa Hammond kindly gave me her perspective on the prospects for Greenlandic Self-Rule and eventual independence. Amongst others, I also spoke to Henrik Leth of the employers' federation, Hans Kristian Olsen of Nunaoil, Inga Markussen, the CEO of Royal Arctic Jens Andersen, Jon Skov Nielsen at the Bureau of Minerals and Petroleum, Klaus Trolle Nedergaard and Jens Bisgaard at Royal Greenland, and Axel at the local small fishermen's association in Nuuk.

In Iceland, I am grateful to Jake Siewert of Alcoa for facilitating my visit to that company's smelter in Reydafjordur, where Erna Indridadottir kindly arranged discussions with Tomas Sigurdsson, the plant's manager, and Gudmnundur Gislason. In Reykjavik, member of parliament Steingrimur Sigfusson explained his view of Iceland's future. I spent a couple of fascinating hours with Andri Snær Magnason discussing the principles of Icelandic economic development, as well as talking to Valur Ingimundarsson – Iceland's leading contemporary historian – and Thorir Ibsen at the foreign ministry. Even as the financial crisis began to take hold, Arni Magnusson, Tryggvi Herbertsson and Thorstein Hilmarsson guided me through their view of Iceland's economic and energy potential.

Across the Atlantic in Washington D.C., I spoke with a number of people involved directly or indirectly in considering American policy on the Arctic, and how changes in the Arctic will affect US priorities in the future, including, amongst others, Margaret Hayes and her colleagues at the State Department, Rachel Halpern, Ronald Filadelfo and David Catarious at the Center for Naval Analyses, Congressman Brian Baird, Neil Brown and Arne Fuglov. In Ottawa, I was re-educated on the legal history of the Northwest Passage by Wendell Sanford, Robert Kadas and Jamie Pennell at the Canadian department of foreign affairs. John Crump at the Many Strong Voices initiative explained how indigenous groups in the Arctic are allying themselves with small-island developing nations affected by climate change. In Anchorage, Alaska, my thanks go to Pete Larsen, Lawson Brigham, Steve Amstrup, Ric Wilson, Mark and Susan Lutz (who managed not just to be excellent hosts at the best place to stay in Anchorage – the Oscar Gill House – but to provide me with trenchant insights into Alaskan politics), Deborah Williams, and the Anchorage staff of Shell: Phil Dyer, general manager Pete Slaiby, Travis Purvis and Barbara Bohn. In Deadhorse, John Maketa and Bill Koski and the staff of Shell gave me an excellent overview of their work, and the work of other oil companies on the North Slope.

A somewhat different view of oil and gas development was provided by Taqulik Hepa, Cheryl Rosa, Craig George and Robert Sudyam at the North Slope Borough Wildlife Department in Barrow, and by Pamela Miller in Fairbanks. Officials at the Fish and Wildlife Service in Fairbanks explained the complications of the process of managing Arctic oil and gas development. Tom Douglas of the US Army Cold Regions Research and Engineering Laboratory opened the permafrost tunnel outside Fairbanks. At the University of Alaska Fairbanks, Vladimir Romanovsky explained the more general permafrost problem in the Arctic while Glenn Juday spoke cogently and passionately about how climate change is affecting Alaska above ground, and why that matters globally.

My understanding of the political and economic situation in endlessly fascinating Russia was informed by a series of excellent lunches with John Helmer in London and Moscow. I am indebted to Evgeny Shvarts of the World Wildlife Fund for talking me through his perspective on Russian environmental issues, and for putting me in touch with other colleagues both in Moscow – Alexei Khiznikov in particular – and Murmansk (Vladimir Krasnopolsky and Sergei Zhavoronkin). Maria Goman at the Murmansk branch of the Norwegian Barents Secretariat was tremendously helpful in helping organise meetings with government officials Victoria Shvets and Victor Gorbunov. Vladimir Advukov provided a powerful insight into the hopes for Murmansk's rebirth. Andrei Ponomarenko's work at the Bellona Foundation provided a reminder of the darker side of the Soviet Union's Arctic legacy. I never would have travelled to the Solovetsky Islands without the help of Alissa de Carbonnel, nor would I have met the enigmatic figure of Yuri Brodsky. In St Petersburg, I am indebted to Ivan Frolov and his colleagues at the Russian Arctic and Antarctic Research Institute, and to Victor Boyarsky and Vladimir Vasilyev.

Further thanks are due to Cleo Paskal and Susan Ambler-Edwards in London for feeding a broader interest in climate and security-related issues, Robert Larsson in Stockholm for his perspective on Russian gas politics, Raymond Arnaudo in Moscow, DJ Peterson in New York, Christian de Morliave in Paris, Cary Fowler of the Global Crop Diversity Trust in Rome, Jean Laherrère, Scott Borgerson and, in Brussels, Janos Herman at the External Relations Directorate of the European Commission (responsible for Arctic policy), Iain Shepherd, Martin Fernandez-Piez-Picazo at DG Mare, and Massimo Lombardini, Ioannis Samoulidis and Catherine Sustek at DG Tren.

Many more informal discussions have gone into this book – too many to mention here. Most of the actual work of writing, and the secondary source material research, was done at the British Library, a wonderful institution and an excellent place to work. When it came to editing and preparing the book for publication, Morgen van Vorst at PublicAffairs took much of the strain, improving both the book's structure and content with her apposite remarks. Beth Wright knocked the text into American style and corrected some of my errors; Chris Erichsen produced the wonderful maps for the book; Mary Chamberlain knocked the text back into English style. Meanwhile, Lisa Kaufman and Melissa Raymond at PublicAffairs, and Will Sulkin and Kay Peddle at Random House, have kept the process rolling. Whitney Peeling in New York and Clara Wormersly in London have expertly introduced me to the wonders of marketing. The smoothness of the process of publishing this book is a testament to the professionalism of the staff at the publishers. I am very grateful to them.

In the end, however, my greatest thanks are for my family and above all my mother and father who are my closest advisers, shrewdest editors and the fount of love within our family; Uncle John, Robert and Pirjo Gardiner and the Gardiner clan; my much-loved sister Chloe and husband JB and, as the book moved into copy-editing, the double-trouble wonders that are Eloise and Theodore.

Notes

Introduction

1. An excellent account of the links between the environment and the early history of Greenland can be found in Diamond, Jared, *Collapse: How Societies Choose to Fail or Succeed*, New York, 2006, pp. 178–276. • **2.** There has long been historical debate on the process of depopulation of the Norse Greenland community and the relative importance of a number of different factors. See, for example, Nansen, Fridtjof, *In Northern Mists*, London, 1911. In the past, some had argued that Inuit harrying of Norse communities played a role in their demise, though Nansen largely discounted that. Instead, Nansen suggested a range of factors, including the possibility that a decline in the variety of diet may have contributed. • **3.** See Lopez, Barry, *Arctic Dreams: Imagination and Desire in a Northern Landscape*, London, 1986. • **4.** See, for example, Vaughan, Richard, *The Arctic: A History*, London, 1994 or McGhee, Robert, *The Last Imaginary Place: A Human History of the Arctic World*, Oxford, 2006. • **5.** The distinction is not, in reality, so clear. Writing history involves weaving known facts, though often distorted by the lens of the present, into a narrative structure that is, by its very nature, a simplification or interpretation of what actually happened. Historians may use deductive reasoning to construct their arguments, but their product is, in one sense, necessarily literary. Futurologists have less to go on than historians, and their speculations tend to be correspondingly broader. There is more room for both speculation and error in drawing together a narrative of uncertain trajectories than there is in drawing together a narrative of certain, but limited, facts. • **6.** See, for example, Schama, Simon, *The American Future: A History*, London, 2008.

PART ONE Visions

1 Oracles and Prophets: Rethinking the North

1. 'Letter of Daniel, bishop of Winchester, to Boniface (722–732)', *English Historical Documents I, circa. 500–1042*, Oxford, 1953. • **2.** Larson, Laurence Marcellus, trans., *The King's Mirror – Speculum Regale – Konungs Skuggsá*, New York, 1917, p. 148. • **3.** Huntington, Ellsworth, *The Character of Races, as Influenced by Physical Environment, Natural Selection and Historical Development*, New York, 1925, p. 67. Huntington's view, in brief, was that, up to some limit, northern areas tended to speed evolution and cultural development compared to the south pp. 50–51: 'It appears to be a biological law that a tropical environment, because of its uniformity, tends to perpetuate primitive, unspecialized forms. Since man split off from the apes his specialization has been in the size, complexity, and functioning of the brain . . . In equatorial regions the mental type of specialization has apparently been slow, largely because there have been no really great changes throughout man's history, not even during the severest glacial epochs . . . That, presumably is one of the chief reasons why it is so difficult to impose upon equatorial people anything more than the outer husk of northern government, northern religion, northern ideals, and northern culture'. • **4.** Huntington, Ellsworth, *Civilization and Climate*, New Haven, 1915, p. 2. • **5.** In fact, while mean temperatures do tend to rise as one approaches the equator – and tend to fall as one moves away from it – latitude is only one determining factor in local climate. Temperatures at any particular locality on earth are determined by prevailing air currents and by the proximity of the oceans as much as by latitude alone. For example, as a consequence of the Gulf Stream, north-west Europe is considerably warmer than areas of North America lying on the same line of latitude. The temperature gradient of Russia – falling from west to east as much as from north to south – is an example of the influence of surrounding land mass on temperature. In terms of liveability (and the survivability of crops) temperature range may matter as much as mean temperature. The key feature of the relatively wealthy coastal areas of Europe is not so much their average temperatures, but the relatively small range within which temperature tends to fluctuate over the year. • **6.** See, for example, Fernández-Armesto, Felipe, *Pathfinders: A Global History of Exploration*, Oxford, 2006. • **7.** One writer has made the (widely disputed) claim that a Chinese fleet under Admiral Zhou circumnavigated Greenland and 'discovered' a passage across the top of the world in the fifteenth century, benefitting from the consequences of a period of relatively warm global climate. If true, the Bering Strait between Asia and America was discovered long before Vitus Bering himself was even born. But the aims of Admiral Zhou were surely no different from those of his European counterparts: it was East-West travel that mattered, above all. And whatever the route which Zhou's voyages may have taken, the experience of them did not change the conclusion of Chinese leaders on the merits of engagement with the outside world. Unchallenged at home, China shut itself off from the world, freezing its technological progress and opening the way for Western global domination within a matter of centuries. Menzies, Gavin, *1421: The Year China Discovered the World*, London, 2002, pp. 343–357. • **8.** Ferrer Maldonado, Laurent, *Voyage de la mer atlantique a l'océan pacifique par le nord-ouest dans la mer glaciale* (translated from Spanish by Amoretti, Charles), 1812. The book exists in various other languages, most notably Italian. This is a later French translation from the Spanish. The original voyage was in 1588, and written up after that. The reason for this French edition was principally to defend the veracity of Ferrer Maldonado's assertions, however questionable they had become over the two centuries since the original voyage was alleged to have taken place. • **9.** Dezhnev, who gives his name to the cape on one of the sides of the Bering Strait, passed there in 1648. The name of the strait was given by Vitus Bering, a Danish navigator in the service of Russian Tsar Peter the Great, who crossed the strait in 1728. • **10.** De la Martinière, Pierre Martin, *Voyage des pays septentrionaux, dans lequel se void les moeurs, maniere de vivre, & superstitions des Norweguiens, Lappons, Kiloppes, Borandiens, Syberiens, Samojedes, Zembliens, Islandois*, Paris, 1676, p. 136. • **11.** Ibid., p. 151. • **12.** Riffenburgh, Bruce A., *The Anglo-American Press and the Sensationalization of the Arctic, 1855–1910*, University of Cambridge, 1991, p. 13. • **13.** Diderot, Denis, *Encyclopédie, ou dictionnaire raisonné des sciences, des arts et des métiers*, published between 1751 and 1772 and subsequently with additions and amendments. • **14.** Macgregor, John, *The Resources and Statics of Nations*, London, 1834. Macgregor subsequently became secretary to the British Board of Trade. • **15.** Shelley, Mary, *Frankenstein, or, the Modern Prometheus*, London, 1818. • **16.** Verne, Jules, *The Adventures and Voyages of Captain Hatteras*, Boston, 1876, p. 440. • **17.** See Riffenburgh, op. cit. • **18.** Hanson, E. P., *Stefansson: Prophet of the North*, New York,

1941. • **19.** Letter quoted in Mead, W. R., 'New Light on Nansen: Review', the *Geographical Journal*, Vol. 128, No. 2, June 1962, pp. 208–209. • **20.** Nansen entitled his account of his travels through Siberia in 1913 *Through Siberia, the Land of the Future* (London, 1914). Though he was hardly a Communist, his vision of Russia's future importance was undiminished by the revolution of 1917. 'Russia's civilisation has not yet burst forth into blossom: it still belongs to the future,' he wrote in 1923 (*Russia and Peace*, London, 1923). • **21.** Nansen, Fridtjof, *Adventure: Doctor F. Nansen's Rectorial Address at St Andrew's University*, 1926, p. 14. • **22.** The best biography of Nansen is Huntford, Roland, *Nansen: The Explorer as Hero*, London, 1997. • **23.** Nansen, Fridtjof, *In Northern Mists*, London, 1911. • **24.** Several years later, Sverdrup would make his own name as the discoverer of several islands in the Canadian Arctic. • **25.** See Nansen, Fridtjof, *Across the Polar Region: A Lecture by Dr Fridtjof Nansen, delivered by him in St James' Hall, London, on Feb. 9th, 1897, and afterwards in many of the larger towns in the British Isles, describing his Voyage Across the Polar Region in the Fram during the years 1893–1896*, 1897. Nansen was by no means the first or last explorer to supplement his income with public speaking. • **26.** Roosevelt's views on frontier development were well established by the multi-volume *The Winning of the West* (published from 1889 onwards). As President, Roosevelt was personally involved in the American attempt to conquer the North Pole, lending his name to the boat which Peary used to take him to his setting-off point. The first line of Roosevelt's preface to Peary's account of his successful expedition (Peary, Robert, *The North Pole*, New York, 1910) described the dinner at which Nansen told Roosevelt that 'Peary is your best man'. • **27.** 'The greatest achievements in history have been brought about more by the aid of ideas than of truth,' he wrote in 1911. *In Northern Mists*, p. 291. • **28.** To make matters worse from Nansen's perspective, the boat which Amadeo had used to reach his setting-off point was the *Stella Polare*. Nansen had used the same boat, then called the *Jason*, to reach his own setting-off point for the Greenland crossing of 1888. • **29.** Peary had reconnoitred Greenland in 1886 and believed that, by rights, he should have been first across. • **30.** *In Northern Mists*, p. 234. • **31.** Nansen, Fridtjof, *Norway and the Union with Sweden*, London, 1905. • **32.** Nansen's publishers argued that the resources of Siberia 'may even prove a vital factor in the decision of the struggle which is now absorbing the attention of the world'. *Through Siberia, the Land of the Future*, p. v. • **33.** Ibid., p. 282. • **34.** Ibid., p. 291. • **35.** Ibid., p. 156. • **36.** Nansen, Fridtjof, 'On North Polar Problems', the *Geographical Journal*, Vol. XXX, No. 5, November 1907, pp. 469–487, and its continuation in the *Geographical Journal*, Vol. XXX, No. 6, December 1907, pp. 585–597. • **37.** Nansen, Fridtjof, 'Klimat-Vekslinger i Nordens Historie', *Avhandlinger Utgitt av Det Norske Videnskaps-Akademi i Oslo*, Oslo, 1926. • **38.** In 1929, the year before he died, Nansen was in the United States, trying to raise money for a trip there by airship, as head of the *Aeroarctic* consortium. • **39.** Nansen, Fridtjof, *Russia and Peace*, London, 1923, p. 16. • **40.** After his death, there were some who sought to appropriate Nansen's legacy for their own ends. A decade after Nansen's death, it was his tireless assistant, Vidkun Quisling, who became leader of Nazi-occupied Norway, espousing a brand of Nordic triumphalism which claimed its roots in Nansen's account of northern history, but which would have been far too shrill for Nansen's individualistic sensibilities. • **41.** 'Nansen Dies at 68 of Heart Paralysis; Explorer Succumbs to a Stroke While Alone – Had Been Believed Recovering; Norway is Grief-Stricken', The *New York Times*, 14 May 1930. • **42.** The speaker was Emil Ludwig, the greatest biographer of the age. Whitehouse, J. Howard, *Nansen: A Book of Homage*, London, 1930. • **43.** Diubaldo, Richard, *Stefansson and the Canadian Arctic*, Montreal, 1998, p. 50. • **44.** Ibid., p. 215. • **45.** Stefansson, Vilhjalmur, *The Friendly Arctic: The Story of Five Years in Polar Regions*, London, 1921, p. 687. • **46.** There is some debate about the appropriate use of the terms 'Eskimo' and 'Inuit'. Broadly, the term 'Inuit' encompasses all the native peoples of the North. Historically, however, the term 'Eskimo' was used to describe, in particular, the native populations of Arctic Canada and Greenland – indeed it was the term used by both Nansen and Stefansson. Throughout this book I have used the term 'Inuit' as the broad term describing the native populations of the Arctic, though it should be remembered that there are, in fact, many different native groupings in the North rather than one homogeneous group. In a few cases, where relevant, I have used the specific names for those other groups – particularly the Alaskan Inupiat and Aleut groups. • **47.** Stefansson, Vilhjalmur, *My Life with the Eskimo*, London, 1913. • **48.** Ibid., p. 15. • **49.** Ibid., p. 61. • **50.** Over the decade prior to Stefansson's expedition there had been growing concern that the basis of Canada's sovereignty in the Arctic might be less secure than originally thought. An expedition led by Otto Sverdrup between 1898 and 1902 raised the possibility of Swedish or Norwegian claim to certain islands, now part of Canada, by right of discovery. At the same time, the ruling of a British judge in the Alaska boundary dispute, favouring the American position over the Canadian one, demonstrated, to some minds, that the United Kingdom could not be relied on as a guarantor of Canadian territorial integrity. • **51.** *The Friendly Arctic*, p. xii. • **52.** Borden wrote that, while he had been freezing in a central London hotel in December 1918, a month after the end of the First World War, forced to commandeer wood fuel from the Canadian Corps near Windsor, 'Stefansson and his party, possessing an abundance of fuel, which the [Arctic] country supplied, were sitting in their shirt-sleeves, hundreds of miles within the Arctic Circle'. 'While we shivered in the temperate zone, there was vast comfort in the vicinity of the North Pole,' he wrote. • **53.** Ibid., p. 2. • **54.** Ibid., p. 6. • **55.** Ibid., p. 7. • **56.** Ibid., p. 8. • **57.** Ibid., p. 11. • **58.** Ibid., p. 16. • **59.** Ibid., p. 24. • **60.** Orientalism was given its current meaning by Edward Saïd (*Orientalism*, London, 1978), who argued that Western attitudes towards the East were framed by the misrepresentations and attitudes of imperialism. • **61.** *The Friendly Arctic*, p. 28. • **62.** Ibid., p. 83. Stefansson further argued that while it might be more difficult for a northern population to move south than the other way around: 'our immunology and preventive medicine are still in their infancy and so we cannot guarantee the south-going Eskimo against germ attack. But we can guarantee the negro against his new enemies of the frost, for we have fought cold and storm successfully for thousands of years, and can show him how to protect himself.' (p. 102.) Stefansson cited the story of 'Jim Fiji' (his real name was James Asasela), brought from the Samoan Islands to Chicago as part of the 1893 World Fair as an exhibit of 'native races'. Having been accidentally shipped to the North on a three-year voyage from San Francisco, he subsequently liked it so much he stayed there. • **63.** A former employer of the American mining company Ayer and Longyear was quoted as telling Stefansson that 'coal can be so cheaply mined and transferred from Spitsbergen to Europe that Spitsbergen will drive Newcastle and Wales out of the continental coal markets north of their latitude'. • **64.** Ibid., p. 14. • **65.** Ibid., p. 51. • **66.** Ibid., p. 60. • **67.** Ibid., p. 120. • **68.** Ibid., p. 65. • **69.** Ibid., p. 136. • **70.** Ibid., p. 199. • **71.** Stefansson proposed a number of different ways of surfacing through the ice. Later, in 1931, when Sir Hubert Wilkins attempted to navigate under the North Pole, the proposed methodology for cracking through the ice was simply to use the natural buoyancy of the submarine and to surface like a whale. In 1922, Stefansson proposed an inverted toboggan (allowing the submarine to slide along the bottom of the ice until it found an opening), the use of a depth charge to open up the ice and, finally, 'an electric coil . . . When the boat rests against the ice a current could be passed through the coil heating it as bread toasters are heated on our breakfast tables, melting the way upward for the boat.' (p. 194.) ı **72.** Stefansson, Vilhjalmur, *The Adventure of Wrangel Island*, London, 1925. His biographer got closer to the truth in labelling the affair the 'misadventure of Wrangel Island'. • **73.** See Webb, Melody, 'Arctic Saga: Vilhjalmur's Attempt to Colonize Wrangel Island', the *Pacific Historical Review*, Vol. 61, No. 2, May 1992, pp. 215–239. • **74.** Stefansson himself had not been with the ship when it sank. He had abandoned the ship as early as 1913, when it was locked in ice, leaving to hunt for food with a plan to return a few days later. But when the ice pack broke, Stefansson's return path was made impossible. The *Karluk* did not finally sink until January 1914. • **75.** LeBouardais, D. M., 'Wrangel's New Laird is Unique Character', *Star*, 23 June 1924. Quoted in appendix to *The Adventure* *of Wrangel Island.* • **76.** 'Wrangel Island of Utmost Value, says Russ Chief', *Star*, 19 November 1924. Quoted in appendix to *The Adventure of Wrangel Island.* • **78.** Stefansson, Vilhjalmur, 'The North American Arctic' in Stefansson and Weigert, eds., *Compass of the World: A Symposium on Political Geography*, London, 1946. This was a reworked and extended edition of an article originally published in *Foreign Affairs* in 1939. • **79.** Nansen described Kibirov as 'a quiet, really pleasant and good-natured-looking man . . . not a trace in him of the savage cruelty that one expects to find in so powerful a man in this country of convicts and exiles'. (*Through Siberia, Land of the Future*, p. 181.) In 1916, as the result either of a bribe or of some Caucasian fellow feeling, Kibirov signed a health certificate without which Djugashvili would have no chance of being conscripted into the Russian army, one of the few means of escaping Kureika. (Sebag Montefiore, Simon, *Young Stalin*, London, 2007, p. 259). • **79.** The 'colleague' was Vyacheslav Molotov, the Commissar for Foreign Affairs of the Soviet Union for seventeen years between 1939 and 1956. • **80.** Sebag Montefiore, op. cit., p. 259.

2 Through a Glass Darkly: The Soviet Arctic

1. Brodsky's work was finally published as *Solovki: Le Isole de Martirio*, Rome, 1998. • **2.** *The Gulag Archipelago* was published in Paris in 1973. Solzhenitsyn was forced into exile from the Soviet Union a few weeks later. • **3.** According the NKVD's statistics the number of registered inmates rose rapidly in the mid- to late 1930s during the period of 'The Terror', rising from 965,742 in 1935 to 1,881,570 in 1938. In general, the numbers fell during the war years (from 1,929,729 in 1941 to 1,460,677 in 1945) and rose in the immediate post-war years, jumping over 2 million for the first time in 1948. But these figures are subject to considerable caveats, however. When she cites this number in *Gulag: A History* (London, 2003), Anne Applebaum points out that the numbers counted at the beginning of each year take no account of the extraordinary turnover of prisoners over the course of a year, particularly the war years – when a very large number of prisoners were released into the Red Army, but an almost equal number of citizens were incarcerated. Applebaum cites estimates that, over the period from 1929 to Stalin's death in 1953, a total of between 17 and 18 million passed through the labour camps. • **4.** Nansen, Fridtjof, *Through Siberia, The Land of the Future*, London, 1913, p. 285. • **5.** For an excellent discussion of the long-term consequences on Russia's economic geography of Soviet planning, see Gaddy, Clifford and Hill, Fiona, *The Siberian Curse: How Communist Planners Left Russia Out in the Cold*, Washington D.C., 2006. • **6.** Mikhailov, N., *Soviet Geography: The New Industrial and Economic Distributions of the U.S.S.R.* (translated Nathalie Rothenstein), 2nd edition, London, 1937, p. 18. • **7.** Quoted in McCannon, John, *Red Arctic: Polar Exploration and the Myth of the North in the Soviet Union, 1932–1939*, Oxford, 1998, p. 15. • **8.** The Trans-Siberian railway was only completed, at great expense, in 1913. • **9.** Hartwig, Georg, *The Polar World: A Popular Description of Man and Nature in the Arctic and Antarctic Regions of the World*, London, 1869, p. 238. • **10.** Mikhailov, N., op. cit., p. 42. • **11.** Ibid., p. xiv. • **12.** Building on Tsar Michael's 1619 interdiction on shipping in the Kara sea – more respected in the breach than in the observance – the Governor of Arkhangelsk province issued an edict forbidding the storage of Norwegian goods on the Russian coastline in 1866. This was only backed up by a physical presence in 1885, with the introduction of the *Murman*. • **13.** Murmansk, which later became by far the largest port on the Kola Peninsula, the home of the Soviet Northern Fleet and the largest city above the Arctic Circle, was only founded in 1916. • **14.** Russians say that the name of Dalian derives from an old Russian word – *'dalny'* – meaning 'far away'. After the Second World War Soviet troops were stationed in Port Arthur until 1955. When I travelled to Dalian in September 2007, the Lushun naval base was closed to foreigners – with the exception of Russian citizens visiting the Russian cemetery. Modern Dalian prides itself on its Arctic links, housing a polar zoo. • **15.** Solzhenitsyn, Alexander, *The Russian Question at the End of the Twentieth Century*, London, 1993, p. 78. • **16.** These terms were supposedly offered to an unmandated American diplomat in Moscow in February 1919, William Bullitt. The Allies, meeting in Paris, failed to approve the deal. When Bullitt returned to Moscow in 1934 as the first American ambassador to the Soviet Union his experiences were sharply different and he became strongly anti-Communist. For a full account of this episode and the attempts to reach peace with Russia in 1919, see MacMillan, Margaret, *Peacemakers: Six Months that Changed the World*, London, 2001, pp. 71–91. • **17.** A Foreign Office note sent to the Undersecretary of State for Foreign Affairs directly after the armistice of 11 November 1918, expressed one side of the British debate: 'Leaving aside the fact that we shall be deserting those whom we have encouraged to expect assistance against the excesses of Bolshevism, we should be in danger of losing for an unknown period the resources of Siberia, which are indispensable for reconstruction after the war. It is unnecessary to emphasise the importance of maintaining our hold on the resources, both from the point of view of denying them to the Bolsheviks and as a guarantee for the acknowledgement of their financial obligations to us by whatever Russian government ultimately assumes control.' Director of Military Operations to the Undersecretary of State for Foreign Affairs, 13 November 1918 (F.O. 371:3365). • **18.** Urquhart had been the director of several Anglo-Russian mining ventures before the First World War. For a full account, see Kolz, Arno W. F., 'British Economic Interests in Siberia during the Russian Civil War, 1918–1920', the *Journal of Modern History*, Vol. 48, No. 3, September 1976, pp. 483–491. • **19.** Stalin, having fought against Marshal Pilsudski's Poles who were swept from the gates of Moscow to the gates of Warsaw and then back into Russia, was keenly aware of Russia's vulnerability in the west. • **20.** That proximity would also have its advantages: later that year Lenin would be forced back across the same border, driven by the same Finnish train driver. • **21.** Mikhailov, N. op. cit., p. 158. • **22.** Komarov, Boris, *The Destruction of Nature in the Soviet Union*, New York, 1980, p. 60. • **23.** Mikhailov, N., op. cit. • **24.** Komarov, Boris, op. cit. p. 17. • **25.** Ibid., p. 135. • **26.** Gorky subsequently wrote a number of essays refuting the notion of 'compulsory labour' in the Soviet Union. In one article from 1931, 'The Legend of Compulsory Labour', he writes, 'In the Soviet State, compulsory labour is not practised even in houses of detention; there illiterate criminals are obliged to learn to read and write, and peasants are allowed leave to go home and work their land and see their families.' That said, Gorky did allude to 'labour communes', but argued that these were inhabited by the 'former "socially dangerous"'. He argued that the West was only interested in the notion of 'compulsory labour' so as to find an excuse to exclude Soviet goods from their markets. Gorky, Maxim, *Articles and Pamphlets*, Moscow, 1950, pp. 233–246. • **27.** This has been doubted by a large number of historians. Anne Applebaum, for one, notes that 'prisoners in earlier, pre-Solovetsky Bolshevik camps also mention being given extra food for extra work, and in any case the idea is in some sense obvious, and need not necessarily have been invented by one man'. Applebaum, Anne, op. cit., p. 51. • **28.** The *Critique of the Gotha Programme*, 1875, stated the principle that, ultimately, Communism would provide for workers based on their need. However, it accepted that, in a transitional period, incentives would still be required and that, therefore, there would still be some relationship in an early-phase Communist society between work and reward. • **29.** Applebaum, Anne, op. cit., p. 53. • **30.** Gorky, Maxim, et al., *The White Sea Canal: Being an Account of the Construction of the New Canal between the White Sea and the Baltic Sea*, London, 1935. The fact that prison workers were used was not hidden – indeed, one of the themes of the book was the reforming power of Soviet labour. • **31.** 'Not everyone worked conscientiously. There were among the prisoners incorrigible idlers who refused to work, and shirked for months. There were malingerers who simulated illness, people who pleaded "rupture" so as to not to go to the diggings. There were recusants who agitated against work. And there were wreckers.' Ibid., p. 168. • **32.** Expectations for the White Sea Canal's economic and strategic importance were impossible to fulfill: 'The

White Sea–Baltic Canal will be the most important artery of the Northern regions. If it can be built it will open up new economic prospects for Karelia, for the Soviet North, for the entire Soviet Union. Grain, salt, oil, metals, machinery, lumber, fish, all sorts of goods will pass through the Canal . . . All the rivers of Soviet Asia flow northward and many are navigable. They open onto this Northern sea route . . . The White Sea–Baltic Canal will be the key to these roads.' Ibid., p. 6. • **33.** Conquest, Robert, *Kolyma: The Arctic Death Camps*, London, 1978, p. 37. • **34.** Applebaum, Anne, op. cit. p. 97. • **35.** The OGPU was incorporated into the NKVD in 1934. • **36.** Taracouzio, T. A., *Soviets in the Arctic: An Historical, Economic and Political Study of the Soviet Advance into the Arctic*, New York, 1938, p. 245. • **37.** For example, Glavsevmorput shipped 271,000 tons of marine cargo in 1936, a vast increase on the levels of the 1920s. But the 1937 quota was 351,800 tons; that for 1942 was 758,000 tons. This was unachieveable without major new investments, yet these were not forthcoming. The agency had long requested six new ice-breaking and ice-forcing ships to add to its total of nine – at a total cost of 99 million rubles. But the ships only began to become available in 1939, two years behind schedule. • **38.** Zenzikov, Vladimir, 'The Soviet Arctic', *Russian Review*, Vol. 3, No. 2, 1944, p. 68 • **39.** Mikhailov, N., *Soviet Geography: The New Industrial and Economic Distributions of the U.S.S.R.* (translated Nathalie Rothenstein), 2nd edition, London 1937 (first edition, 1935), p. V. • **40.** Smolka, H. P., 'Soviet Development of the Arctic: New Industries and Strategical Possibilities', *International Affairs*, Vol. 16, No. 4, July 1937, p. 567. • **41.** Smolka, H. P., *Forty Thousand Against the Arctic: Russia's Polar Empire*, London, 1937, p. 152. • **42.** Smolka, also known as Smollet, later served as the post-war correspondent for *The Times* in Vienna, and provided ideas and experience for Graham Greene's screenplay for *The Third Man*. • **43.** Kitchin, George, *Prisoner of the OGPU*, London, 1935, pp. 267–270, quoted in Applebaum, Anne, op. cit. • **44.** Stefansson, Vilhjalmur, 'The North American Arctic' in Stefansson and Weigert, eds., *Compass of the World: A Symposium on Political Geography*, London, 1946. This was an extended edition of an article originally published in *Foreign Affairs* in 1939. • **45.** In 1991, a decree from President Yeltsin allowed religious institutions the right to reclaim buildings confiscated under the Soviet Union. • **46.** 'Arctic Museum to Make Room for Church', *St Petersburg Times*, 24 October 2000. • **47.** Trotsky, Leon, *The Revolution Betrayed*, trans. Eastman, Max, London, 1937, p. 195. • **48.** Many of the names given to ships and geographical features drew on a history of Russian exploration of the Arctic which predated the Soviet Union, frequently from the perceived glories of Russia's expansive phases of the seventeenth and eighteenth centuries. The name *Cheliuskin* referred to one of the members of the 'Great Northern Expedition' of 1733–45 which mapped, for the first time, a large stretch of the northern Russian coastline from the White Sea to Kolyma. • **49.** In contrast, much of the equipment they used was built outside the Soviet Union. The ice-breaker *Sibiriakov* had previously been a humble Scottish whaling vessel, the *Bellaventura*. The *Cheliuskin* had originally been a Danish transport vessel, while the ice-breakers *Sedov*, *Krasin*, *Malygin* and *Lenin* were built in Britain in 1916–17. Zenzikov, op. cit., p. 69. • **50.** Foreword to Boris Gromov's *Gibel' Arktiki*, Moscow, 1932, pp. 3–4. • **51.** The honour of Hero of the Soviet Union was first given to the pilots who rescued the so-called 'Cheliuskinites' in 1934. • **52.** Including Evgeny Feodorov's unimaginatively titled *On Drifting Ice*. • **53.** In 'Positive Heroes at the Pole: Celebrity Status, Socialist–Realist Ideals and the Soviet Myth of the Arctic, 1932–1939', *Russian Review*, Vol. 56, No. 3, 1997, John McCannon reports that Mikhail Vodopianov, himself a borderline alcoholic with a tendency towards violence, apparently denounced his fellow pilot Mavriki Slepnev as being politically unreliable, on the basis that he had become close to Charles Lindbergh, during the American pilot's trip to Moscow in 1938. • **54.** The supposed reason for Chkalov's arranged death was his public opinion that Nikolai Bukharin and Alexei Rykov were innocent of the charges against them in the third Moscow show trial, in 1938. McCannon argues that human error – Chkalov was known for his reckless flying – or technical failure of his Polikarpov I-180 are more likely explanations. • **55.** McCannon, John, *Red Arctic:*

Polar Exploration and the Myth of the North in the Soviet Union, 1932–1939, Oxford, 1998, p. 177. • **56.** It was probably celebrity which allowed Schmidt to avoid the gulag when he was demoted from his position as head of the Glavsevmorput (the Main Administration of the Northern Sea Route) in 1937–38. Schmidt retained his job as editor-in-chief of the *Bolshaya sovetskaia entsiklopediia* (the Soviet Encyclopedia) until 1941 (a post he had held since 1924) and was made a vice-president of the Academy of Sciences of the USSR. • **57.** Whatever Soviet suspicions there were of Nazi Germany in 1940– and Stalin, it appears, thought until the last minute that a German invasion of the USSR would never happen – the Soviet Union and Germany were nominally allies in 1940, bound by the Molotov–Ribbentrop Pact of 1939. • **58.** Up until its collapse in 1990, the USSR basked in the glory of its military victory in 1945, using its greatest generation as its bedrock of political support and its victory as a moral justification not only for the existence of the Soviet Union, but for the existence of its empire. • **59.** The Great Patriotic War and, above all, the generation that fought it, provided at least one of the fundamental elements of what Ernest Renan had considered vital to create a nation: 'Where national memories are concerned, griefs are of more value than triumphs, for they impose duties, and require a common effort. A nation is therefore a large-scale solidarity, constituted by the feeling of the sacrifices that one has made in the past and of those that one is prepared to make in the future.' *Qu'est-ce qu'une nation?*, Paris, 1882. More than this, from the point of view of the sociability of the elite, military service in the war provided strong links which served to cement loyalties across the national boundaries within the Soviet Union.

PART TWO Power

3 Northern Designs: The Making of the American Arctic

1. The original French lyrics of 'O Canada', the current Canadian national anthem, written by Sir Adolphe Basile Routhier in 1880, did not mention the North at all. 'The True North strong and free!' only appears in the subsequent English version, written in 1908. Francophone Canadians were never as keen on their country's territorial expansion as Anglophone Canadians, sensing – correctly – that it would further diminish their weight in national politics. • **2.** Stephen Harper's focus on the Arctic has allowed him to proclaim a specifically Conservative tradition of Canadian patriotism – rather than the Liberal tradition which has predominated over the last few decades. The symbolism is significant. In August 2008, a few months before the federal general election, Harper travelled to Inuvik to announce the name of Canada's newest ice-breaker, due to enter service in 2017, the *CCGS John G. Diefenbaker*, named after a Conservative Prime Minister of the 1950s and author of the 'Roads to Riches' programme. Diefenbaker himself had named a previous ice-breaker *CCGS John A. Macdonald*, after the first Conservative Prime Minister of Canada and the man credited with expanding Canada west and north in the 1870s. • **3.** As part of the celebrations of the centenary of its purchase, in 1967, Alaska adopted a forward-looking motto of which the true prophets of the North would have been proud: 'North to the Future'. • **4.** Over 2,000 bowhead whales were killed in the peak year of 1852. • **5.** Subsequent historians argued over whether Seward's motivations for the purchase were principally economic, strategic or simply as a means to pay the Russians off for their loyalty to the Union during the Civil War. • **6.** The phrase was coined the following year by Democratic essayist John O'Sullivan. • **7.** Seward, William H., *The Elements of Empire in America*, New York, 1844, p. 7. • **8.** Adams, John, 'A Defence of the Constitutions of Government of the United States of America, against the Attack of M. Turgot, in his Letter to Dr Price, Dated the Twenty-Second Day of March, 1778'. • **9.** Seward argued that one great advantage of the republican form of government was that it provided a template, easily replicated, which could be applied to any new territory.

In other words, the republic would not be diluted by new territories, only strengthened. • **10.** Seward viewed republican institutions – 'the heart's desire of mankind' – as a magnetic force which would make the use of military force superfluous. He reminded his audience that 'Frederick the Great impoverished and almost depopulated his little kingdom by glorious conquests'. But the United States would be different, expanding by 'unnoticed acquisitions'. • **11.** In 1846, Seward reaffirmed his belief in America's northern destiny, writing to Chautauqua, New York, 'Our population is destined to roll its resistless waves to the icy barriers of the north,' in Baker, George E., ed., *The Works of William H. Seward*, New York, 1853, vol. III, p. 409. • **12.** Thomas Roys had become the first American to sail through the strait, aboard his whaling ship the *Superior*, a few years earlier. • **13.** Writing back to the *Albany Evening Journal*, Seward stated that his earlier opinion of the inevitability of the accession of British colonies had been misplaced. Excerpts from these letters were reprinted in an article in the *New York Times* in January 1862 entitled 'Mr. Seward's Views on Canadian Annexation'. By this time, Seward had become President Lincoln's Secretary of State. The purpose of the article was to defuse British criticism that Seward intended the conquest of Canada. Whether or not the 1857 trip had truly changed Seward's mind on the question of inevitability, the desirability of the North remained as strong. In 1860, Seward told an audience in St Paul, Minnesota, a city that would benefit greatly from any northward expansion of the United States, that the Canadians 'are building excellent states to be hereafter admitted to the American Union'. Minnesota adopted the motto 'L'Etoile du Nord' ('The Star of the North') the following year. • **14.** As Seward put it, 'all Southern political stars must set, though many times they rise with diminished splendor. But those which illuminate the Pole remain forever shining, forever increasing in splendor,' 'Mr Seward's Views on Canadian Annexation', *New York Times*, January 1862. • **15.** La Feber, Walter, *The Cambridge History of American Foreign Relations: The American Search for Opportunity, 1865–1913*, Cambridge 1995, p. 16. • **16.** In Nicaragua, the United States negotiated a treaty providing for non-exclusive rights to build a canal. Although this passed the Senate, a second and more ambitious deal, to construct an American-controlled canal through the Colombian province of Panama, failed. In the Caribbean, an initial American offer of $5 million to buy all three of the Danish Virgin Islands had to be raised to $7.5 million – more than the United States would ultimately pay for Alaska – for just two of them. Without Senate approval, the deal fell through. • **17.** The two colonies were united in 1866. • **18.** These were the so-called 'Alabama Claims' made against Britain for alleged violations of neutrality during the Civil War. • **19.** 'Secretary Seward's Speech in St. Louis', the *New York Times*, 12 September 1866. • **20.** For more on American attempts to annex British Columbia in these years see Shi, David E., 'Seward's Attempt to Annex British Columbia, 1865–1869', the *Pacific Historical Review*, Vol. 47, No. 2, May 1978, pp. 217–238. • **21.** Alekseev, A. I., *The Destiny of Russian America, 1741– 1867*, Kingston, 1990 (translated from the Russian) p. 290. • **22.** For discussion on this point see Jensen, R. J., *The Alaska Purchase and Russian–American Relations*, London, 1975. • **23.** In *Our New Alaska, or The Seward Purchase Vindicated*, New York, 1886, Charles Hallock reported the tale that a Russian mining engineer had been sent to the territory in 1855 to assess its mineral wealth, but that he spent most of his time 'in Sitka, at "potlatch" [essentially traditional indigenous feasting] and dancing' (p. 127). • **24.** The poem was published in the *Bulletin* in San Francisco on 8 April 1867. With some irony, Harte, born in New York state, died in England. • **25.** Sumner's Senate speech on the purchase of Alaska lasted several hours, citing Russian, English, French and German sources in his support of Seward's acquisition. • **26.** London, Jack, *Gold Hunters of the North*, 1902. London suggested that the United States had bought Alaska 'for its furs and fisheries, without a thought of its treasures underground'. • **27.** In this, Western Union failed – an alternative undersea Atlantic cable arrived at Heart's Content, Newfoundland, in 1866. • **28.** Quoted in Ernest N. Paolino, *The Foundations of American Empire: William Henry Seward and U.S. Foreign Policy*, London 1973, p. 110. • **29.** Davidson described the site of the Exxon-Valdez disaster, warning that it required 'the greatest circumspection to navigate . . . It diverges into many extensive arms, yet none of them can be considered as commodious harbours, on account of the rocks and shoals that obstruct the approach to them.' Davidson, George, *Pacific Coast: Coast Pilot of Alaska from Southern Boundary to Cook's Inlet*, Washington D.C., 1869, p. 153. • **30.** Ibid., p. 38. • **31.** Dall, William, *Alaska and its Resources*, Boston, 1870, p. 476. Davidson had also mentioned the discovery of petroleum on Kodiak Island a few years previously. The discovery was ascribed to an unknown 'teacher in the Russian–American company'. • **32.** Ibid., p. 480 and p. 505. • **33.** This was specifically pointed out by Macdonald in his famous 'On Confederation' speech to parliament in February 1865. • **34.** The text was clear: 'Therefore, in order to give effect to that Agreement [on the construction of an intercolonial railway], it shall be the Duty of the Government and Parliament of Canada to provide for the Commencement, within Six Months after the Union, of a Railway connecting the River St Lawrence with the City of Halifax in Nova Scotia, and for the Construction thereof without Intermission, and the Completion thereof with all practicable Speed.' British North America Act, 29 March 1867, Part X, section 145. • **35.** Newfoundland only became part of Canada in 1949. • **36.** British Parliamentary Debates, Third Series, Vol. CLXXXVI, London, 1867. • **37.** Franklin published *The Interest of Great Britain considered With Regard to her Colonies, And the Acquisitions of Canada and Guadeloupe. To Which are Added, Observations concerning the Increase of Mankind, Peopling of Countries, &c.*, in 1760. Countering those who claimed that maintaining British possession of Canada would be economic folly, provide American colonies the security to ultimately rebel and perpetuate Britain's marginalisation in the sugar market, Franklin argued that the union of American colonies against Britain was 'impossible'. • **38.** In the late eighteenth and early nineteenth century the North-Western Territory had been exploited both by the Hudson's Bay Company and by its Montreal-based rival, the North West Company. In 1821 the two companies were merged. • **39.** For example, in 1863, Sioux had fled American cavalry across the border into the territory of the Hudson's Bay Company. In response to a request from the US commander, the Company allowed the cavalry to continue their pursuit. At best, this was embarrassing evidence of the Company's inability to control its own border. At worst, it was a travesty of sovereignty – as British authorities had not been consulted. Then, in 1867, Charles Francis Adams, the American minister in London, complained that the Sioux were receiving safe haven within the territories of the Company. Seward requested – and was granted – a right of limited pursuit. • **40.** See Galbraith, John S., *The Hudson's Bay Company as an Imperial Factor, 1821-1869*, Berkeley, 1957. A Bill was introduced to the House of Representatives entitled 'An Act for the Admission of the States of Nova Scotia, New Brunswick, Canada East, Canada West and for the organisation of the territories of Selkirk, Saskatchewan and Columbia'. Seward caused the Bill to be withdrawn. • **41.** *Address to Her Majesty the Queen from The Senate and House of Commons of the Dominion of Canada*, 17 December 1867. • **42.** The British government warned the Company that it would have to accept a lower offer or face drawn-out legal uncertainty over its Charter status. • **43.** In the meantime, the government had been forced to put down a rebellion by Louis Riel, a Francophone Métis. Riel would come back to haunt Macdonald and the Conservatives. In 1885, following a second rebellion, Riel was sentenced to death. Macdonald refused to commute the sentence. The result was a collapse of support for the Conservative party in Francophone Quebec which would last for generations. • **44.** Seward's speech in Victoria, British Columbia, was included as a part of Seward, William H., *Alaska Speech of William H. Seward at Sitka*, Washington, 1869. • **45.** Musgrave quoted in Howay, F. W., 'British Columbia's Entry into Confederation', *The Canadian Historical Association: Report of Annual Meeting Held in the City of Toronto*, Ottawa, 1927. • **46.** Carnarvon initially suggested resolving the problem by making an expansive statement of Canada's boundaries, but finding an appropriate wording which would neither claim more than Britain could justifiably pretend to own, nor undermine possible future British claims, was difficult: 'To the East, the British Territories might perhaps be defined to be bounded by the

Atlantic Ocean, Davis Straits, Baffin Bay, Smith Sound and Kennedy Channel. But even this definition wld' exclude the extreme North West of Greenland, which is marked in some maps as British territory, from having been discovered probably by British subjects. To the North, to use the words of the Hudson's Bay Co. in 1750, the boundaries might perhaps be, "the utmost limits of the lands towards the North Pole".' Quoted in Smith, Gordon, W., 'The Transfer of Arctic Territories from Great Britain to Canada in 1880, and Some Related Matters, as Seen in Official Correspondence', *Arctic*, Vol. 14, No. 1, 1961, p. 56. • **47.** 'Another Arctic Expedition', the *New York Times*, 31 May 1876. • **48.** 'From the Arctic Regions', the *New York Times*, 12 October 1876. • **49.** Macdonald, John A., *Debates of the House of Commons of the Dominion of Canada: fifth session, third parliament*, Ottawa, 1878, vol. 5, p. 2390. • **50.** The occasion was the opening of parliament, in Stephen Harper's speech in reply to the speech of the Governor-General. • **51.** Smith, Gordon W., *Transfer of Arctic Territories*, p. 60. • **52.** Imperial Order in Council, 31 July 1880. • **53.** A plan for a line had been set out in 1872 by a consortium headed by Sir Hugh Allan. However, allegations of bribery of Sir John A. Macdonald brought the plan – and Macdonald's government – crashing down in 1873. • **54.** Macdonald, John A., *Debates of the House of Commons of the Dominion of Canada: third session, fourth parliament*, Ottawa, 1881, vol. 10, p. 488. • **54.** Macdonald's biographer put it succinctly: 'He [Macdonald] had begun as a politician, he was to end as an institution.' Lower, Arthur R. M., *Colony to Nation: A History of Canada*, Toronto, 1946, p. 365. • **56.** Both Iceland and Greenland were possessions of the Kingdom of Denmark (though the precise extent of Danish rule over Greenland was not yet clear). • **57.** Benjamin Mills Peirce came from a strong scientific and political family, closely linked to the Untied States Coast Survey. His father was the superintendent of the Survey from 1867 to 1874. His brother, Charles Saunders Peirce, was an employee of the Survey from 1859 to 1861. His grandfather on his father's side was the librarian at Harvard University. His grandfather on his mother's side was the United States Senator from Massachusetts from 1820 to 1827, Elijah Hunt Mills. • **58.** Peirce noted that Icelandic sulphur was cheaper than potential European competitors and that therefore, 'with improved means of transportation it would control the market'. • **59.** Peirce viewed the cod fishery as particularly valuable (though described as being in the hands of French fishermen). As for the grassland of Iceland: 'With proper care, draining, and so forth, much of the land now covered with heath (nearly half of Iceland) could be made fertile enough for capital grazing land . . . some of the more sanguine of the agriculturists believe that grain even could be grown in Iceland were the soil, naturally excellent, properly prepared by draining and ploughing.' Peirce estimated the population of sheep on Iceland as 600,000, some ten times the human population, and pointed out that horses which would cost $150 or $200 in Boston or New York fetched just $10 in Reykjavik. • **60.** Stefansson, Vilhjalmur, *Iceland: The First American Republic*, New York, 1939. • **61.** Peirce's report was quoted by Robert E. Peary in 1916, in his criticism of American plans to give up claims to Greenland as part of the purchase price for the Danish Virgin Islands (Peary, Robert E., 'Greenland as an American Naval Base', the *New York Times*, 11 September 1916). After the Second World War, the US reportedly offered Denmark $100 million for Greenland ('Deepfreeze Defense', *Time*, 27 January 1947). • **62.** This and subsequent quotations come from Seward, William H., *Alaska Speech of William H. Seward at Sitka*, Washington, 1869.

4 Scramble: Dividing the Arctic

1. 'Notifies the New York Times that he reached it on April 6, 1909', the *New York Times*, 7 September, 1909. Peary actually sent several telegraphs. The one quoted here was sent to the Associated Press. • **2.** The media had a field day, not least when it emerged that some of the footage which Reuters had circulated of the *Arktika* expedition turned out to be Russian television footage, itself cut from a major Hollywood film, *Titanic*.

Holmwood, Leigh, 'Reuters Gets That Sinking Feeling', the *Guardian*, 10 August 2007. • **3.** Galloway, Gloria and Freeman, Alan, 'Ottawa Assails Russia's Arctic Ambition', the *Globe and Mail*, 3 August 2007. • **4.** 'Taft has Faith in Peary', the *New York Times*, 9 September 1909. • **5.** See Smith, Mark, A. and Giles, Keir, *Russia and the Arctic: The 'Last Dash North'*, Defence Academy of the United Kingdom, 2007. • **6.** 'Russia's Message from the North Pole: We're a Force to be Reckoned With', *International Herald Tribune*, 7 August 2007. • **7.** In 1878–79 Swedish explorer Adolf Erik Nordenskjöld completed the first voyage through the Northeast Passage in an expedition financed by Russian businessman Alexander Sibiryakov. In 2007, it would be a Swede financing the Russians, to the tune of several million dollars. • **8.** McDowell and Paulsen managed to keep their places on the expedition. McLaren did not. • **9.** Strictly speaking no nation is making a 'claim' for the seabed, as the territory is granted under UNCLOS. The only question is defining the geographic extent of ownership, not determining whether ownership exists or not. • **10.** The United Nations Convention on the Law of the Sea was opened for signature and ratification on 10 December 1982, with 119 countries signing the Convention on the first day. Since then a total of 157 countries have ratified UNCLOS. In 1994, with the ratification of the sixtieth state – Bosnia-Herzegovina – the Convention legally came into force. All the Arctic coastal states bar the United States have ratified UNCLOS: Iceland (1985), Norway (1996), Russia (1997), Canada (2003) and Denmark (2004). • **11.** In 2008, the International Boundaries Research Unit of Durham University, in the United Kingdom, produced a map which marked out the areas under discussion, the possible areas of overlap, and areas where special agreements have already been made. The map is available at www.dur.ac.uk/ibru/resources/arctic/. • **12.** In May 2008 the coastal states met at the Greenlandic settlement of Ilulissat. The declaration that they issued made clear, as one Danish diplomat put it to me, that 'an actual treaty for the Arctic is not necessary since we already have a well functioning legal regime in place'. • **13.** Hans Island, between Greenland and the Canadian Arctic Archipelago, is a case in point. Although there have been exchanges of diplomatic notes on the island, the dispute over the island has not prevented Canada and Denmark from agreeing on the border between their respective continental shelves, nor has it prevented the two countries from cooperating on a project to ascertain the outward northern extent of their continental shelves. • **14.** In the late nineteenth and early twentieth centuries, 'unclaimed' essentially meant unclaimed by those of European descent representing modern nation states. • **15.** The term 'scramble for Africa' was coined during the 1884 Berlin Conference, where European states decided how they were going to divide the spoils. Between 1876 and 1912 nearly the entire African continent was partitioned between just six European states: Belgium, France, Germany, Italy, Portugal and the United Kingdom. See Pakenham, Thomas, *The Scramble for Africa*, 1876–1912, London, 1991. • **16.** The instrument of this was the Indian Citizenship Act, granting citizenship to all Native Americans born in the United States. • **17.** See, for example, Vaughan, Richard, *The Arctic: A History*, London, 1994, pp. 267–295. • **18.** Native claims became heated points of law and politics across the Arctic in the 1970s and 1980s. The Alaska Native Claims Act was signed in 1971, as the pre-condition for the construction of the Trans-Alaska Pipeline. In Canada, the Nunavut Territory was established in April 1999. Greenland, meanwhile, acquired Home Rule in 1979 and enhanced Self-Rule in 2009. In many cases, however, the basis on which claims have been 'settled' has subsequently been criticised. • **19.** See Timtchenko, Leonid, 'The Russian Arctic Sectoral Concept: Past And Present', *Arctic*, Vol. 50, No. 1, March 1997, pp. 29–35. • **20.** The Treaty of Tordesillas was itself based on a papal bull issued the previous year by Pope Alexander VI, one of the most corrupt of all popes. His original proposal had been extremely favourable to Spain. The final treaty moved the demarcation line several degrees of longitude to the west, thus advantaging Portugal. By any standards, the Treaty of Tordesillas was a statement of breathtaking ambition, and, of course, it took absolutely no account whatsoever of native populations. • **21.** Quoted in Timtchenko, *The Russian Arctic*, p. 30. • **22.** Lakhtine referred to the area over which a coastal state could exercise sovereignty as the 'region of attraction'. Lakhtine, W., 'Rights over the Arctic', the *American Journal of International Law*, Vol. 24, No. 4., October 1930, pp. 703–717. The Russian version of Lakhtine's article had been published in 1928, exciting discussion and debate in the Soviet Union. • **23.** The Soviet Union was not about to just rely on the 'sector theory' for its sovereignty. From the late 1920s on, the Soviet Union pursued a number of projects in the Arctic with the aim of asserting effective ownership. • **24.** This was particularly relevant because, if any land were left to be discovered in the Arctic region, it was most likely to lie north of the Soviet Union. • **25.** This point had not been made explicitly in the 1926 declaration. Two years later, when Lakhtine wrote the Russian version of 'Rights over the Arctic' it had become an issue of some moment, as it became increasingly obvious that airship travel in the Arctic was technically possible. • **26.** In the Antarctic, though the origins of most territorial claims were initially by right of discovery, they subsequently took on the character of a southern version of the 'sector theory' in terms of their extent over the continent of Antarctica. For example, the claims of Argentina, Australia, Britain, Chile and France all correspond to adjacent sovereign territory. The United States never made an Antarctic claim. The Soviet Union, meanwhile, rejected the notion that the Antarctic could be divided by the 'sector theory' – in which case it would clearly own none of Antarctica – which necessarily undermined its advocacy of the 'sector theory' in the Arctic. (Of course, by that time Soviet ownership of its Arctic territories was unquestioned so, in a sense, the 'sector theory' could be laid to rest.) After several models of international cooperation over the Antarctic were rejected in the late 1940s, the Antarctic Treaty was finally signed in 1959. This treaty did not explicitly reject existing Antarctic claims, it simply put them into abeyance for the length of application of the treaty, and it stated that no action undertaken during the period of application of the treaty could be used to assert, support or deny claims of sovereignty thereafter. See Rothwell, Donald R., *The Polar Regions and the Development of International Law*, Cambridge, 1996. • **27.** Since the Dutch jurist Hugo Grotius had formulated the notion of a 'Mare Liberum' in the early seventeenth century, the idea had become, bar a few dissenting voices, a fixed element of customary international law. (Grotius, Hugo, *The Free Sea* (translated from the Latin by Richard Hakluyt), Indianapolis, 2004 (work originally published as Mare Liberum in 1609)). The English jurist John Selden had taken issue with Grotius's view, writing a treatise entitled *Mare Clausum*. However, maritime states in particular – England, Portugal, Spain – were keen to protect the principle of open seas. Over time, state practice turned the notion of the 'freedom of the high seas' into a generally accepted principle of international law. • **28.** Balch, Thomas Willing, 'The Arctic and Antarctic Regions and the Law of Nations', the *American Journal of International Law*, Vol. 4, No. 2, April 1910, pp. 265–275. • **29.** Were ice to be assimilated to land, a number of significant issues would arise from the fact that the boundaries of a territory might shift as ice melted and reformed, or as icebergs calved into the sea. For example, '. . . it has been said that ships bound for the Franz Josef Archipelago frequently encounter an ice barrier as much as two hundred miles away from the islands, but that between this barrier and the land – sometimes even during winter – there is a stretch of open water often from thirty to forty miles wide. Would a Norwegian war vessel be violating the sovereignty of the USSR if it were cruising in that open water at a distance of more than twelve miles from the shore? . . . Would a Japanese cruiser be considered to be in Soviet territory if it were found blocked in a field of ice of "considerable" size, "more or less" drifting in the East Siberian Sea forty miles from the Siberian mainland, but separated from the coastal ice by a "polyn'ia" fifteen miles wide?' Taracouzio, T. A., *Soviets in the Arctic: An Historical, Economic and Political Study of the Soviet Advance into the Arctic*, New York, 1938, p. 358. • **30.** The dispute had been settled by arbitration by a panel of six judges, three from the United States, two from Canada and the British Lord Chief Justice, Lord Alverstone. The British judge ruled in favour of the US position on most points, leading to claims that imperial interests had been put before Canadian interests. The Canadian Prime Minister, Wilfrid Laurier, told the House of Commons that 'we are living beside a great neighbour who, I believe I can say without being deemed unfriendly to them, are very grasping in their national acts, and who are determined upon every occasion to get the best in every agreement which they make'. Bothwell, Robert, *The Penguin History of Canada*, Toronto, 2006, p. 272. • **31.** King, W. F., *Report upon the Title of Canada to the Islands North of the Mainland of Canada*, Ottawa, 1905 (Government Printing Bureau). • **32.** In 1895 Canada issued a public Order-in-Council defining its Arctic borders and claiming sovereignty over the entire Arctic archipelago west of the strait which separated the archipelago from Greenland, but only up to 83¼° North. Subsequent expeditions led by Low and Bernier deposited further claims and were intended to demonstrate Canadian sovereignty. Bernier extended the Canadian claim to 90° North. Otto Sverdrup's travels through the Far North between 1898 and 1902 alerted some Canadians to potential challenges to their sovereignty. Sverdrup's discoveries were later used unsuccessfully by Norway as the basis for an official claim over what Norway called the 'Sverdrup Islands' (Axel Heiberg Island, and the Ringnes Islands). The United States made claims to parts of Ellesmere Island. For a full account, see Johnston, V. Kenneth, 'Canada's Title to the Arctic Islands', *Canadian Historical Review*, Vol. XIV, 1933, pp. 24– 41. • **33.** In 1871, Charles Francis Hall's *Polaris* expedition, largely funded by the United States government, had reached up to 82°11′ North. Wintering on the north coast of Greenland, at a place then called Thank God Harbor, Hall fell ill and died. • **34.** *Convention Between the United States and Denmark: Cession of the Danish West Indies*, United States Treaty Series No. 269, 39 Stat. 1706. • **35.** Johnston, V. Kenneth, op. cit., pp. 36–37. • **36.** Of course, the first settlers of Greenland were not in fact the Norse at all but the Dorset culture Inuit who had arrived on the island around 800 BC. • **37.** Subsequent treaties between Sweden (to which Norway was joined from 1814 to 1905) and Denmark implicitly accepted Danish sovereignty over Greenland. • **38.** Established in 1776, the so-called 'prohibited area', over which Denmark exercised control over who could enter, initially covered only a quite small portion of Greenland: the west coast from 60° North to 73° North. This was extended over time, but, even so, it did not cover all of Greenland until 1921. • **39.** After all, some argued, if Denmark truly believed the 1814 Treaty of Kiel clearly entitled it to the whole of Greenland, why had it spent diplomatic capital trying to extract declarations to that effect from Britain, France, Italy, Japan and the United States? Norway rejected Denmark's claim that Norwegian Foreign Minister Nils Ihlen had verbally accepted the extension of Danish sovereignty in 1919. See Preuss, Lawrence, 'The Dispute between Denmark and Norway over the Sovereignty of East Greenland', the *American Journal of International Law*, Vol. 26, No. 3, July 1932, pp. 469–487. • **40.** The legal argument came in the form of the Soviet Union's 1926 declaration of ownership over all land (discovered and undiscovered) lying in a sector starting from the Soviet Union's land borders and running to the North Pole. The argument of effective authority, which would perhaps be more valuable, was that the American and Canadian colonists placed on the island by Stefansson's expedition, were removed from it by sailors from the Soviet ice-breaker *Krasny Oktyabr* (*Red October*) in the summer of 1924. • **41.** The name 'Spitsbergen' is often used interchangeably with Svalbard, though strictly it refers to just the main island of the archipelago. • **42.** Vaughan, Richard, op. cit., p. 267. • **43.** The treaty was signed on 9 February 1920 by Britain, Denmark, France, Italy, Japan, the Netherlands, Norway, Sweden and the United States. • **44.** Treaty Concerning the Archipelago of Spitsbergen, and Protocol, Paris, 9 February 1920. Article I recognises Norwegian sovereignty. Article II protects the rights of citizens of the contracting parties to hunt and fish on the archipelago 'and in their territorial waters', while also granting Norway the right to enact measures of environmental protection, as long as they apply 'equally to the nationals of all the High Contracting Parties without any exemption, privilege or favour whatsoever, direct or indirect to the advantage of any one of them'. Article III guarantees rights of access to citizens of the contracting parties, subject to local laws, including the 'practice of all maritime, industrial, mining or commercial enterprises both on land and in the territorial waters', with limitations on the restrictions to trade which Norway is allowed to implement. Article IV guarantees equal access to any wireless stations which Norway may set up on the archipelago. Article V recommended establishment of an international meteorological station. Article VI protected the acquired rights of nationals of the contracting parties. Article VII granted equality in terms of the maintenance and acquisition of land rights (including mineral rights). Article VIII set the conditions under which mining regulations would be set up and the tax arrangements under which mining would operate. Article IX disallowed the construction of any naval bases and 'any fortification in the said territories, which may never be used for warlike purposes'. Finally, in Article X, the parties accepted Russian rights, in the absence of a recognised Russian government. • **45.** Governor of Svalbard, *Tourism Statistics for Svalbard*, 2006. • **46.** For example, under the Svalbard Environmental Protection Act, 'no person may possess or initiate anything that may entail a risk of pollution unless this is lawful pursuant to this Act'. • **47.** In 2007, Moscow set up a Svalbard commission under the chairmanship of Deputy Prime Minister Sergei Naryshkin, who visited the archipelago in October 2007 at the head of a forty-strong delegation. Since then, development of Svalbard has remained an issue of interest in Russian Arctic planning. Proposals for an airfield at Barentsburg caused particular concern in Oslo. • **48.** One Norwegian government official I spoke to called Britain's actions 'pure ego interests' and warned 'if the UK crosses us on this one, why should we land [future natural] gas in the UK?'. • **49.** First, the existing customary law of the sea was codified in a series of conventions, known as the Geneva Conventions on the Law of the Sea, in 1958. Then, in 1970, a process was begun to negotiate a comprehensive agreement on the law of the sea, which would also cover the question of rights to the potential economic resources of the seabed. This was not just a theoretical exercise: twice, in the 1970s, Britain and Iceland faced off over access to fishing grounds off the coast of Iceland. • **50.** That provides a strong incentive for any remaining hold-outs to sign and ratify the convention so that they can then better influence its proceedings. • **51.** I have used the words 'mile' and 'nautical mile' interchangeably in this section. All measurements under UNCLOS are measured in 'nautical miles', which are about 1.15 standard miles and about 1.85 kilometres. • **52.** This was an issue of particular importance to the United States, which feared that the encroachment of territorial seas on narrow but economically important stretches of water – such as the Straits of Hormuz or the Malacca Straits – might ultimately give the coastal states too much power in the international system. The fact that UNCLOS provided for legal certainty over this issue is one reason US defence community tends to favour American ratification of UNCLOS, though there are some who still view the convention as an unnecessary and dangerous infringement of state sovereignty. • **53.** Article 82 provides for payments from the exploitation of non-living resources, to be made to the International Seabed Authority and then distributed, amounting to 1 per cent in the fifth year of production and then rising in one-point increments to a total of 7 per cent in the twelfth. Developing countries are exempt. • **54.** The prospects here are 'manganese nodules' or 'polymetallic nodules', which tend to lie at considerable depths, and offer resources of several metals, including nickel. In the 1960s and 1970s, as UNCLOS was being negotiated, several commercial organisations were set up to engage in deep-sea mining. Production proved uneconomic in the 1980s and 1990s. But if production costs fall and the price of metals rise then, over time, it may become lucrative. • **55.** The United States and the Soviet Union signed an agreement on their maritime border in the Bering Strait in 1990. Some in Russia thought that the deal struck by the Soviet Union did not sufficiently protect Russian interests. At the time of writing the Russian Duma had not ratified the agreement. • **56.** Canada claims that the maritime border was set by the 1825 Anglo-Russian treaty, and therefore lies along a line of longitude of 141° West. The United States contends the maritime border has not yet been delimited, and that the 'median' line should be used, which would give the US a greater share of the Beaufort Sea.

• **57.** Norway claims a 'median' line for the maritime border. Russia, on the other hand, claims a sector line – as claimed in the 1926 declaration of the Central Executive Committee of the USSR – which runs due north from the land border (deviating slightly around the area covered by the Spitsbergen Treaty). • **58.** In 1957, Norway and the Soviet Union reached agreement on the maritime border in the Varangerfjord area, but this extends only a few miles out to sea from the coast. In 1978 the two countries agreed on a fishing regime to cover the so-called 'Grey Zone' which straddles the maritime claims of both states. • **59.** At various times diplomats and analysts have suggested that a final settlement of the border is just around the corner, and that most elements of a final settlement have already been decided. However, it is likely that there will be no final settlement of the Barents Sea dispute until there is also agreement between Russia and Norway over the Svalbard area. There is a strong incentive for the two issues to be linked. • **60.** The best overall in-depth account of the history of the law behind the Northwest Passage is Pharand, Donat, *Canada's Arctic Waters in International Law*, Cambridge, 1988. • **61.** In order to prevent any international legal challenge to the Arctic Waters Pollution Prevention Act (AWPPA), Canada notified the International Court of Justice that it no longer recognised ICJ jurisdiction in this particular area. With the signing of UNCLOS in 1982, and particularly Article 234, which deals with management of ice-covered areas, Canada succeeded in eliminating doubts over the legality under international law of the AWPPA. • **62.** United States Department of State statement, 15 April 1970, reproduced in International Legal Materials 1970, 605. • **63.** *Information Memorandun for Mr Kissinger, The White House. Subject: Imminent Canadian Legislation on the Arctic*, 12 March 1970. E.O. 12958. • **64.** The United States had informed Canada of the planned voyage of the *USCG Polar Sea* but had neither sought nor received official permission for the transit. • **65.** Canada was entitled to redraw its baselines under the provisions of UNCLOS. But there are plenty of open legal questions around what 'historic rights' to passage other states may be able to claim. • **66.** *National Security Presidential Directive 66*, January 2009. • **67.** It is, in fact, more complicated than that. Measurement is made from the 'foot of the slope', which is the slope from the edge of the continental shelf down to something called the 'continental rise'. All of these terms are open to some level of debate and discussion, though all are defined by the manuals prepared by the Commission on the Limits of the Continental Shelf. • **68.** Where there is a 'submarine ridge', the constraint line of 350 nautical miles must be applied. Where there is a 'submarine plateau', however, the state may decide which rule to apply. • **69.** Extended to 2009 for countries which ratified UNCLOS before 1999. • **70.** This claim was submitted in 2004. The CLCS ruled on the submission in 2008. The number 14.2 million square kilometres includes the area of seabed underneath the 200-mile EEZ, and the extended continental shelf. It also includes the EEZ and seabed which Australia attaches to its dependencies, including its claims in Antarctica (though these are held in abeyance by the Antarctic Treaty) which are substantial. See *Continental Shelf Submission of Australia*, Canberra, 2004. • **71.** Presentation at the International Geological Congress, 6–14 August, 2008. • **72.** United States, *Presidential Proclamation No. 2667*, 28 September 1945. • **73.** Presentation at the International Geological Congress, 6–14 August 2008. • **74.** Division for Ocean Affairs and the Law of the Sea, United Nations, *Training Manual for Delineation of the Outer Limits of the Continental Shelf Beyond 200 Nautical Miles and for Preparation of Submissions to the Commission on the Limits of the Continental Shelf*, New York, 2006. • **75.** R. Macnab, quoted in Potts, Tavis and Schofield, Clive, 'Current Legal Developments in the Arctic', the *International Journal of Marine and Coastal Law*, No. 23, 2008, p. 163. • **76.** In February 2002, John D. Negroponte, then the United States Ambassador to the United Nations, wrote to the Secretary-General telling him that 'the United States believes that the [Russian] claim has major flaws as it relates to the continental shelf claim in the Arctic'. He asked that the United States' position be circulated to all members of the United Nations. The United States disagreed that the Alpha-Mendeleev Ridge could be taken as a prolongation of the land mass of

Russia. Instead, the US submission argued that it was the result of volcanic activity rising from the sea floor 120 to 130 million years ago, arguing that its rough edges and its magnetic properties tended to bear out that idea. The US argued that the Lomonosov Ridge was not a 'natural component of the continental margins of either Russia or any other State'. Norway stated that it was acceptable for the Commission to consider the continental shelf in the 'area in dispute' (i.e. areas between Russian and Norwegian ideas of borders in the Barents Sea). Denmark reserved its position, pointing out the possibility of an overlapping claim with Denmark/Greenland. Canada did much the same. • **77.** The first chairman elected to the CLCS was a Russian, Yuri Borisovitch Kazmin. Whatever might have motivated such an early submission by Russia to the CLCS, several years before their deadline, there is no evidence to suggest any corruption or undue influence whatsoever. The subcommission which actually looked at the Russian claim did not have a Russian member and was chaired by a Mexican expert, Mr Galo Carrera Hurtado. • **78.** In 2006 an English artist, Alex Hartley, wrote to the Norwegian government claiming Nymark, an island off Svalbard recently revealed by a retreating glacier and previously unknown. Kilner, James, 'Englishman Claims Sovereignty over Norwegian Island', *Reuters*, 11 May 2006. • **79.** Chircop, Aldo, *Laws & Conventions: Questions for the Regulation of International Shipping Through the Arctic, Impacts of Climate Change on the Maritime Industry*, World Maritime University, Malmö, June 2008. • **80.** This applies, however, only within the Exclusive Economic Zone. Canada was particularly keen on this article in UNCLOS, as a retrospective acceptance of their own, unilateral, Arctic Waters Pollution Prevention Act.

5 Parade Ground: War and Peace in the North

1. The treaty in question is the Treaty on Conventional Armed Forces in Europe (CFE) suspended on the Russian side in July 2007. Who is to blame for the chilling of the strategic relationship between Russia and NATO depends on one's perspective, and not all NATO members share the same assessment. In general, NATO members view the invasion of Georgia (a NATO aspirant), the use of energy as a geopolitical weapon (though Russia debates this), the proposal of an alternative Euro-Asian security structure and the reconstruction of Russia's armed forces as being to blame for the chilling of Russia's relationship with the West. Some states view Russia's moves as largely symbolic; others are acutely alarmed by them. From the Russian perspective, NATO's expansion into Eastern Europe (particularly the Baltic States) represents an attempt at strategic encirclement and is in direct contravention of oral assurances given in 1990 that the United States did not necessarily seek NATO's eastward expansion. The establishment of missile defence systems in Eastern Europe is viewed as an attempt to negate the effectiveness of Russia's nuclear deterrent. In the Russian narrative, therefore, it is Western expansionism, riding roughshod over the geopolitical rights of Russia in her own 'sphere of influence' which is the main problem. • **2.** For a discussion of the issues regarding any possible NATO expansion to Finland, see Salonius-Pasternak, Charly, *From Protecting Some to Securing Many: NATO's Journey from a Military Alliance to a Security Manager*, Finnish Institute of International Affairs, 2007. • **3.** The Treaty of Friendship, Cooperation and Assistance required Finland to repulse any encroachment on her sovereignty (with Soviet assistance if necessary). It was a cornerstone of Finnish security policy up until 1990, when the treaty lapsed. In effect, the Soviet Union neutralised the threat of Finland being used as a base to invade and, at the same time, secured a major point of leverage over Finnish foreign policy. • **4.** The comments were made in the context of an event hosted by the Center for Strategic and International Studies (CSIS), on 6 September 2007. • **5.** The report was prepared by former Norwegian Foreign Minister Thorvald Stoltenberg. • **6.** Although Norway is not a member of the European Union, the country has signed up to the Schengen agreement, abolishing permanent border controls with all the countries of the European Union and the European Economic Area, with the exception of the Republic of Ireland and the United Kingdom. • **7.** Mackinder, Halford, 'The Geographical Pivot of History', *Geographical Journal*, Vol. 23, 1904, pp. 421–437. Mackinder was already a well-known geographer and one of the founders, in 1895, of the London School of Economics. But, in posterity, it was 'The Geographical Pivot of History' which gave him the mantle of geopolitical theorist. • **8.** In 'The Geographical Pivot of History' Mackinder noted that 'she [Russia] can strike on all sides and be struck on all sides, save the north'. Forty years later his position remained broadly the same. While 'the Arctic shore is no longer inaccessible in the absolute sense that held until a few years ago', invasion through northern Siberia was still 'impossible'. ('The Round World and the Winning of the Peace', *Foreign Affairs*, July 1943, pp. 595–605.) At the same time, he reinforced his view of the North's economic importance, presenting 'Lenaland' – essentially Siberia – as a mine of 'rich natural resources . . . as yet practically untouched' and source of Russia's future power. • **9.** Quoted in Bartlett, Robert A., 'Greenland from 1898 to Now', *National Geographic*, Vol. LXXVIII, No. 1, July 1940, pp. 111–140. • **10.** See Fogelson, Nancy, 'Greenland: Strategic Base on a Northern Defense Line', the *Journal of Military History*, Vol. 53, No. 1, January 1989, pp. 51–63. The *USS Shenandoah*, named after the confederate ship which, in the Arctic, fired the last shots of the Civil War, was destroyed in an accident in 1925. Mitchell criticised his superiors, and was court-martialled as a result. • **11.** Hurley, Alfred F., *Billy Mitchell: Crusader for Air Power*, Bloomington, 1975, p. 118. • **12.** See, for example, Mikhailov, N., *Soviet Geography: The New Industrial and Economic Distributions of the U.S.S.R.* (trans. Rothenstein, N.), 2nd edition, London, 1937 (first edition, 1935). • **13.** Arnold, H. H., 'Our Air Frontier in Alaska', the *National Geographic*, Vol. LXXVIII, No. 4., October 1940, pp. 487–504. Arnold had been threatened with court martial in the 1920s for his support of Mitchell in the *Shenandoah* affair. • **14.** British troops were first to set foot in Iceland, landing there in May 1940. They were subsequently replaced by American forces. • **15.** In some cases, German and Allied weather stations operated from the same territory. On Spitsbergen, a fixed weather station at Barentsburg provided information to the Allies from 1942 onwards, while the German army set up weather stations at different places around the archipelago in different years, trying to keep as far away as possible from Allied forces, so as to avoid detection and destruction. The eleven crew members of the last German weather station, *Haudegen*, at the far north-eastern end of the Spitsbergen archipelago, were not evacuated until September 1945, several months after the fall of Berlin had ended the war in Europe. • **16.** The Soviet border officially moved as a result of the Paris peace treaty of 1946, by which Finland confirmed the cession of the Petsamo district to the Soviet Union. • **17.** An earlier aim, to establish a Communist government in Helsinki, was abandoned. • **18.** 'Red Army Reaches Norway Border', the *New York Times*, 23 October 1944. • **19.** There were some who initially thought the Soviet Union would be forced to concentrate on domestic reconstruction rather than territorial expansion, and there were Americans keen for US disengagement from Europe. Two things prevented that from happening. First, the Soviet Union failed to honour the promise of free elections in Eastern Europe. Second, American officials won the argument that their own security depended on forward engagement. In March 1946, former British Prime Minister Winston Churchill speaking at Fulton, Missouri, warned his audience, which included President Harry Truman, that 'an iron curtain has descended across the [European] continent', acknowledging an emerging geopolitical reality. • **20.** Pearson, Lester B., 'Canada Looks "Down North"', *Foreign Affairs*, Vol. 24, No. 4, July 1946. • **21.** The Maginot Line was a reference to the line of defensive installations on the French border with Germany, Luxembourg and Belgium. This line proved unequal to the decision of the German army to invade France through neutral Belgium. • **22.** Opposition was strongest in Iceland, leading to riots in Rekjavik. Opposition to the basing of American forces provided the background to Halldór Laxnes's novel, *The Atom Station*. There were discussions in 1948–49 of a possible Nordic alternative to NATO, but this collapsed when Norway pulled out, warning that, without an implicit or explicit American guarantee a purely Nordic option would not provide sufficient security against the Soviet Union. • **23.** For a discussion of Icelandic politics and defence, see Ingimundarson, Valur, 'Buttressing the North: The Atlantic Alliance, Economic Warfare, and the Soviet Challenge in Iceland, 1956–59', the *International History Review*, Vol. XXI, No. 1, March 1999, pp. 80–103. In a reflection of Iceland's continuing strategic importance, and the difficulties with its government, the National Security Council of the Eisenhower administration produced no fewer than three papers on Iceland, more than for most other countries, including major US allies. • **24.** Indeed, all the territory above the Arctic Circle was either Soviet or NATO-held, with the exceptions of Spitsbergen, Arctic Sweden (which retained a policy of Westward leaning neutrality) and Arctic Finland (which was in essence neutered by its peace treaty with the Soviet Union). • **25.** *NSC 68: United States Objectives and Programs for National Security*, 14 April 1950. • **26.** Nesbitt, Paul H., 'A Brief History of the Arctic, Desert and Tropic Information Center and its Arctic Research Activities', in Friis, Herman R. and Bale, Shelby G., *United States Polar Exploration*, Athens (Ohio), 1970, pp. 135–145. ADTIC was actually temporarily shut down in 1945, only to be reopened in 1947. • **27.** *Arctic Manual, Prepared Under Direction of the Chief of the Air Corps*, United States Army Air Corps, Washington, 1940. • **28.** Office of Naval Research, Navy Department, *Across the Top of the World: A Discussion of the Arctic*, Shelesnyak, M.C. and Stefansson, V., August 1947. • **29.** CRREL was set up in 1961, in Dartmouth, New Hampshire, as the result of a merger between the Snow, Ice, and Permafrost Research Establishment (SIPRE) and ACFEL (which itself combined the Frost Effects Laboratory, established in Boston in 1944, and the Permafrost Division of the St Paul (Minnesota) District, Corps of Engineers). Northway Fields was the field station of the Permafrost Division. Two booklets on the CRREL and its projects have been produced: Edmund A. Wright, *CRREL's First 25 Years*, June 1986 and a follow-up report *Retrospective*, 1990. • **30.** Three previous expeditions – *Eskimo*, *Polar Bear* and *Lemming* – had been smaller. • **31.** Zimmerman, John S., 'Arctic Airborne Operations', *Military Review*, Vol. XXXII, No. 5, August 1952, United States Command and General Staff College, Fort Leavenworth, Kansas, p. 26. • **32.** The first US fusion bomb was tested in November 1952; the first Soviet device was tested in November 1955. • **33.** See, for example, 'A Voyage of Importance', *Time*, 18 August 1958. The tone of media coverage of both American and Soviet exploits was one of breathless enthusiasm. 'It was Aug. 3, 1958. Time: 11:15 p.m. E.D.T. That day in Peking the Kremlin's Khrushchev had wound up four days of secret conferences with Red China's Mao. In Washington US officials were again on tenterhooks about a parley at the summit. In the quivering Middle East more US ground troops were pouring ashore. But there beneath the peaceful, sunlit icecap, the 116 US Navymen were making more pages for the history books than anybody else. They were setting a new sea tradition for their countrymen, to rate alongside Jones, Farragut, Peary, Byrd. The submarine was blunt-bowed Nautilus, world's first nuclear-powered ship. Nautilus' position: under the ice at the North Pole.' • **34.** Wilkins wrote a book about his 1931 attempt, *Under the North Pole: The Wilkins-Ellsworth Submarine Expedition*, London, 1931. • **35.** McLaren, Alfred S., *Unknown Waters: A First-Hand Account of the Historic Under-Ice Survey of the Siberian Continental Shelf by USS Queenfish (SSN–651)*, Tuscaloosa, 2008. • **36.** Overall, whereas the Northern Fleet had made up just 10 per cent of the complement of Soviet submarines in the 1950s it made up 50 per cent by the 1980s. The numbers tell only part of the story. In the late 1970s the proportion of SSBNs assigned to the Northern Fleet versus the Pacific Fleet actually fell, but the new submarines being assigned to the Northern Fleet, the *Typhoon* class, in the 1980s represented a further strengthening of the Northern Fleet and placed an even greater emphasis on its strategic importance. See Skogan, John Kristen, 'The Evolution of the Four Soviet Fleets 1968–1987', in Skogan, John K., and Brudtland, Arne O., *Soviet Seapower in Northern Waters: Facts, Motivation, Impact and Responses*, London, 1990, pp. 18–33. • **37.** Ries, Tomas and Skorve,

Johnny, *Investigating Kola: A Study of Military Bases using Satellite Photography*, Norwegian Institute for International Affairs, 1987. • **38.** In 1989–90 Skogan served as Norway's deputy minister of defence. • **39.** In April 1970, the Soviet navy had involved over 200 ships worldwide in a naval exercise known as *Okean*. • **40.** Gorshkov, S. G., *The Sea Power of the State*, London, 1979 (translated from Russian edition, 1976), p. 60. Gorshkov thought the balance of naval forces was turning in the Soviet Union's favour: 'America, separated from Europe by the vast expanses of the Atlantic Ocean, has for centuries been safe and not experienced the horrors of war. It grew used to its safety and impunity, finding itself protected by its powerful fleet. Today, the position has changed and oceanic areas are now the least secure in the US defence system.' (P. 61.) He argued that a strong Soviet navy was necessary as a a defensive measure: 'The imperialists are turning the World Ocean into an extensive launching-pad . . . our navy must be capable of standing up to that very real threat.' (P. 280.) • **41.** Brome, Adam and Nossal, Kim Richard, 'Tensions in Canada's Foreign Policy', *Foreign Affairs*, Vol. 62, No. 2, Winter 1983/84, pp. 335–353. • **42.** Lamm, Manfred R., *Ten Steps to Counter Moscow's Threat to Northern Europe*, Heritage Foundation Backgrounder, No. 356, 30 May 1984. • **43.** He tried to force a wedge between Western allies by drawing a distinction between Scandinavian and American attitudes. Gorbachev approvingly noted the various ways in which Scandinavian countries had supported East–West rapprochement. The Conference on Cooperation and Security in Europe had held its crucial meetings in Helsinki. A confidence-building treaty had been signed in Stockholm. Reagan and Gorbachev had met in Reykjavik to discuss arms limitations in 1986. • **44.** A rather sceptical analysis of Gorbachev's speech was produced by a Canadian expert in 1988. See Hayward, Dan, 'Gorbachev's Murmansk Initiative: New Prospects for Arms Control in the Arctic?', *Northern Perspectives*, Vol. 16, No. 4, Canadian Arctic Resources Committee, July–August 1988. • **45.** Previous Soviet leaders had dealt with problems in Eastern Europe quite differently, bolstering forces in East Germany in 1948, crushing rebellion in Hungary in 1956 and invading Czechoslovakia in 1968. In 1980–81 Poland avoided Soviet invasion by imposing Martial Law and cracking down on dissidents itself. Although troops were used to quell dissent in the Baltic states and in the Caucasus in the last years of the Soviet Union, Moscow ultimately refrained from full-scale military intervention. • **46.** The United States would have liked to have left even earlier. • **47.** President Putin was criticised for failing to return immediately from his holiday on the Black Sea. Follow-up investigations were widely considered inadequate, as were the punishments for senior naval officers and the compensation to the families of dead crewmen. In 2003, Moscow radio reported that an aquapark was being built in Vidyayevo, the home base of the *Kursk* (*Ekho Moskvy*, 12 August 2003). • **48.** Nikitin, Aleksandr; Kudrik, Igor; Nilsen, Thomas, *The Russian Northern Fleet: Sources of Radioactive Contamination*, Bellona Foundation, 1996. • **49.** Bøhmer, Nils; Nikitin, Aleksandr; Kudrik, Igor; Nilsen, Thomas, McGovern, Michael H., Zolotkov, Andrey, *The Arctic Nuclear Challenge*, Bellona Foundation, 2001. • **50.** Some argue that Russia is now ideally placed to turn the processing of international nuclear waste into a commercial business (Smith, Mark, A., *Russian Environmental Problems*, Defence Academy of the United Kingdom, 2006). In 2008, the Zvezdochka shipyard was reported as having offered to decommission Britain's submarines as well. • **51.** Skorve believes a test took place on 12 September 1973 yielding two megatons – over a hundred times more powerful than the Hiroshima bomb in August 1945. Skorve, Johnny and Skogan, John Kristen, *The NUPI Satellite Study of the Northern Underground Nuclear Test Area on Novaya Zemlya: A Summary Report of Preliminary Results*, Research Report No. 164, Norwegian Institute of International Affairs, December 1992. • **52.** Interview with John Skogan and Johnny Skorve, Oslo, April 2008. The Russian official was Viktor Mikhaylov. • **53.** Peterson, D. J., *Troubled Lands: The Legacy of Soviet Environmental Destruction*, Boulder, 1993, pp. 5–6. • **54.** Velikhov, Evgeny, *Use of Russian Nuclear Shipbuilding Technologies for Developing Energy and Supporting Global Energy Security*, Arctic Energy Summit, Anchorage, October 2007. • **55.** NC-401 provides for a role for NATO in energy security issues. Some have argued that the disruption of energy supplies (such as natural gas from Russia) to NATO member states (such as any of the Baltic states) should be taken up, as a matter of course, by the North Atlantic Council as a whole. • **56.** Interview with Bård Bredrup Knudsen, Norwegian Ministry of Defence, April 2008. • **57.** At least one of the five new Fridtjof Nansen-class frigates built by Norway is intended to be permanently attached to NATO's Atlantic fleet. The other four will be on domestic operations, used for training or undergoing maintenance in port. • **58.** Struzik, Ed, 'Who's Guarding our Back Door?', the *Star*, 18 November 2007. • **59.** At the time, Leblanc ran Canada's Northern Area Command headquarters in Yellowknife. For a history of the role of Canada's military in the Arctic, see Huebert, Rob, 'The Rise and Fall (and Rise?) of Canadian Arctic Security', in MacDonald, Brian, ed., *Defence Requirements for Canada's Arctic*, Vimy Paper, Conference of Defence Associations Institute, 2007, pp. 8–23. • **60.** The choice of Arctic/Offshore patrol ships has been criticised in Canada as not offering the flexibility required for ice operations. Cost considerations have meant a downgrading of the ships' military specifications. • **61.** In May 2008 the Canadian government took the highly unusual step of blocking the sale of Radarsat's operator – MacDonald Dettwiler and Associates of Vancouver, British Columbia – to an American company, Alliant Techsystems. • **62.** *Final Report: Naval Operations in an Ice-Free Arctic Symposium 17–18 April 2001*, Office of Naval Research, Naval Ice Center, Oceanographer of the Navy, and the Arctic Research Commission. Similar reports were produced for the 2007 and 2009 symposia. • **63.** Evidence to US Senate Subcommittee on Oceans, Atmosphere, Fisheries, 7 July 2009. • **64.** At the time of writing, the USCG only had two functional ice-breakers, both close to the end of their service life, and a third ice-breaker out of action and confined to port. • **65.** Rice, Gary, 'Four Selected Intrusion Scenarios', in MacDonald, Brian, *Defence Requirements for Canada's Arctic*, Vimy Paper, the Conference of Defence Associations Institute, 2007, pp. 65–78. In 2007 the Canadian military ran an exercise involving an attempted terrorist attack, with two terrorists from Edmonton, attempting to sabotage oil infrastructure around Norman Wells, just below the Arctic Circle. • **66.** Brooks, Gene, 'Arctic Journal', *Coast Guard Journal*, 15 April 2008.

PART THREE Nature

6 Signs: Nature's Front Line

1. The extent of the maximum sea-ice cover in March is no clear predictor of the extent of minimum sea-ice cover in September. The two are, of course, linked. But one does not determine the other. If it did, then sea-ice cover from year to year would follow a straight line, rather than exhibiting variation around a trend line. • **2.** That is still a huge area. But it is far less than even in 2005, taking away an area the size of Texas, California and Florida combined. • **3.** See, for example, Kay, Jennifer E.; L'Ecuyer, Tristan; Gettelman, Andrew; Stephens, Graeme and O'Dell, Chris, 'The Contribution of Cloud and Radiation Anomalies to the 2007 Arctic Sea Ice Extent Minimum', *Geophysical Research Letters*, Vol. 35, 2008. The study pointed out that the particularly warm and sunny conditions of the 2007 summer were important in explaining the exceptional reduction in sea-ice extent. But the study also argued that this indicated a more general point: 'In a warmer world with thinner ice, natural summertime circulation and cloud variability is an increasingly important control on sea-ice extent minima.' See also Overland J. E.; Wang, M., and Salo, S., 'The Recent Arctic Warm Period', *Tellus A*, No. 60, 2008. • **4.** In general, the surface area of Arctic ice has been used as the key figure, for two main reasons. First, it is far easier to measure – satellite data can accurately measure the extent of sea ice as it changes from minute to minute, and from month to month. The thickness of the ice is far more difficult to measure in a consistent and comprehensive way, though data on ice thickness have existed in some parts of the Arctic (particularly the Russian Arctic) for several decades. Second, from a climatological point of view, the surface area of Arctic ice was considered more important than its volume because the main effect of the whiteness of the Arctic is to reflect sunlight back into the atmosphere. Other natural factors might depend on the volume of melting ice – the proportion of fresh water to salt water in the northern oceans, for example – but the surface area was and is considered to have the most important impact on potential changes in global climate. • **5.** See Frolov, Ivan E.; Gudkovich, Zalman M.; Radionov, Vladimir F.; Shirochkov, Alexander V. and Timokhov, Leonid A., *The Arctic Basin: Results from Russian Drifting Stations*. Chichester, 2005. • **6.** Hegerl, G. C.; Zwiers, F. W.; Braconnot, P.; Gillett, N. P.; Luo, Y.; Marengo Orsini, J. A.; Nicholls, N.; Penner, J. E. and Stott, P. A., '2007: Understanding and Attributing Climate Change', *Climate Change 2007: The Physical Science Basis. Contribution of Working Group I to the Fourth Assessment Report of the Intergovernmental Panel on Climate Change*, Cambridge, 2007, p. 716. • **7.** It does not necessarily follow that more abrupt changes are changes which will happen sooner. Abruptness simply means that, when things do change, they will change quickly. The proximate cause of that change – a particularly warm summer, for example – will not be brought forward in time simply because the consequences of the change are now considered to be greater. • **8.** Holland, Marika M.; Bitz, Cecilia M.; Tremblay, Bruno, 'Future Abrupt Reductions in the Summer Arctic Sea Ice', *Geophysical Research Letters*, Vol. 33, 2006. • **9.** The idea of an open polar sea was a staple of thinking about the Arctic until conclusively disproved by expeditions in the late nineteenth and early twentieth centuries. • **10.** Ewing, Maurice and Donn, William L., 'A Theory of Ice Ages', *Science*, Vol. 123, No. 3207, June 1956. • **11.** Wexler, H., 'Modifying Weather on a Large Scale', *Science*, Vol. 128, No. 3331, October 1958. The idea of geo-engineering, fashionable in the middle of the twentieth century, became unpopular in the second half of the twentieth century as concern over environmental protection increased and as fears over the unintentional consequences of geo-engineering also increased. Now, however, some of the ideas of geo-engineering have returned. But whereas in the 1950s the main objective was likely to be to melt Arctic ice – the main objective now is to use geo-engineering to prevent or retard the melting of Arctic ice. • **12.** In 2008 the US Department of the Interior accepted the recommendations of the Fish and Wildlife Service (FWS) that the polar bear should be listed as an endangered species under the Endangered Species Act (1973). In the run-up to the decision on whether or not to list, opponents of listing argued that it would solve nothing: 'While America sleeps better at night, falsely believing they have assisted this iconic species, they will still fly planes, drive cars, and power their homes,' argued one witness before a US Senate committee investigating the possibility of a listing. See testimony of Richard Glenn to the US Senate, 30 January 2008. • **13.** See Durner, George M.; Douglas, David C.; Nielson, Ryan M.; Amstrup, Steven C., and McDonald, Trent L., *Predicting the Future Distribution of Polar Bear Habitat in the Polar Basin from Resource Selection Functions Applied to 21st Century General Circulation Model Projections of Sea Ice*, United States Geological Survey, 2007. Other factors which are of key importance include pollution – a major problem for polar bears, because their position at the apex of the food chain tends to mean that they integrate a large portion of the persistent pollutants eaten or absorbed by animals and organisms which are lower on the food chain – and local conditions. Variations between Arctic regions is highly likely. Polar bears may survive longer in the southern Hudson Bay area, where ice tends to be pushed from the north. There may also be variations dependent on changes in the size and distribution of seal and walrus populations, some of which may make them more vulnerable to predatory polar bear activity. • **14.** Polar bear populations are thought to be falling in five areas (Southern Beaufort Sea, Norwegian Bay, Western Hudson Bay, Baffin Bay and Kane Basin). They are thought to be stable in five others (Northern Beaufort Sea, Lancaster Sound, Gulf of Boothia, Foxe Basin and Southern Hudson Bay). In two areas, populations are rising (Viscount Melville Sound and McClintock Channel), but in both cases populations are counted in the low hundreds rather than in the thousands (as is the case for some of the areas where populations are known to be falling). In all the other areas of the Arctic, data are not sufficient to tell whether populations are rising, falling or stable. See Simpkins, M., 'Marine Mammals', *Arctic Report Card 2008*, October 2008. • **15.** In 2007, the United Nations Environment Programme (UNEP) launched World Environment Day with a seminar in Tromsø entitled 'Melting Ice – A Hot Topic?'. In preparation for that theme, UNEP published a book on the most recent science on snow and ice and how it influences the global environment. See *Global Outlook for Ice & Snow*, Nairobi, 2007. • **16.** *Arctic Climate Impact Assessment – Scientific Report*, Cambridge, 2005. The original report was presented at the Fourth Arctic Council Ministerial Meeting, held in Reykjavik on 24 November 2004. • **17.** The albedo feedback mechanism describes a process whereby the declining whiteness of the earth's surface – as snow and ice melt – tends to increase the absorption of heat from the sun, thus further accelerating the process of melting snow and ice, and so on. This is called a positive feedback loop, because the consequences of the process tend to re-enforce the process itself – in the same way that leaving a microphone near to a speaker can result in extremely loud 'feedback' even in an apparently quiet room. There can also be negative feedback, where the consequences of a particular process tend to be attenuated by a countervailing process which it sets off, in the same way that a thermostat turns down heating when a certain temperature is reached. Whereas positive feedback tends to lead to acceleration of a process, negative feedback tends to lead to a return to equilibrium. • **18.** Despite the difficulties of constructing numeric models to describe and predict climate changes, there is basically little choice. 'We can try and make predictions based on extrapolation, or we can try to look for analogues in the earth's past history when we had the same sort of mix of situations,' Holmen says, 'but the first is unhealthy – because we are moving into an area we haven't been in before – and the second is impossible – because analogues are impossible to find, we've never had this mix of boundary conditions before.' • **19.** Melting sea ice has no direct impact on sea level as it displaces as much when liquid as when frozen solid. Warming water does have some impact on sea level simply as a result of thermal expansion. But by far the largest potential contributor to rising sea level is the water frozen in the ice caps of Greenland and Antarctica. Together, these amount to 98 or 99 per cent of all ice on earth. Unlike sea ice, the melting of the ice caps would contribute considerably to sea-level rise because melting would imply that water which is currently stored on land as ice, would now contribute to the volume of the oceans. • **20.** See, for example, Appenzeller, Tim, 'The Big Thaw', the *National Geographic*, Vol. 211, No. 6, June 2007. • **21.** The standard definition of 'permafrost' is soil that has been kept below freezing for at least two years. Not all the land in the Arctic is permafrost, and not all the permafrost on earth is in the Arctic, though it tends to be concentrated in the higher latitudes of the northern hemisphere. Permafrost in mountainous areas is known as 'alpine permafrost'. There is an additional distinction in permafrost between 'continuous' and 'discontinuous' permafrost. While the former describes areas where all of the land below the active layer is frozen, the latter describes areas where patches of frozen soil coexist with areas which are not frozen. • **22.** It is more likely that there will be increased releases of methane from marine methane clathrates, particularly those in the Arctic Ocean (where the residues tend to be closer to the surface of the sea floor). It is also likely that release will be steady rather than sudden, though some scientists have hypothesised that large releases of methane may have been responsible for substantial and rapid periods of climate change in the past, referred to as 'abrupt'. See, for example, Brook, Edward; Archer, David; Dlugokencky, Ed; Frolking, Steve; and Lawrence, David, 'Potential for Abrupt Changes in Atmospheric Methane', *Abrupt Climate Change*, U.S. Climate Change Science Program and the Subcommittee on Global Change Research (lead agency United States Geological Survey), December 2008. • **23.** See Barber, Valerie A.; Juday, Glenn, P. and Finney, Bruce P., 'Reduced Growth of Alaskan White Spruce in the

Twentieth Century from Temperature-Induced Drought Stress', *Nature*, No. 405, June 2000. • **24.** Individual trees may grow more quickly as a result of rising temperatures, of course. They are 'positive responders'. But the question is not whether some trees grow more quickly but the overall numbers of 'positive' versus 'negative' responders and their overall carbon dioxide uptake. • **25.** Higuera, P. E.; Brubaker, L. B.; Anderson, P. M.; Brown, T. A.; Kennedy, A. T.; and Hu, F. S., 'Frequent Fires in Ancient Shrub Tundra: Implications of Paleo-records for Arctic Environmental Change', *PLoS One*, No. 3, March 2008. • **26.** There is natural annual variation in carbon dioxide concentrations as a result of the different levels of uptake and production by land and sea, and by the radically different proportions of the northern and southern hemispheres covered by each. Thus global carbon dioxide concentrations always fall in the northern summer, when the photosynthetic processes of northern hemisphere vegetation reach its peak. But they always rise over the northern hemisphere winter. • **27.** Occasionally, the variety of nations and scientific projects can lead to problems of coordination. Trond Svenøe, the station manager for Norway's Svedrup Station, responsible for overall scientific coordination at Ny-Ålesund, is crystal clear: 'If scientists are coming up to do work which is already being done then they can be turned away – it's a waste of money.' In reality, however, it is hard to fully monitor what is happening in Ny-Ålesund and still harder to refuse access to international researchers. • **28.** The goal of German scientists Professor Hans-Otto Poertner and Kathleen Walther is to understand the processes and extent of crab acclimatisation to different future climate scenarios (which include not only increases in ambient temperature but also, as a result, increases in ocean acidification, which tends to dissolve the exoskeleton of crustaceans in much the same way that it damages both warm-water and fresh-water corals). • **29.** Poland has its own scientific research station on Spitsbergen, at Hornsund. The Russians have a research establishment at Barentsburg. • **30.** The French and German stations were merged in 2003, on the occasion of the fortieth anniversary of the 1963 Franco–German Elysée treaty. • **31.** Japan came to Ny-Ålesund first, opening its research station in 1990. A decade later, in 2002, the Korean Polar Research Institute set up its 'Dasan'. Then, in 2004, the Shanghai-based Chinese Arctic and Antarctic Administration opened the 'Yellow River' station, now one of the most active research stations at Ny-Ålesund. Finally, in 2008, India opened the 'Himdari' research base. • **32.** Wei Luo believes that her country has a lot to offer culturally, as well: 'In the West you are more individualist, in the East we are more focused on the group – we believe in teamwork.' • **33.** Of those forty-three, most were American or Canadian and a significant further portion were European (mostly from the European Arctic states and the United Kingdom). None were Russian. However, Chinese, Korean and Indian scientists have all been involved in elements of bioprospecting in the Arctic. See Leary, David, *Bioprospecting in the Arctic*, United Nations University, Yokohama, 2008. • **34.** See Battisti, David S. and Naylor, Rosamond L., 'Historical Warnings of Future Food Security with Unprecedented Seasonal Heat', *Science*, Vol. 323, January 2009, pp. 240–244. • **35.** The International Treaty on Plant Genetic Resources for Food and Agriculture was agreed in 2001. • **36.** The economics of managing climate change were discussed by the Stern review, published by the British government in 2006 (Stern, Nicholas, *The Economics of Climate Change: The Stern Review*, Cambridge 2006). However, while this report looks at climate change from a global perspective – though some might argue with some Western biases – and forcefully makes the case for mitigation measures on that basis, it is inevitable that there will be disagreement between states on whether mitigation or adaptation is the more appropriate strategy, depending on a whole host of factors, including current wealth, expectations of future wealth and estimation of vulnerability to the consequences of climate change. • **37.** The politician is Winston Churchill. The notion of 'Age of Consequences' in terms of climate change is drawn from the title of a report on the security implications of global warming, published in 2007. Campbell, Kurt M.; Gulledge, Jay; McNeill, J. R.; Podesta, John; Ogden, Peter; Fuerth, Leon; Woolsey, James R.; Lennon, Alexander T. J.;

Smith, Julianne; Weitz, Richard; and Mix, Derek, *The Age of Consequences: The Foreign Policy and National Security Implications of Global Climate Change*, Center for Strategic and International Studies and the Center for a New American Security, November 2007. • **38.** Reports on the political consequences of climate change agree on the potential stresses which climate change may add to the global system – increased migration, disease, flooding, concern over natural resources. But there are different views on whether climate change can be considered a security issue in itself, or as a factor which may tend to exacerbate existing security issues. See, for example, Campbell et al., op. cit.; Smith, Dan, and Vivekanada, Janani, *A Climate of Conflict: The Links between Climate Change, Peace and War*, International Alert, November 2007; or, Goodman, Sherri W.; Catarious, David M.; Filadelfo, Ronald; Gaffney, Henry; Maybee, Sean; and Morehouse, Thomas, *National Security and the Threat of Climate Change*, The CNA Corporation, 2007.

7 Consequences: Reworking Geography

1. See, for example, Jones, B. M; Arp, C. D.; Jorgenson, M. T.; Hinkel, K. M.; Schmutz, J. A.; and Flint, P. L., 'Increase in the Rate and Uniformity of Coastline Erosion in Arctic Alaska', *Geophysical Research Letters*, Vol. 36, 14 February 2009. • **2.** US Army Corps of Engineers, *An Examination of Erosion Issues in the Communities of Bethel, Dillingham, Kaktovik, Kivalina, Newtok, Shishmaref, and Unalakleet*, April 2006. • **3.** Lemmen, Donald S.; Warren, Fiona J.; Lacroix, Jacinthe; and Bush, Elizabeth, eds., *From Impacts to Adaptation: Canada in a Changing Climate 2007*, Government of Canada, Ottawa, 2008. Chapter three, covering northern Canada and edited by C. Furgal and T. D. Prowse, is of particular relevance. • **4.** Larsen, Peter; Goldsmith, Scott; Smith, Orson; Wilson, Meghan; Strzepek, Ken; Chinowsky, Paul; and Saylor, Ben, *Estimating Future Costs for Alaska Public Infrastructure at Risk from Climate Change*, Institute of Social and Economic Research, University of Alaska Anchorage, June 2007. A revised version was subsequently published as an article in *Global Environmental Change*, 2008. The report looked at two time frames for climate change costs: from 2006 to 2030 and from 2006 to 2080. The report considered several different climate models, and scenarios with and without adaptation. Over the period 2006 to 2030 the report estimated that adaptation made virtually no difference to the net additional costs to public infrastructure incurred as a result of global warming. Over the period 2006 to 2080, however, net costs without adaptation were considerably higher. For example, in the 'warmest model' used by the researchers, costs with adaptation amounted to $6.7 billion while costs without adaptation were estimated at $12.3 billion. • **5.** The project involves a number of different institutions in Canada, Finland, Iceland, Norway, Russia, Sweden and the United States with CICERO sharing the lead with the University of Guelph, in Ontario, Canada. • **6.** See Wohlforth, Charles, *The Whale and the Supercomputer: On the Northern Front of Climate Change*, New York, 2004. • **7.** Sheila-Watt Cloutier, chair of the Inuit Circumpolar Conference, in 2005. • **8.** The Stockholm Convention on Persistent Organic Pollutants entered into force in 2004. The particular vulnerability of Arctic communities to POPs is due to the slow decomposition of POPs in Arctic conditions and the build-up of POPs in Arctic ecosystems (concentrations rising through the food chain). The diet of many Inuit communities is rich in marine mammals – themselves carrying high concentrations of POPs – with serious consequences for human health. See Cone, Martha, *Silent Snow: The Slow Poisoning of the Arctic*, New York, 2005. • **9.** The *Arctic Report Card 2008* provides an overview of several impacts of climate change on marine ecosystems. See Overland, J., 'Fisheries in the Bering Sea'; Overland, J., 'Fisheries in the Barents Sea'; and, Sawatzky, C.D. and Reist, J.D., 'The State of Char in the Arctic'. • **10.** This study was conducted by the Canadian Ministry of Agriculture. • **11.** The Northwest Passage refers to a series of possible routes through the Canadian Arctic archipelago between Baffin Bay and the Beaufort Sea. The Northeast Passage refers, in general terms, to any sea route between Europe and Asia across the northern shores of Europe and Asia. The Russian term, the Northern Sea Route, is generally used in the more limited sense of transportation along Russia's Arctic coastline. Here, the terms are used more or less interchangeably. There are a number of variations on the Northern Sea Route. One, known as the 'inner' Northern Sea Route involves travel south of Novaya Zemlya and Severnya Zemlya. Another, known as the 'outer' Northern Sea Route follows a path north of these island groups. While the first route has the advantage of being ice free for a longer part of the year, it also involves the navigation of a number of narrow and shallow straits. The second route is both deeper and broader – but, being further north, it tends to be ice covered for longer. • **12.** Freight on the the Northern Sea Route peaked at 6,579,000 tons in 1987. Freight carried between Europe and Asia peaked some years later, in 1993, but only amounted to some 226,000 tons. • **13.** See, for example, Brigham, Lawson W., ed., *The Soviet Maritime Arctic*, London, 1991. • **14.** 'German ships successfully make "Arctic Passage",' *Reuters*, 12 September 2009. • **15.** Icelandic Ministry of Foreign Affairs working group, *North Meets North: Navigation and the Future of the Arctic*, 2006 (originally published in Icelandic in 2005 as *Fyrir stafni haf*). • **16.** See Brigham, Lawson W., 'Thinking about the Arctic's Future: Scenarios for 2040', the *Futurist*, September–October 2007. • **17.** Hakkinen, Sirpa; Proshutinsky, Andrey; Ashik, Igor, 'Sea ice drift in the Arctic since the 1950s', *Geophysical Research Letters*, Vol. 35, October 2008. • **18.** The company has a long Arctic heritage, building the first Finnish ice-breaker in 1954 and converting the *SS Manhattan* oil tanker for its historic voyage through the Northwest Passage back in 1969. • **19.** See, for example, *Governance of Arctic Marine Shipping*, Dalhousie University, October 2008. • **20.** In 2007, the Econ Pöyry group produced several scenarios for Arctic shipping to 2030, commissioned by the Norwegian shipping industry. (*Arctic Shipping 2030: From Russia with Oil, Stormy Passage or Arctic Great Game?*, Econ, 2007.) In 2008, a body working with the Arctic Council produced the *Arctic Marine Shipping Assessment*, overview of four possible futures for Arctic shipping to 2050, comprising 'Arctic Race', 'Arctic Saga', 'Polar Lows' and 'Polar Preserve' with very different outcomes depending on whether demand for Arctic resources is high or low, and whether Arctic governance is stable and rules based or unstable and ad hoc. (*The Future of Arctic Marine Navigation in Mid-Century*, Global Business Network, May 2008.)

PART FOUR Riches

8 The (Slow) Rush for Northern Resources

1. 'Wealth of the Klondike', the *New York Times*, 18 July 1897. • **2.** 'By Air Ship to Klondike: A Hoboken Inventor to Start in Three Weeks for a Quick Trip through the Atmosphere', the *New York Times*, 30 August 1897. • **3.** Vaughan, Richard, op. cit., p. 249. • **4.** In contrast, artisanal mining is still very much alive in the developing world. • **5.** Gold is still mined in the Arctic. So are diamonds and rubies. • **6.** On occasion, gold could be of strategic importance. For example, in the 1920s and 1930s, exporting gold was the only way the Soviet Union could obtain the hard currency necessary to import foreign industrial equipment and to fund external political objectives through Comintern. • **7.** Interview with Frederic H. Wilson, United States Geological Survey Alaska Office, Anchorage, August 2008. • **8.** Vaughan, Richard, op. cit., pp. 78–97. • **9.** See Dolin, Eric Jay, *Leviathan: The History of Whaling in America*, New York, 2007. • **10.** Cited in Bone, Robert M. and Mahnic, Robert J., 'Norman Wells: The Oil Center of the Northwest Territories', *Arctic*, Vol. 37, No. 1, March 1984, pp. 53–96. • **11.** See, for example, Dall, William, *Alaska and its Resources*, Boston, 1870. • **12.** The area was named the Naval Petroleum Reserve 4 (NPR-4). It is now known as the National Petroleum Reserve – Alaska, or NPR-A. • **13.** Though Norman Wells is generally accepted as the first producing Arctic oil well, it lies a degree below the Arctic Circle. • **14.** The account of the oil find at Norman Wells, and the immediate press and political response to it, is derived from Fumoleau, René, *As Long as This Land Shall Last: A History of Treaty 8 and Treaty 11, 1870–1939*, Calgary, 2004 (originally published in 1975). • **15.** In the United States, the Teapot Dome field opened in Wyoming, while oil production took off in California. Canadian oil was still being produced in Ontario, close to the main domestic markets of Toronto and Montreal. Alberta remained the more immediate prospect for future Canadian oil production, indeed oil from the Athabasca tar sands was already being mined. • **16.** From a high of $3 in 1920, oil prices in the 1920s fluctuated between $1 and $2 per barrel (around $20 in inflation-adjusted terms), with huge regional variations. • **17.** Sikstrom, C. B., 'Theodore August Link (1897–1980)', *Arctic*, Vol. 48, No. 1, March 1995, pp. 96–98, and Bone, Robert and Mahnic, Robert, 'Norman Wells', appear to disagree on the exact number of wells drilled, and their results. • **18.** The Canol project was strongly criticised during construction ('Demand US Halt Canol Oil Project', the *New York Times*, 23 November 1943). In 1945, a Senate committee chaired by future President Harry S. Truman called the $134 million cost of building Canol 'inexcusable' ('End of Canol', *Time*, 5 March 1945). • **19.** In essence, production in each of the oil-producing states was pro-rationed by a state administration. By enforcing production quotas below capacity, the intention was to manage the market to avoid any disastrous collapse in domestic oil prices. The result was to create what Dan Yergin has called a 'surge capacity' for the world. In 1957–63 this surplus capacity amounted to 4 million barrels per day; by 1970, it had been trimmed to 1 million bpd or less. Yergin, Daniel, *The Prize: The Epic Quest for Oil Money and Power*, New York, 1991, p. 567. • **20.** The introduction of import quotas represented a political victory for small domestic producers against the larger international oil companies (also mostly American) which sourced their oil from outside the US. (Aramco, the producing company in Saudi Arabia, was owned outright by a consortium of four American companies.) Eisenhower originally opposed quotas on the basis that they distorted free markets. But the political alliances of the oil-producing states won through, with House Senate Leader Lyndon Johnson leading the charge. • **21.** Yergin, Daniel, *The Prize*, p. 589. • **22.** In 1970, US oil production reached 11,297,000 barrels per day. The following year it fell to 11,156,000 bpd. In Canada, production has not yet peaked, rising from 920,000 bpd in 1965 to 3,309,000 bpd in 2007. • **23.** The man responsible for Winter Harbor #1 was Jack Gallagher, of Dome Petroleum. Working for the Geological Survey of Canada in the 1930s, Gallagher had come to the view that the Mackenzie Delta was potentially as prolific as the deltas of the Mississippi, the Tigris-Euphrates or Venezuela's Orinoco. In 1950, he was hired to run the petroleum branch of Dome Minerals. In the 1970s and 1980s, Jack Gallagher would be one of the leading advocates in Canada for Arctic oil and gas development, staking – and losing – a fortune in the process. • **24.** Reed, John C., *Exploration of Naval Petroleum Reserve No. 4 and Adjacent Areas, Northern Alaska, 1944–1953*, Geological Survey Professional Paper 301, Washington D.C. 1958. • **25.** BP had originally produced exclusively in Iran, operating under the now defunct name of Anglo-Persian (and then Anglo-Iranian) from 1908 until nationalisation in 1951. • **26.** Previous lease sales in 1964, 1965 and 1967 had raised a total of $12 million. • **27.** Amerada/Getty paid a total of $72.3 million. The difference, of course, was that whereas BP had bought on the edge of an area where considerable exploration might be needed before oil was struck in sufficient quantity to be produced, Amerada/Getty was buying into the middle of an area where the presence of oil had been conclusively proven. • **28.** The government held a 45 per cent share in the company. • **29.** McKenzie Brown, Peter; Jaremko, Gordon; and Finch, David, *The Great Oil Age: The Petroleum Industry in Canada*, Calgary, Alberta, 1993, p. 97. • **30.** In 1970, the United States government launched a programme to investigate the consequences of a colder Arctic. Sullivan, Walter, 'US and Soviet Press Studies of a Colder Arctic', the *New York Times*, 18 July 1970. • **31.** Meneley, Bob, *Exploration Results in the Canadian Arctic Islands*, Arctic Energy Summit, Anchorage, October 2007. • **32.** Rockefeller obtained large discounts on rail transport of oil, guaranteed by supply which his competitors could not match. This conferred a price advantage deemed anti-competitive and struck down in a number of court and legislative rulings culminating in the US Sherman Antitrust Act of 1890. • **33.** 'Alaska's New Strike', *Time*, 13 December 1968. • **34.** The Trans-Alaska Pipeline consortium, consisting of Arco, BP and Humble, was organised in October 1968. In order to benefit from tax breaks available to municipally financed projects, the debt required to finance the construction of the pipeline was issued by the city of Valdez itself, as tax exempt municipal bonds, underwritten by the consortium. • **35.** Though the transit was achieved with US and Canadian coastguard support, there was no prior authorisation from the Canadian government. In 1970, the Arctic Waters Pollution Prevention Act, essentially extending State control to 100 miles north of the Canadian coast, was contested by the United States which feared this could allow the Soviet Union to do likewise. • **36.** An inhabitant of the Anuktuvuk area of the Brooks mountain range named Jesse Ahgook, recalled the passage of the first white men in 1901 – a reconnaissance party from the United States Geological Survey sent to survey the mineral potential of the North Slope. • **37.** 'Old Eskimo Pass Becomes a Truck Route', the *New York Times*, 22 April 1969. • **38.** The first $462.5 million was to be paid out over eleven years. A further $500 million was to be paid out from future royalties. The mechanism for dispensing these funds was the establishment of regional and village corporations, with all the (non-transferable) shares held by the local population. An odd mixture of socialism and capitalism, the corporation system remains a defining feature of Alaskan politics in the twenty-first century. • **39.** The lawsuits claimed that the Department of the Interior had failed to comply with the environmental impact statement requirements of NEPA in its initial approval of the pipeline. Moreover, the request for a 100-foot right of way for the pipeline – made in June 1969 – was said to be in violation of the 50-foot limit under the Mineral Leasing Act of 1920. Whereas the first environmental impact statement produced by the Department of the Interior in January 1971 ran to barely 200 pages, with a further 60 pages of revised plans, the final environmental impact statement released just over a year later ran to six volumes, covering everything from impacts on caribou to earthquake risk. The report stated that a major earthquake along the southern portion of the Trans-Alaska Pipeline was 'almost a certainty . . . during the lifetime of the pipeline'. US Department of the Interior, *Final Environmental Impact Statement: Proposed Trans-Alaska Pipeline*, March 1972, Vol. 1. • **40.** The lead on this was taken by government scientist Art Lachenbruch, who had been working on permafrost issues since the 1940s. See 'Some Effects of a Heated Pipeline in Permafrost', USGS Circular 632, 1970. • **41.** The *Torrey Canyon* disaster in the English Channel, just a few years earlier, highlighted the risk of sea transport of oil. • **42.** It was a point taken up in the US Department of the Interior's final environmental impact statement. (US Department of the Interior, *Final Environmental Impact Statement: Proposed Trans-Alaska Pipeline*, March 1972, Vol. 5.) A single route for both Alaskan and Canadian oil and gas would be cheaper in the long run, and cause less environmental damage. A report supportive of a trans-Canadian option warned presciently that 'the consequences of a TAP-tanker system for transporting Alaskan oil could haunt an entire continent for years to come'. Cichetti, Charles J., *Alaskan Oil: Alternative Routes and Markets*, Resources of the Future, Inc., Baltimore, 1972. • **43.** The 1920 Merchant Marine Act, commonly referred to as the Jones Act, restricts the carrying of goods – including oil – from one US port to another to US vessels. Exporting oil to East Asia would allow oil companies to use cheaper, foreign vessels. • **44.** Discussion of the Canadian route is described in Christopher Kirkey, 'Moving Alaskan Oil to Market: Canadian National Interests and the Trans-Alaska Pipeline', *American Review of Canadian Studies*, Vol. 27, 1997. Ottawa's support for a Canadian pipeline seemed to waver. Prime Minister Pierre Trudeau told the House of Commons, 'If it were in the Canadian interest, if it were desirable to have such a pipeline built we would make the conditions available whereby it could be built. This is not a clear invitation.' House of Commons Debates IV (1971): 4292. • **45.** United States Department of the

Interior, *An Analysis of the Economic and Security Aspects of the Trans-Alaska Pipeline*, 1971, p. 5. • **46.** US Secretary of the Interior Rogers Morton, Joint Economic Committee of Congress, *Natural Gas Regulation and the Trans-Alaska Pipeline*, 1972, p. 300. • **47.** In July 1973 the US Senate attempted to sweep aside legal challenges to construction by voting that the conditions of NEPA had been met. Further, the amendment introduced by Alaska Senator Mike Gravel attempted to bar future litigation on the issue by what his colleague Ted Stevens called 'environmental extremists'. The amendment passed by fifty votes to forty-nine, with Vice-President Spiro Agnew casting the deciding ballot. • **48.** In inflation-adjusted 2007 dollars the price of oil averaged $15.42 in 1973, and $48.92 in 1974. After 1973, the lowest year-average price of oil was $16.69, in 1998. *BP Statistical Review of World Energy 2008*. • **49.** In 2007 dollars, this is equivalent to a rise from $44.77 to $93.08. *BP Statistical Review of World Energy 2008*. • **50.** In 1938, Mexico's President Cárdenas, after a bruising labour dispute between foreign-owned oil companies and Mexican workers, had nationalised the entire oil industry, forming Petóleos Mexicanos (PEMEX) as a result. In 1943, Venezuela, initially one of the strongest fiefs of the American oil industry outside the United States, signed an agreement with Standard Oil of New Jersey which split the economic rents of oil production equally between the state and the producer, far more advantageous to the state than previous agreements. After 1945, when the Venezuelan interim government was overthrown by a more left-leaning junta, the arrangements between Venezuela and companies were revised again, such that by 1948 government income was six times greater than in 1942. • **51.** Bronson, Rachel, *Thicker than Oil: America's Uneasy Partnership with Saudi Arabia*, Oxford, 2006, p. 55. • **52.** *The Oil Import Question: A Report on the Relationship of Oil Imports to National Security*, Washington D.C., February 1970. • **53.** Akins, James E., 'The Oil Crisis: This Time the Wolf is Here', *Foreign Affairs*, April 1973. • **54.** European countries had, after all, experienced shortages during the Suez Crisis in 1956. Moreover, unlike the United States, most European countries had never been major oil producers and were therefore used to the notion of import dependence. • **55.** For a history of the interrelationship between oil supply and the geopolitical outlook of the United States, see Klare, Michael, *Blood and Oil: The Dangers and Consequences of America's Growing Petroleum Dependency*, London, 2004. • **56.** 'Results of Beaufort Sea Drilling Crucial Factor in Dome's Future', the *Globe and Mail*, 26 October 1981. • **57.** Some inroads had been made by Soviet oil production before 1973, but they were the exception not the rule. The Italian oil major ENI had first breached the exclusion of Soviet oil in 1960, as ENI tried to break the stranglehold of the larger oil companies by circumventing the system they had set up. In the twenty-first century ENI would recreate the reputation Mattei had given it, by following a strategy consistently more accommodating to producer countries than its larger rivals. • **58.** 'The Great Arctic Energy Rush', *Business Week*, 24 January 1983. • **59.** Berger, Thomas R., *Northern Frontier, Northern Homeland: The Report of the Mackenzie Valley Pipeline Enquiry*, Ottawa, 1977. Berger suggested that the economic benefits of the pipeline were dubious and environmental harm was certain. Moreover, he argued that 'the future of the North ought not to be determined only by our southern ideas of frontier development. It should also reflect the ideas of the people who call it their homeland.' Settling native land claims would have to take priority. • **60.** For example, the Frontier Exploration Allowance allowed for generous tax relief on wells that cost more than C$5 million to drill. • **61.** Some investors bucked the trend. The Reichmann brothers invested heavily in Canadian Arctic oil in the mid-1980s. In 1987, Paul Reichmann told a journalist, 'Unless we meet some nuts on the way, oil will be flowing from the Beaufort next year and there will be full and permanent production by 1993.' See Francis, Diane, 'Reichmanns Gamble Billions on Arctic Oil', the *Toronto Star*, 27 January 1987. • **62.** In 2007 dollars the fall was from $53.21 in 1985 to $27.22 in 1986. *BP Statistical Review of World Energy 2008*. • **63.** Saudi production fell from 10.2 million bpd in 1980 to 3.6 million bpd in 1985 (at one point in 1985 reaching as low as just 2.2 million bpd). Some suggest the US persuaded Saudi Arabia to boost production so as to cause prices to fall, not only helping the American economy but also hurting the Soviet Union. But the Saudis had their own reasons for wanting to raise production. Having borne most of OPEC's costs of reducing production, thus reducing revenues unless the oil price rose to compensate (which it did not), the Saudis were fed up. A fair market share would now replace a high oil price as the guiding strategy of OPEC. • **64.** The companies involved are Imperial Oil, Exxon Mobil, Royal Dutch Shell and ConocoPhillips. • **65.** Such peaks are local, not global. Gasoline demand may peak in the US – there is a limit to the number of cars any country needs, and the distance they are driven – but it does not follow that global demand is anywhere near its peak. China and India account for the bulk of current demand growth, where per-capital utilisation is much lower. • **66.** See, for example, Lovins, Amory B., *Winning the Oil Endgame: Innovation for Profits, Jobs and Security*, Snowmass, Co, 2004 or Roberts, Paul, *The End of Oil: The Decline of the Petroleum Economy and the Rise of a New Energy Order*, London, 2004. • **67.** See Colin Campbell and Jean Laherrère in 'The End of Cheap Oil', *Scientific American*, March 1998. This argument is contested, but a few central points are beyond debate, even if their causes and consequences are not: the rate and size of new discoveries of oil has declined for several decades; the secrecy around the claimed oil reserves of some countries makes them impossible to verify; there are relatively few places where new 'conventional' oil could be found in any quantity. • **68.** See Cohen, David, 'We are Using Up Minerals at an Alarming Rate. How Long Before They Run Out?', *New Scientist*, 26 May 2007. • **69.** Many executives accept that the world is running out of oil which is easy to produce, but argue that enhanced recovery, deep-water production and production in remote areas can replace it (up to a point). Beyond that, technology exists to convert other hydrocarbons (mainly gas and tar sands, but also coal) into the liquid fuels most of us associate with oil. In short, 'unconventional oil' can prolong the oil-based economy. But as Jean Laherrère puts it, 'It's not about the size of the oil tank, but the size of the oil tap.' Even if there is plenty of 'unconventional oil', there may be constraints to production. In the case of tar sands, the availability of water is a constraint, because huge quantities of steam are needed to free the flow of oil from the tar sands. If the water is unavailable, the quantity of hydrocarbons available is irrelevant. • **70.** International Energy Agency, *World Energy Outlook 2008*, p. 56. • **71.** Maugeri, Leonardo, *The Age of Oil: The Mythology, History and Future of the World's Most Controversial Resource*, London, 2006. • **72.** National oil companies make up an increasing share of global production, far outstripping that of the IOCs, a reversal of their relative strengths forty years ago. • **73.** The *BP Statistical Review of World Energy 2008* put the proven oil reserves of Saudi Arabia at 264.2 billion barrels of oil. • **74.** United States Geological Survey, *Circum-Arctic Resource Appraisal: Estimates of Undiscovered Oil and Gas North of the Arctic Circle*, USGS Fact Sheet 2008–3049, 2008. • **75.** United States Geological Survey, *Assessment of Undiscovered Oil and Gas Resources of the West Siberian Basin Province, Russia, 2008*, USGS Fact Sheet 2008–3064, August 2008. • **76.** United States Geological Survey, *Assessment of the Undiscovered Petroleum Resources of the Laptev Sea Shelf Province, Russian Federation*, USGS Fact Sheet 2007–3096, November 2007. • **77.** Extrapolating from historical exploration data to 'undiscovered' may make sense in a well-known and well-researched geological province but it may not be valid for a province such as the Arctic, where the exploration data points are either insufficiently numerous or insufficiently distributed geographically to be viewed as being statistically representative of the province as a whole. • **78.** Interview with Jean Laherrère, June 2008. • **79.** Jean Laherrère, 'Arctic Oil and Gas Ultimates', the Oil Drum blog, March 2008 (www.theoildrum.com/node/3666). His figures were 50 billion barrels of oil and 1,000 trillion cubic feet of natural gas, which I have calculated back to barrels of oil equivalent using an approximate factor of 6,000 cubic feet per barrel of oil equivalent. • **80.** Scott Borgerson cites a Russian figure of 586 billion barrels of oil ('Arctic Meltdown: The Economic and Security Implications of Global Warming', *Foreign Affairs*, March/April 2008). But this figure was used ambiguously. It is actually derived from a Russian figure of 80 billion tons of hydrocarbons in the Russian part of the Arctic, amounting to 586 billion barrels of oil equivalent, not oil itself. Most of the forecast resource is natural gas. See Yenikeyeff, Shamil Midkhatovich and Krysiek, Timothy Fenton, 'The Battle for the Next Energy Frontier: The Russian Polar Expedition and the Future of Arctic Hydrocarbons', *Oxford Energy Comment*, Oxford Institute for Energy Studies, August 2007. • **81.** Of a total of 412.16 billion barrels of oil equivalent undiscovered in the Arctic, the USGS projected that 89.98 billion would in the form of oil itself and 265.29 billion would be in the form of natural gas, with the remainder in the form of natural gas liquids (NGL). United States Geological Survey, *Circum-Arctic Resource Appraisal: Estimates of Undiscovered Oil and Gas North of the Arctic Circle*, USGS Fact Sheet 2008–3049, 2008. • **82.** From under 50 per cent in what is termed 'Arctic Alaska' to over 80 per cent in what is described as the 'East Barents Basin'. • **83.** For a basic overview of methane hydrates, and their Arctic occurrence, see either Collett, Timothy S., 'Energy Resource Potential of Natural Gas Hydrates', *AAPG Bulletin*, Vol. 86, No. 11, November 2002, pp. 1971–1992 or, more recently, Ruppel, Carolyn, 'Tapping Methane Hydrates for Unconventional Natural Gas', *Elements*, Vol. 3, pp. 193–199. • **84.** Japan has also invested in research into deep-water hydrates off her coast. • **85.** For more information see www.netl.doe.gov. • **86.** Charlie Chaplin's silent movie, *The Gold Rush* (1925), was so successful that it was re-released in 1942 with accompanying music and a voice-over by Chaplin himself. *The Klondike Kid*, starring Mickey Mouse, was released in 1932.

9 Russia's Arctic Dilemma

1. Victor, David G., 'Three Reasons for Getting Scared: Energy Markets and the Tremendous Lack of Political Strategies', *Post*, No.95, January–June 2007. • **2.** Gazprom was formed as a State corporation in 1989, headed by the then USSR Minister for the Gas Industry, Viktor Chernomyrdin. In 1992–93 the corporation was transformed into a joint stock company. For a detailed description of Gazprom's role in Russia, see Victor, Nadejda Makarova, *Gazprom: Gas Giant Under Strain*, Working Paper #71, Program on Energy and Sustainable Development, Stanford University, January 2008. • **3.** Russia claims to be 'objectively' the world's most reliable gas supplier, pointing out that, even under the Soviet Union, supply to Western Europe was never interrupted for political reasons. However, the potential use of the 'energy weapon' remains. A number of incidents – though many are defended in Russia as either technical mishaps or realignments based on objective price criteria – suggest that Russia's willingness to use its energy position as a bargaining chip is particularly great with the countries of the former Soviet Union, some of which are now members of the European Union. And even if Western Europe is not the intended victim of supply cut-offs, as in the Russia/Ukraine dispute of 2006–07 or in the dispute between the same countries in January 2009, Western Europe may suffer the knock-on effects of Russian policy within the former Soviet Union, though the risk of this will be reduced with the construction of Nord-Stream. The failure of Russia to sign the Energy Charter Treaty, despite pressure from the European Union, does not inspire confidence. See Larsson, Robert, *Russia's Energy Policy: Security Dimensions and Russia's Reliability as an Energy Supplier*, Swedish Defence Research Agency, March 2006. • **4.** The 2007 revenue figures of the companies whose chief executives were present, either on the panel or in the room, were as follows: ExxonMobil ($390 billion), Shell ($355 billion), BP ($284 billion), Total ($233 billion), Conoco Phillips ($193 billion), ENI ($127 billion), Lukoil ($82 billion). • **5.** In 2006 Shell was essentially forced to cede control of the Sakhalin-2 project to Gazprom. In June 2009, however, it appeared that Shell was on the brink of being invited to play a major role in the development of further gas fields at Sakhalin. • **6.** Putin himself had already been highly critical of the influx of Western managers before the TNK-BP dispute blew up. BP's attempts to strengthen its political and commercial position in Russia have included taking part in an auction of Yukos assets which would have been rendered legally invalid without a second bidder, and the purchase of $1 billion of Rosneft shares at that company's London Initial Public Offering in 2006. BP's success has been mixed. In 2007 the company was forced to relinquish control of the East Siberian Kovytka gas field to Gazprom. Nonetheless, in September 2008 BP reached a deal with Russian shareholders on the future of TNK-BP, leading to the appointment of a Russian-speaking CEO. • **7.** For Browne, Russia was the future of BP, the place where his company stole a march on his rivals, cementing BP's access to Russian hydrocarbon resources with the TNK-BP deal. In the 1990s, in a world where equity oil was harder and harder to find, Russia seemed the great prize, and Browne put BP in on the ground floor – indirectly with the acquisition of Amoco's Sidanko assets in 1998, then directly with the TNK-BP fifty-fifty partnership agreed to in 2003. • **8.** This allegation, impossible to verify without the original dissertation, was reported as having been made by Clifford Gaddy, a respected American researcher and academic, in 2006. The allegation was that sixteen pages of one of the dissertation sections were essentially plagiarised from a 1978 textbook by two American professors, William King and David Cleland, entitled *Strategic Planning and Policy*. • **9.** A translation of Putin's 1999 article was published in 2006 in an article by Harley Balzer, 'Vladimir Putin's Academic Writings and Russian Natural Resource Policy', *Problems of Post-Communism*, vol. 53, no. 1, January/February 2006, pp. 48–54. The quotations in this chapter are from that translation. • **10.** And this figure is under the relatively low oil and gas prices of 1999. However, he estimated that the profitable portion of those resources was much lower: $1.5 trillion. • **11.** Viktor Chernomyrdin, long-time Chairman of Gazprom, was removed in 2000, and his handpicked successor, Rem Vyakhirev, was removed a year later. Boris Berezovsky, owner of Sibneft, was forced to leave the country in 2000, yielding control of the company to his business partner Roman Abramovich, who subsequently sold the company to Gazprom for $13 billion and left Russia. • **12.** Some had predicted that the energy relationship would not last, citing a fundamental difference of interests. See Victor, David G. and Victor, Nadejda M., 'Axis of Oil?', *Foreign Affairs*, March/April 2003. • **13.** Dutch gas production peaked at 75.8 billion cubic metres (bcm) in 1996, declining to 64.5 bcm in 2007. The British decline has been more dramatic, from a peak of 108.4 bcm in 2000 to 72.4 bcm in 2007. • **14.** *Europe's Current and Future Energy Position: Demand – Resources – Investments*, November 2008. However, both the past trajectory and future direction of the relationship between Russia and the EU are debated. In *Beyond Dependence: How to Deal with Russian Gas* (European Council on Foreign Relations, November 2008), Pierre Noël argues that EU-27 dependence on Russian natural gas has fallen as a share of imports since 1990, and remained constant as a share of primary energy consumption. He is sceptical of the idea that increasing EU demand is a given, pointing out that global recession and strategic decisions made in some Western European countries after the January 2009 Ukraine/Russia dispute may reduce demand in both the short and the long term. Noël further argues that Russia's ability to supply any future demand is deeply questionable. • **15.** Russia dismisses these concerns as exaggerated – and in some cases politically motivated. Instead, it emphasises its need for 'security of demand' as the necessary counterpart to 'security of supply'. • **16.** The complexity of the relationship, and the key role of the lack of an integrated European gas market, is underlined in Pierre Noël's paper, *Europe's Current and Future Energy Position*. • **17.** Environmental concerns, political opposition and fears that the gas line could serve a secondary intelligence-gathering function in the Baltic Sea have led to numerous delays. The exact routing of the pipeline, from Vyborg in Russia to Greifswald in eastern Germany, was not confirmed at the time of writing. The gas line, originally due to open in 2010, is now more likely to open after 2011. The costs of the gas line, originally estimated at €5 billion, were officially raised to €7.4 billion in 2008, though some experts suggested the final cost might be as high as €12 billion. Certain technical questions on the pipeline – for example, how to fix an undersea pipeline should there be a problem – were not fully answered at the time of writing. For an analysis of the political and security implications of the Nord-Stream project, see Larsson, Robert, *Nord Stream, Sweden and Baltic Sea Security*, Swedish Defence Research Agency, March 2007 or Whist, Bendik Solum, *Nord Stream, Not Just a Pipeline: An Analysis of the Political Debates in the Baltic Sea Region Regarding the Planned Gas Pipeline from Russia to Germany*, Fridtjof Nansen Institute, November 2008. • **18.** Gazprom's partner in South-Stream is ENI, the same Italian company which, under the leadership of Enrico Mattei, breached the post-war exclusion of Soviet oil from global markets in 1960. Before that, despite obvious political differences, Mussolini's Italy was the Soviet Union's largest foreign market for oil. • **19.** In July 2009, Turkey, Romania, Bulgaria, Hungary and Austria signed a deal to begin construction on the Nabucco pipeline, due to be finished in 2014. • **20.** Pierre Noël, presentation of European Council on Foreign Relations report, 3 February 2009. • **21.** *BP Statistical Review of World Energy 2008*. • **22.** Rodova, Nadia, 'Russian output to fall after 2020: National Energy Ministry', *Platts*, 8 December 2008. • **23.** *World Energy Outlook 2008*, International Energy Agency, December 2008. • **24.** See, for example, 'Trouble in the Pipeline', the *Economist*, 8 May 2008. • **25.** Quoted in Yenikeyeff, Shamil Midkhatovich and Krysiek, Timothy Fenton, 'The Battle for the Next Energy Frontier: The Russian Polar Expedition and the Future of Arctic Hydrocarbons', *Oxford Energy Comment*, Oxford Institute for Energy Studies, August 2007. • **26.** Viktor Chernomyrdin was subsequently appointed Prime Minister by President Yeltsin in 1992, serving until 1998. In 2001 President Putin named him Moscow's ambassador to the Ukraine, a post he held until June 2009. • **27.** In fact, Gazprom's production of natural gas was almost unchanged between 2003 and 2007. Official production figures are as follows: 512.0 bcm (2001), 525.6 bcm (2002), 547.6 bcm (2003), 552.5 bcm (2004), 555.0 bcm (2005), 556.0 bcm (2006), 548.6 bcm (2007). • **28.** Approximately 90 per cent of Russia's natural gas production comes from the Nadym Pur Taz region, particularly from three super-giant fields: Medvezhe, Urengoy and Yamburg. • **29.** Pierre Noël calls this the 'gas bridge'. In recognition of Turkmenistan's vital supply role, the price that Gazprom paid for Turkmen gas has increased, though it remains far below the onward sale price of natural gas in Western Europe. • **30.** Gazprom forecasts overall production to reach 570 bcm in 2010, 610–615 bcm in 2015 and 650–670 bcm in 2020, broadly in line with Russia's national energy strategy which projects a 33 per cent growth in gas production between 2010 and 2030. These numbers depend on a hugely optimistic timeline for Yamal and Shtokman. • **31.** At the time of writing, Yamal was forecast to come onstream by 2011, though many analysts suggested a start date of 2015 was more likely. Gas deliveries from Shtokman are forecast to start in 2013, but many view that as overly optimistic. • **32.** Gazprom's financial position deteriorated rapidly as the world economy turned down from late 2008. Investment in 2009 was estimated at 920 billion rubles ($33 billion), down from previous estimates of over 1 trillion rubles. The impact of falling oil prices – which have a knock-on impact on Russian gas prices through the existing price formula – may delay further investment decisions. The Shtokman field is considered to be profitable at an oil price of $50–$60 per barrel. • **33.** *Yamal Megaproject*, Gazprom, 2008. (www.gazprom.com/documents/Book_MY_Eng_1.pdf) • **34.** Tchurilov, Lev, *Lifeblood of Empire: A Personal History of the Rise and Fall of the Soviet Oil Industry*, PIW Publications, New York, 1996. • **35.** Ibid., pp. 125–126. • **36.** Ibid., p. 19. • **37.** Tchurilov identified what amounted to a distracting bureacratic war of targets: 'Soviet geologists tended to overestimate reserves so as to increase their standing at the Ministry. In contrast, oilmen had the unfortunate reputation for underestimating potential production in the hope that less demanding annual targets would be set.' Ibid., p. 150. • **38.** *The Impending Soviet Oil Crisis*, Intelligence Memorandum ER77-10147, March 1977. • **39.** The negotiations themselves had actually begun in 1989, before the collapse of the Soviet Union. • **40.** 'Joint US–Russian Statement in Support of the Timan–Pechora Project', White House Press Release, 28 September 1994. In an unfortunate echo of the exploitative Turkish Petroleum Company – the consortium of Western oil companies which produced all of Iraq's oil from the late 1920s to 1961 – the Timan–Pechora Company was publicly referred to by its three-letter acronym: TPC. • **41.** The Production Sharing Agreement was signed in 1995 and came into force in 1999, with Total as the lead partner with 50 per cent of the project, NorskHydro with 40 per cent and a Russian Nenets Oil Company with 10 per cent. In 2008 it became clear that Total would lose its majority position through the sale of 20 per cent of the project to the state-owned Zarubezhneft company. • **42.** The BP Statistical Review of World Energy 2008 put the proved natural gas reserves of the United States in 2007 at 5.98 trillion cubic metres. The combined reserves of the United Kingdom, the Netherlands, Italy, Germany, Poland and Romania were put at 2.75 trillion cubic metres. • **43.** Up to now, unwilling to buy in the technology, Gazprom has sought to acquire it in other ways. In 2006, on the margins of a state visit by President Putin to Algeria – where Russia wrote off some $5 billion in Algerian debts and agreed to a $7.5 billion contract for Russian military hardware – Gazprom forged an alliance with Sonatrach, the state gas company of Algeria and one of the world's LNG pioneers. Elsewhere, Shell is considering building an LNG plant to export from Sakhalin. But given the primary importance attached to gas exports by pipeline to Europe, and the risk of a substantial gap between Gazprom's production and its existing export obligations, some argue that Shtokman's planned LNG plant, to be built in the derelict Russian fishing village of Teriberka, may yet be cancelled. • **44.** Foreign companies had been involved in discussions over Shtokman, on and off, for nearly twenty years. In 2005, after substantial submissions from eleven oil and gas companies, Gazprom had drawn up a shortlist of five suitors for the Shtokman development: Chevron, ConocoPhillips, NorskHydro, Statoil and Total. Negotiations with these companies were expected to end up with Gazprom choosing one of these companies or, more likely, a consortium made up of a number of the group. • **45.** The rights to the Shtokman field are owned by a Russian company called Sevmorneftegaz, owned 100 per cent by Gazprom. • **46.** 'Total Wins Share in Shtokman: Chat Between Putin and Sarkozy Gives French Company Rights to Shtokman Development', *Kommersant*, 13 July 2007. • **47.** See Øverland, Indra, 'Shtokman and Russia's Arctic Petroleum Frontier', *Russian Analytical Digest*, No. 33, January 2008. • **48.** Total is involved in liquefaction projects in Abu Dhabi, Indonesia, Nigeria, Norway, Qatar and Oman. • **49.** In September 2009, Putin invited western executives (including the CEOs of Shell, Total, ConocoPhillips, StatoilHydro, E.ON, ENI, Mitsui and Mitsubishi) to Salekhard, well above the Arctic circle, to explain the terms on which they would be allowed to participate in the Yamal project. Access, he explained, would depend on a willingness to transfer technology to Russia, and to swap upstream assets in Russia with downstream assets (such as refineries) in developed markets. Though attempting to negotiate from a position of strength, the Russians know that if their Arctic ambitions are to be fulfilled – particularly with regard to LNG – they need both western technology and capital. • **50.** The new port management company is controlled by the Murmansk Sea Commercial Port (40 per cent), the Russian Railways (25 per cent), the Rosmorport (15 per cent), Rosneft (15 per cent) and the Murmansk regional administration (5 per cent). • **51.** Helmer, John, 'Russia Opening Up the Arctic Market: Icebreaking is the Key to the Expansion of Energy and Mineral Shipments in the Arctic', *Mineweb*, 17 March 2008. • **52.** Recent problems of the Sevmash yard include cost overruns and delays on the refurbishment of the Admiral Gorshkov aircraft carrier leased to India, delays on the construction of Russia's fourth-generation nuclear submarines and delays on the construction of controversial floating nuclear power stations. • **53.** The ten listed by the Blacksmith Institute were Sumgayit (Azerbaijan), Tinfen (China), Tianying (China), Sukinda (India), Vapi (India), La Oroya (Peru), Dzerzinsk (Russia), Norilsk (Russia), Chernobyl (Ukraine) and Kabwe (Zambia). *The World's Most Polluted Places*, Blacksmith Institute, September 2007. • **54.** Kramer, Andrew E., 'Shell Cedes Control of Sakhalin-2 to Gazprom', *International Herald Tribune*, 21 December 2006. If only indirectly, were environmental protesters against Shell essentially co-opted into acting as the instruments of the Kremlin's policy of resource control? Certainly, new laws on NGO registration have made it far more difficult for environmental NGOs to maintain their independence. • **55.** The fact that TNK-BP's Russian and British shareholders have different visions for the future of the company doesn't help. • **56.** One example is horizontal drilling. In some cases it is a necessity for year-round production, avoiding the risk of potential ice damage to floating infrastructure by drilling horizontally from an onshore drill site. But it also tends to reduce the ecological footprint of oil and gas extraction. • **57.** Vinocur, John, 'Europe Looks at Putin with Prudence and Respect, and at Bush with Indifference', *International Herald Tribune*, 9 June 2008.

10 End of Empire in Alaska

1. In 2001, an independent task force convened by Vice President Dick Cheney concluded the 'central dilemma' for future US energy policy was that 'the American people continue to demand plentiful and cheap energy without sacrifice or inconvenience'. Morse, Edward L. and Jaffe, Amy Myers, *Strategic Energy Policy Challenges for the 21st Century: Report of an Independent Task Force Cosponsored by the James A. Baker III Institute for Public Policy of Rice University and the Council on Foreign Relations*, 2001, p. vi. • **2.** Approximately 270,000 gallons of oil were spilled in one incident. A former BP employee told me that, in his view, pressure from other oil companies with a stake in Alaskan oil was at least as much to blame as the culture of financial – as opposed to operational – management within BP itself. 'The order from head office was always: "Cut costs by 20 per cent,"' he said, 'even if there wasn't 20 per cent to cut. And the other companies around the table were pushing just as hard.' • **3.** Oil brands have always been important, but increasingly they are designed to associate the company with broader qualities. BP, originally British Petroleum, has become Beyond Petroleum. Shell advertises itself as 'responsible energy'. Chevron has run a highly successful campaign on energy security, with the byline, 'Will you join us?' and the slogan, 'The power of human energy.' ExxonMobil has adopted, 'Taking on the world's toughest energy challenges.' For a view of the world from the point of view of corporate oil, see Clarke, Duncan, *Empires of Oil: Corporate Oil in Barbarian Worlds*, London, 2007. • **4.** BP owns 46.93 per cent of the pipeline, ConocoPhillips owns 28.29 per cent, ExxonMobil has a 20.34 per cent stake, with the remaining shares being held by Unocal (1.36 per cent) and Koch Alaska (3.08 per cent). • **5.** For Conoco Phillips, production from Alaska amounted to 263,000 bpd from a core group production figure of 856,000 (excluding equity affiliates), while Alaskan reserves amounted to 1.36 billion barrels of oil, out of 3.20 billion barrels of oil for the core group (excluding equity affiliates.) (ConocoPhillips, *Factbook 2007*.) BP's Alaskan operations produced 209,000 bpd, out of a group total of 1.3 million bpd. This latter figure excludes major equity-accounted entities, the major part of which was BP's share of TNK-BP's production. (BP, *Annual Report and Accounts 2007*.) • **6.** *Alaska's North Slope: Requirements for Restoring Land after Production Ceases*, General Accounting Office, GAO-02-357, June 2002. • **7.** The 5.9 billion barrel Kuparuk field came on-stream in 1981; the Alpine field entered production in 2000 and, after three capacity expansions, was producing 74,000 bpd by 2006. Lisburne started producing in 1986 from an estimated in-place reserve of 1.8 billion; the smaller Point McIntyre field, thought to contain around 900 million barrels of oil, came onstream in 1993. The 1.1 billion barrel Endicott field was brought into production in 1987, requiring the construction of a causeway and a 45-acre artificial island three miles into the Beaufort Sea; in November 2001 BP opened the controversial Northstar field, the first stand-alone offshore field in the Arctic, connected to the Alaskan coast by a six-mile subsea pipeline. BP faced down the opposition of the local Inupiat community in the process and defeated a legal petition which claimed the Minerals Management Service had not taken full account of potential environmental impacts when granting a permit for the field to be developed. • **8.** In the late 1990s, after exhaustive environmental assessments, the Clinton administration directed the Bureau of Land Management to prepare for leasing of lands in NPR-A.

(See Bourne, Joel, K., 'Fall of the Wild: Our Appetite for Oil Threatens to Devour Alaska's North Slope', *National Geographic*, May 2006.) The Bush administration pursued an active leasing programme in NPR-A, often in the face of local and environmental opposition, with one major reversal around Teshepuk Lake. • **9.** National Energy Technology Laboratory, *Alaska North Slope Oil and Gas: A Promising Future or an Area in Decline?*, DOE/NETL-2007/1279, August 2007, pp. 3–125. • **10.** The pipeline has already undergone a 'strategic reconfiguration' to take account of the fall in throughput from a design capacity of 2 million bpd to just over one third of that now. The permits for the pipeline were renewed for a further thirty years in 2004, but some argue that the owners do not believe the pipeline will last that long. One campaigner told me, 'BP is milking [the pipeline] at the moment – they are running it as a depreciating asset; they are not running it as if it is going to be used for fifty years.' • **11.** The USGS estimate is a probability – the true volume of oil and gas may be considerably greater or lower than the headline figure of 30 billion. The techniques used by USGS were geologic synthesis and analogue modelling, rather than an analysis of discovery history, prospect counting or reservoir modelling – principally because there are too few existing wells or discoveries to be statistically significant for the whole area. Some are highly critical of such methodology. United States Geological Survey, *Circum-Arctic Resource Appraisal: Estimates of Undiscovered Oil and Gas North of the Arctic Circle*, USGS Fact Sheet 2008–3049, 2008. The ANWR figures are based on a 1998 USGS assessment *Arctic National Wildlife Refuge, 1002 Area, Petroleum Assessment*, USGS Fact Sheet FS-028-01, 1998. The USGS estimated that most of that oil was onshore, on federal – rather than native – lands. In the probability distribution of oil, USGS estimated a 5 per cent probability of either 5.7 billion or 16.0 billion barrels of oil. • **12.** Speech made by Senator Lisa Murkowski to the Arctic Energy Summit, Anchorage, 15 October 2007. • **13.** The original song, 'North to Alaska' was written by country singer Johnny Horton. • **14.** Alaskans receive annual dividends from the multi-billion dollar Alaska Permanent Fund. In 2008, anyone who had been an Alaska resident for the entire previous calendar year could claim a total of $3,269, including a one-off $1,200 payment. • **15.** Marshall, Bob, 'The Problem of the Wilderness', *Scientific Monthly*, 30/2, February 1930, pp. 141–148. Amongst other things, Marshall argued that 'this theorizing is justified empirically by the number of America's most virile minds, including Thomas Jefferson, Henry Thoreau, Louis Agassiz, Herman Melville, Mark Twain, John Muir and William James, who have felt the compulsion of periodical retirements into the solitudes.' A father of the wilderness movement in the United States, Marshall was one of the inspirations behind the founding of the Arctic National Wildlife Refuge. • **16.** Amongst the loudest advocates of the national security argument for developing Arctic oil is the industry itself. The following extract from a speech by BP America's Robert Malone is a case in point: 'The US accounts for just 5 per cent of world oil production. We purchase and consume 25 per cent. And rather than open new areas to exploration and production, US presidents periodically swallow their pride and ask foreign oil ministers to increase oil production in foreign lands . . . The resource estimates for the places now off limits [including, but not limited to parts of the offshore American Arctic] exceed 100 billion barrels, with 30 billion barrels recoverable. There could be more. There could be less. We can't know unless we explore.' (Speech to the National Governors' Association, July 2008.) In 2006, a blue-ribbon report into US dependency on imports of foreign oil, directed by David G. Victor and chaired by John Deutch (former Director of the CIA) and James Schlesinger (former Director of Central Intelligence, Secretary of Defense and Secretary of Energy) stated that 'increased supply from US sources, just as increased supply from any source, places a downward pressure on oil prices. But increased supply from US sources has the additional foreign policy benefit of directly reducing imports of oil into the United States.' (*National Security Consequences of US Oil Dependency*, Independent Task Force Report No. 58, Council on Foreign Relations, New York, 2006.) • **17.** These and subsequent figures for petroleum imports are derived from the Energy

Information Administration statistics database. • **18.** Net imports peaked at 8.6 million bpd in 1977, falling to 4.3 million bpd by 1985. The 1977 figure remained the highest recorded figure for net imports of petroleum until 1997. But the reasons for the decline in imports were not all related to the existence of new domestic supply from the Arctic. Total consumption of crude oil and petroleum peaked at 18.9 million bpd in 1978, then fell rapidly during the harsh recession of the early 1980s, bottoming out at 15.2 million in 1983. Consumption then rose for most of the 1980s and 1990s, with a falling off in the recession of the early 1990s. Consumption peaked at 20.8 million barrels in 2005, falling to 20.7 million in 2007. In effect, nearly three quarters of the decline in net oil imports in the late 1970s and early 1980s came from demand decline rather than supply increase. • **19.** In 2007, the Department of Energy reported that Alaska North Slope Production 'still represents about 17 per cent of the U.S. domestic production'. National Energy Technology Laboratory, *Alaska North Slope Oil and Gas: A Promising Future or an Area in Decline?*, DOE/NETL-2007/1279, August 2007, p. vii. This figure, however, was derived from production figures from 2003. The preliminary figures from the EIA's *Annual Energy Review 2007* put Alaskan production at 719,000 bpd out of a US domestic total of 5,103,000. But there are lies, damn lies and statistics. The use of the word 'energy' as interchangeable with 'oil' further inflates the apparent importance of Alaska to US energy supplies. • **20.** Former House Leader Newt Gingrich, Chairman of American Solutions for Winning the Future, wrote a book entitled *Drill Here, Drill Now, Pay Less: A Handbook for Slashing Gas Prices and Solving Our Energy Crisis*, (Washington D.C., 2008). The title of the book was subsequently turned into a country song. • **21.** Energy Information Administration, *Analysis of Crude Oil Production in the Arctic National Wildlife Refuge*, SR/OIAF/2008–03, May 2008. • **22.** One can point to several ways in which the conclusions of the EIA report need to be contextualised. First, the report only covered one area from which Arctic oil and gas might be produced. It did not look at the impact on national import dependence and gasoline price of, say, opening the entire North Slope to oil and gas development, plus the entire Beaufort and Chukchi Seas. Second, the report assumed future levels of domestic demand which could prove too high. If some future level of Arctic oil production was closing an import gap which had already been narrowed by the widespread adoption of domestic energy efficiency measures, for example, reducing demand at home, then the impact of additional domestic supply on dependence and price would be all the greater. Third, the report looked at crude oil – whereas some argue that the reason that US Arctic development should be encouraged is to boost the supply of natural gas. • **23.** State jurisdiction applies to waters less than three nautical miles offshore. Federal jurisdiction applies from that point to the edge of the Outer Continental Shelf. • **24.** Shell bid $105,304,581 for tract 6763, adjacent to the Burger well. The next-highest bid on the same tract was ConocoPhillips's bid of $36 million. • **25.** The discovery was led by Unocal, the California-based oil company now part of Chevron. The lease was relinquished in 1999. • **26.** Shell Offshore Inc., *2008 Beaufort Sea Exploration Drilling Program*, 15211-02.03-07-017A/07-470A, November 2007. • **27.** Sarah Palin was reported as saying, 'I am very disappointed in the 9th Circuit Court of Appeals ruling blocking Shell Oil Company from drilling in the Beaufort Sea . . . This is the second development project with costs exceeding $200 million to be blocked by an action by this court. Decisions such as these pose a threat to our economic future. Nevertheless, I remain committed to help responsible parties develop Alaska's resources in a manner that protects our way of life.', 'Shell's Beaufort Drilling Plans Dealt Costly Setback by Court', *Anchorage Daily News*, 16 August 2007. • **28.** See Mouawad, Jad, 'US Court Says Shell Can't Drill Near Alaska', *International Herald Tribune*, 21 November 2008. • **29.** Personal communication. • **30.** Personal communication. • **31.** See, for example, Minerals Management Service, Alaska Outer Continental Shelf region, *Alaska Annual Studies Plan: Final FY 2009*, September 2008. • **32.** The Arctic National Wildlife Refuge is much larger, covering 19 million acres, of which 8 million are designated as a wilderness area. • **33.** Miller, Pamela A., *Protecting* *the Arctic: An Exciting New Chapter, The Northern Line*, Winter 2008. • **34.** An excellent book has been written on the background to the refuge, and the conflicting ideas of land use that it brought out. Kaye, Roger, *Last Great Wilderness: The Campaign to Establish the Arctic National Wildlife Refuge*, Fairbanks, 2006. • **35.** *In This Place: A Guide for Those Who Would Work in The Country of The Kaktovikmiut, An Unfinished and On-going Work Of the People of Kaktovik, Alaska*. The document is available at www.kaktovik.com. • **36.** See Wohlforth, Charles, *The Whale and the Supercomputer: On the Northern Front of Climate Change*, New York, 2004. • **37.** Cheryl draws a distinction between 'grey' (semi-polluted) and 'black' (polluted) water. The standard to which any discharges will be treated is the key factor. • **38.** This is one of several responses highlighted in a document entitled *Shell's Beaufort Sea Exploratory Drilling Program Oil Spill Response*.

11 Balance: Norway and the Arctic Model

1. In December 2008, the United Nations listed Iceland's Human Development Index (HDI) score at 0.968, with Norway on the same level. Canada came third (0.967) and Australia fourth (0.965). For comparison, the United States was fifteenth (0.950), the United Kingdom was twenty-first (0.942), Germany was twenty-third (0.940) and Russia was seventy-third (0.806). Sierra Leone was last (0.329). • **2.** In reality, however, the World Wildlife Fund puts Norway eighth globally in terms of its ecological footprint on a per capita basis, just behind Canada, suggesting it is actually one of the least sustainable countries in the world. See *Living Planet Report 2008*, p. 22. • **3.** The last Norwegian recipient was Fridtjof Nansen, the Arctic explorer turned diplomat, in 1922. • **4.** The World Commission on Environment and Development was convened in 1983, reporting that year. The report was published by the Oxford University Press as *Our Common Future* in 1987. • **5.** Most of the companies blacklisted are in the defence sector, from BAE Systems to Lockheed Martin and Northrop Grumann. But, from 2008, the list included mining company Rio Tinto, entailing the divestment of an investment valued at approximately 4.8 billion Norwegian crowns ($550 million). • **6.** One of the people who works on these issues for StatoilHydro, Olav Kaarstad, has also worked on the Intergovernmental Panel on Climate Change on carbon dioxide capture and storage and co-authored a book on the future of fossil fuel use. Freund, Paul and Kaarstad, Olav, *Keeping the Lights On: Fossil Fuels in the Century of Climate Change*, Oslo, 2007. The argument of whether CCS used to enhance oil recovery slows or accelerates climate change hinges on whether one believes the additional oil produced is replacing production that would have occurred elsewhere in the world without it, and whether the amount of carbon dioxide sequestered is greater than that which will be released by burning the hydrocarbons produced. • **7.** In 1905, partly to snub the Swedish royal family, the Danish Prince Carl became Haakon VII, taking the name of a fourteenth-century Norwegian king. The independence of the country is traditionally viewed as dating from 7 June 1905 but the country celebrates its national day on 17 May, the anniversary of the proclamation of Norway's 1814 constitution. • **8.** See Kindingstad, Torbjørn and Hagemann, Fredrik, eds., *Norwegian Oil History*, Stavanger, 2002. • **9.** Detailed information on all of Norway's oilfields can be obtained from the Norwegian Petroleum Directorate. • **10.** See Gordon, Richard and Stenvoll, Thomas, *Statoil: A Study in Political Entrepreneurship*, James A. Baker III Institute for Public Policy of Rice University, March 2007. • **11.** In 1972, Statoil's first task was to administer the state's equity interest in oilfields – observing and learning, but not operating any fields itself. This changed in 1981, when the company became the operator of the Gullfaks field – halfway between the Shetland Isles and the south-west coast of Norway – now expected to be producing oil up until 2030. • **12.** Discriminatory provisions in domestic law in favour of Norwegian companies were removed in 1991. The coming into force of the European Economic Area treaty in 1995 means that, additionally, Norway is bound by European Union rules on state aid. • **13.** Personal communication. • **14.** *BP Statistical Review of World Energy 2008.* • **15.** In 2007 Norway's consumption of oil amounted to 221,000 bpd giving net exports of 2,335,000 bpd. In Kuwait, production was 2,626,000 bpd but consumption was higher than in Norway, at 276,000 bpd giving a net figure of 2,350,000 bpd. At the same time, while Norway's proved reserves of oil were estimated at 8.5 billion barrels of oil, Kuwait's were estimated at 101.5 billion barrels. *BP Statistical Review of World Energy 2008.* • **16.** The Barents Sea is estimated to contain 6.2 billion barrels of oil equivalent (boe) of which more than half is natural gas, compared to 7.5 billion boe in the North Sea. However, because the Barents Sea is relatively less well known, the upward boundary of probability (representing a 10 per cent probability) is much higher for the Barents Sea. The 'high estimate' for the Barents Sea is 10.7 billion boe, compared to 8.7 billion boe for the North Sea. The Norwegian Sea is estimated to be larger than both, with an expected resource of 7.7 billion boe and a 'high estimate' of 11.1 billion boe. The Royal Norwegian Ministry for the Environment, Report No. 8 to the Storting: *Integrated Management of the Marine Environment of the Barents Sea and the Sea Areas off the Lofoten Islands*, (*Helhetlig forvaltning av det marine miljø i Barentshavet og havområdene utenfor Lofoten [forvaltningsplan]*) 2006, p. 38. • **17.** Ministry of Petroleum and Energy/Norwegian Petroleum Directorate, *Facts 2008*, p. 14. • **18.** Interview, European Commission, Brussels, 2008. • **19.** The government held approximately 71 per cent of Statoil stock and 44 per cent of NorskHydro stock ahead of the October 2007 merger. After the merger of NorskHydro's petroleum business with Statoil, the state's ownership of the combined StatoilHydro group was diluted to 62.5 per cent. However, the state has announced the intention to increase this to 67 per cent over time. • **20.** 'Mr Lund's claims of very high environmental and safety standards are not supported by StatoilHydro's recent track record,' argues Rasmus Hansson, head of the World Wildlife Fund in Norway. Hansson, Rasmus, 'StatoilHydro's Lack of Vision', *Arctic Bulletin*, 2/08, 2008. • **21.** The Goliath field (known as Goliat in Norwegian) was discovered in 2000, after ENI had been awarded the block as lead company in the 1997 Barents Sea round. The licensees are ENI with 65 per cent, StatoilHydro with 20 per cent and Det norske oljeselskap with 15 per cent. • **22.** *Global Climate will be the Loser if Norwegian Oil is Left in the Ground*, Press release from the Oljeindustriens Landsforening (Norwegian Oil Industry Association), 16 October 2008. • **23.** *Production Development on the Norwegian Continental Shelf*, Konkraft, December 2008. • **24.** That strategy was put on hold in the late 1980s after the Mongstad refinery scandal undercut the company's strategy, unseated Johnsen and renewed fears that Statoil was getting out of hand. Cost overruns on the re-equipping of the Mongstad refinery had led to questions over the competence of Statoil to manage large-scale infrastructure projects. Arve Johnsen was forced to resign in 1987. Problems with the Mongstad refinery, coupled with a collapse in oil prices in the mid-1980s, cut the state's revenues from oil and gas significantly. Whereas in 1986, the state had received 1.2 billion Norwegian crowns in dividends alone from Statoil, the company paid no dividends at all in 1988 and 1989. The overall share of state revenues coming from the oil and gas sector declined from 20 per cent in 1984 to just 1 per cent in 1988. • **25.** The Royal Norwegian Ministry for the Environment, trans. Report No. 8 to the Storting: *Integrated Management of the Marine Environment of the Barents Sea.* For a description of how the plan was put together see Olsen, Erik; Gjøsæter, Harald; Røttingen, Ingolf; Dommasnes, Are; Fossum, Petter; and Sandberg, Per, 'The Norwegian ecosystem-based management plan for the Barents Sea', *Journal of Marine Science*, 64 (4), pp. 599–602. • **26.** The Norwegian–Russian Environmental Commission was set up 1988. The management of the Barents Sea is only one of the issues it covers. Historically, it has been air and water pollution that have been more prominent in the commission's work. • **27.** Gazprom projects Shtokman to begin producing natural gas by 2013, with an initial annual production level of 23.7 bcm. Current plans are for LNG exports to begin in 2014. • **28.** StatoilHydro owns 33.53 per cent of the project. The other owners are Petoro (30 per cent), Total (18.4 per cent), Gaz de France (12 per cent),

Hess (3.26 per cent) and RWE Dea (2.81 per cent). • **29.** The compressor was designed using proprietary technology owned jointly by StatoilHydro and Linde, the German engineering company. • **30.** The plant itself was built in Spain, in Cádiz, and then transported by ship to Hammerfest in an eleven-day trip. But it wasn't entirely finished when it arrived, 'so more work than expected had to be done up here'. • **31.** See, in particular, Huebert, Robert and Yeager, Brooks, *A New Sea: The Need for a Regional Agreement on Management and Conservation of the Arctic Marine Environment*, World Wildlife Fund, January 2008. • **32.** Stokke, Olav Schram, 'A Legal Regime for the Arctic? Interplay with the Law of the Sea Convention', *Marine Policy*, Vol. 31, No. 4, 2007, pp. 402–408. • **33.** Two other Arctic states, Iceland and Finland, were excluded on the basis that they are not coastal states, though Iceland – just below the Arctic Circle – would certainly be heavily affected by any decisions on the status or management of the Arctic marine environment. The conference, called by the Danish Foreign Ministry, took place in Ilulissat on 27–29 May 2008.

PART FIVE Freedom

12 Greenland's Search for Independence

1. Legally, the Act became law on 12 June in Denmark, having received royal assent from Queen Margarethe II at Amalienborg palace, Copenhagen. • **2.** The *Self-Government Act* provides for Greenland to enter into some international agreements, and to have some right of influence over international agreements entered by Denmark which may affect Greenland, but these are limited. Greenland's powers 'shall not limit the Danish authorities' constitutional responsibility and power in international affairs, as foreign and security policy matters are affairs of the [Danish] Realm' (Chapter 4). • **3.** Chapter 8 of the *Self-Government Act* includes the important caveat, 'An agreement between Naalakkersuisut [the Greenlandic administration] and the [Danish] Government regarding the introduction of independence for Greenland shall be concluded with the consent of the Inatsisartut [Greenlandic parliament] and shall be endorsed by a referendum in Greenland. The agreement shall, furthermore, be concluded with the consent of the Folketing [the Danish parliament in Copenhagen].' Independence for Greenland is therefore not a right, but one possible outcome of a future constitutional process. • **4.** There is a minimum amount of 75 million Danish crowns (around $15 million) which the Greenlandic authorities will receive from resource extraction before it will count against the subsidy. • **5.** Some argue that this was in fact a step back from the Home Rule agreement, in that it reintroduces uncertainty around the future division of income from resource exploitation on Greenland. • **6.** As of 2008, the Greenland Home Rule government owned 37.1 per cent of Nunaminerals, with the intention for this holding to reduce over time. • **7.** Revenues for BHP Billiton in 2008 were $59.5 billion, with $5.8 billion budgeted for capital expenditure on projects approved in 2008 alone. Group revenue at Anglo-American (including associates) amounted to $30.6 billion in 2007, with $12 billion of projects under development. • **8.** The alliance agreement was signed in December 2007. • **9.** This complements LKAB's major sources of iron ore. • **10.** Alcoa suggests that the final full-time employment at the plant, once it is operating, would be six hundred. After a Memorandum of Understanding signed between Alcoa and the Home Rule government in May 2007 the project moved into a preparatory phase in 2009, with the expectation that the smelter could come online as early as 2014–15. Greenland's government will decide on its level of financial participation in the project in 2010. • **11.** Much later, when European geologists came to Disko Bay in the eighteenth century, they disagreed about the origin of the iron deposits. Some believed that the iron came from meteors, brought down by ancient meteor showers. • **12.** The Bergen Greenland Company (1723), from Bergen, Norway; the Greenland Trading Company (1734); the General Trading Company (1747). • **13.** In a February 1906 answer to a question from the ministry regarding an application from a number of private Danish companies to become involved in fishing and commercial rearing of sheep, reindeer, eider ducks and foxes. Quoted in Sorensen, Axel Kjaer, *Denmark–Greenland in the Twentieth Century*, Copenhagen, 2006, p. 27. • **14.** When the mine closed in 1987, some 3.7 million tons of cryolite ore had been produced. • **15.** Aluminium and silver both cost around $1 an ounce. The pyramid weighed approximately 2.8 kilograms; total US production, all from the Frishmuth Foundry in Philadelphia, was estimated at 51 kilograms. Production in France and in the United Kingdom was considerably more than this. • **16.** Aviation became a major user in the 1930s, with aluminium consumption doubling between 1937 and 1939. Given the strategic importance of aluminium the US government imposed price restrictions on aluminium during the war at 15 cents per pound. • **17.** Commercial agreements were also in place with the Pennsylvania Salt Manufacturing Company of Philadelphia (Pennsalt) and with the Aluminium Company of Canada (Alcan). • **18.** For example, a Dutch trading ship visited Nuuk in 1656, returning to Holland with narwhal tusks, 900 seal skins, whalebone, fur clothing, weapons and Inuit kayaks. • **19.** Icelanders argued that the Act of Union was originally intended only for twenty-five years, to 1943. With renegotiations impossible under conditions of German occupation of Denmark, renunciation of the Act of Union therefore merely confirmed the expiry of the original act. Danes viewed the vote as a betrayal. But Icelanders' votes were unequivocal: 97 per cent in favour of abolishing the 1918 Act of Union and 95 per cent in favour of a new republican constitution. • **20.** The Danish parliament voted to retrospectively accept the base agreement which had been made without Copenhagen's specific approval in 1941 by the then Danish ambassador to the United States, Henrik Kauffmann. Later, there would be moves to rescind the agreement as both being inconsistent with full Danish sovereignty and potentially offering a pretext for Soviet forces to remain on the strategically important Bornholm Island, deep in the Baltic Sea, due north of the Soviet sector of occupied Germany. • **21.** An article from *Time* magazine in January 1947 put the strategic attractions of Greenland to the US succinctly: 'Greenland's 800,000 square miles make it the world's largest island and stationary aircraft carrier. It would be as valuable as Alaska during the next few years, before bombers with a 10,000-mile range are in general use. It would be invaluable, in either conventional or push-button war, as an advance radar outpost. It would be a forward position for future rocket-launching sites. In peace or war it is the weather factory for northwest Europe'. 'Deepfreeze Defense', *Time*, 27 January 1947. • **22.** Denmark hesitated to join NATO in 1948. It ultimately agreed to do so with the understanding, as with Norway, that foreign troops would not be stationed on its territory in peacetime. This did not apply to Greenland, where basing agreements with the US were already in place, confirmed in 1951. Greenland was considered part of the Western Hemisphere under the 1948 Rio Treaty, though a non-signatory. • **23.** Under the United Nations charter, member states were required to list their colonial possessions. Despite some doubts in the Danish foreign ministry, Denmark registered Greenland as a 'non-self-governing territory' in 1946. The one hold-out in the 1953 vote at the UN, Belgium, voted against on a matter of principle: deeply troubled by the attitude taken towards her possessions in Central Africa, she refused to recognise UN authority in colonial matters. • **24.** By 1962 fully 45 per cent of Greenlanders were on the public payroll. The proportion of Danes in Greenland's population increased fourfold between 1950 and 1970. • **25.** For example, the Inuit Circumpolar Council was founded at Barrow in 1977. • **26.** In 1926 the Danish government, under pressure from the Faeroese who were being pushed out of their traditional fishing grounds by British and German trawlers, granted them limited fishing rights in Greenlandic waters amounting to 97 nautical miles of coastline. By 1939 this had been extended to 425 nautical miles, despite Greenlandic opposition. • **27.** Broadcast interview on 4 November 1976. Quoted in Sorensen, Axel Kjaer, op. cit., p. 150. • **28.** In 1984 the EEC consisted of Belgium, Denmark, France, Greece, Ireland, Italy, Luxembourg, the Netherlands, the United Kingdom and West Germany, totalling 1.7 million square kilometres. Greenland covers 2.1 million square kilometres. • **29.** Calouste Gulbenkian acquired the name 'Mr Five Per Cent' from the 5 per cent financial interest that he retained, in 1914, from negotiations which principally brought Anglo-Persian (the forerunner of BP), Shell and Deutsche Bank to agree to cooperate within a consortium on the development of oil ventures in the territory of the Ottoman Empire. • **30.** DONG had revenues of 41.6 billion Danish crowns ($8 billion) in 2007. • **31.** StatoilHydro had revenues of 397.9 billion Norwegian crowns ($68 billion) in 2007. • **32.** Kanumas is the acronym for Kalaallit Nunaat Marine Seismic Project. The project covered some 7,000 kilometres of seismic lines. In return for the financial outlay, the sponsors of Kanumas received a preferential exploration position. • **33.** PGS Geophysical, CGG Veritas, Wavefield and Western Geco were all contracted to cover all or part of ten offshore leases in Greenland. The most aggressive seismic work was that led by PGS Geophysical for Cairn Energy, expected to cover over 8,000 square kilometres. • **34.** As of the middle of 2008, the following leases were held: West Disko area: Sigguk and Eqqua (87.5 per cent Cairn, 12.5 per cent Nunaoil); Kangerluk and Ikermiut (87.5 per cent Husky, 12.5 per cent Nunaoil); Puilassoq (29.17 per cent Chevron, 29.17 per cent DONG, 29.17 per cent Exxon, 12.5 per cent Nunaoil); Orsivik (43.75 per cent ExxonMobil, 43.75 per cent Husky, 12.5 per cent Nunaoil); Naternaq (PA Resources). West Greenland: Atammik and Lady Franklin (47.5 per cent EnCana, 40 per cent Cairn, 12.5 per cent Nunaoil). Open Door Area: Saqqamiut and Kingittoq (92 per cent Cairn, 8 per cent Nunaoil). • **35.** United States Geological Survey, *Assessment of Undiscovered Oil and Gas Resources of the East Greenland Rift Basins Province*, Fact Sheet 2007–3077, August 2007. The survey estimated 86.1 trillion cubic feet of natural gas (approximately 2,400 billion cubic metres of natural gas, equivalent to 15 billion barrels of oil) and an additional 8.1 billion barrels of Natural Gas Liquids (NGLs). • **36.** United States Geological Survey, *Assessment of Undiscovered Oil and Gas Resources of the West Greenland–East Canada Province*, USGS Fact Sheet 2008–3014, May 2008. • **37.** Freight rates have been reduced by 10 per cent since 2003, and have not been increased since. In 2007, Royal Arctic Line's turnover amounted to 793 million Danish crowns ($155 million). • **38.** This interview was conducted in 2008. • **39.** Potential economic benefits of climate change to Greenland are not restricted to the retreat of ice from the land, or the increase in run-off from Greenlandic glaciers. Sheep farming has long been a marginal part of the economy around Qaqortoq in southern Greenland, the area settled by Norse farmers more than a thousand years ago. But now some believe that, as a result of warming temperatures, Greenland could become self-sufficient in vegetables over the next ten years.

13 Selling Iceland: The Grand Illusion

1. It took another hundred years before the population reached 100,000 in 1926. The next doubling of the population took place in just forty-two years, reaching 200,000 in 1968. In 2008 the Icelandic Statistics Office estimated the country's population at 313,000. • **2.** The island of Flatey gave its name to one of the largest books of Icelandic manuscripts, the fourteenth-century *Flateyjarbók*. In the seventeenth century the book was taken to Copenhagen, and not returned to Iceland until the late twentieth century. • **3.** Newspaper article from *Althydubladid*, December 1959, quoted in Ingimudarson, Valur, 'Immunizing Against the American Other: Racism, Nationalism, and Gender in US–Icelandic Military Relations during the Cold War', *Journal of Cold War Studies*, Vol. 6, No 4., Fall 2004, pp. 65–88. • **4.** In 1918 Iceland gained independence from Denmark (though up until 1944 it remained tied to Denmark for its defence and foreign policy, and remained conjoined to Denmark under the Danish monarchy). The 1918 agreement recognised Iceland's 'permanent neutrality'. • **5.** This despite the fact that Iceland's membership of the European Economic Area means that it must follow most EU law. • **6.** The Åaland Islands, an autonomous territory within Finland, have been officially demilitarised since the Treaty of Paris (1856). • **7.** Magnason, Andri Snær, *Dreamland: A Self-Help Manual for a Frightened Nation*,

London, 2008. • **8.** At several points in the 1950s the government actively contemplated the closure of the American airbase at Keflavik, even though it provided as much as one fifth of Iceland's foreign currency earnings. At the same time, the Soviet Union waged an active economic campaign to lure the country closer to Soviet positions. See Nuechterlein, Donald E., *Iceland: Reluctant Ally*, Ithaca, 1961 and Ingimundarson, Valur, 'Buttressing the North: The Atlantic Alliance, Economic Warfare, and the Soviet Challenge in Iceland, 1956–1959', the *International History Review*, Vol. XXI, No.1, March 1999, pp. 80–103. • **9.** Some view this as absurd. In May 2008, Steingrimur Sigfusson, leader of the Left-Green Party (and Minister of Finance since February 2009) argued that: 'If someone was going to attack us they would just wait till the Frenchmen are gone. We should rather use the money for real security issues – civil defence, police, drugs smuggling, being on alert for volcanoes and floods . . . But having an agreement with others to come here and play? The French like to come here because they can fly low and stay in a hotel at Iceland's expense. In my mind there's no justification. We wouldn't want a group of Hell's Angels taking over the country – but instead of having an army we should have a well-equipped police force. In time of war the Americans still have a responsibility to defend us. But they don't think they need a base for that. So why should we want someone else to have a base here? We need tow boats, we need cooperation, but we don't need military fighter planes.' • **10.** Magnason, *Dreamland*, p. 272. • **11.** Moody's *Global Sovereign Report: Iceland*, January 2009. • **12.** The final vote came in October 2008, just as the Icelandic economy was beginning to collapse. In the end Iceland was defeated by Austria and by Turkey, despite strong Nordic support for Iceland's candidacy. • **13.** At the height of Iceland's financial crisis, in October 2008, the British government used anti-terrorism legislation to freeze Icelandic assets, as it became clear that there would not be guarantees for foreign depositors in Icelandic banks. • **14.** Zarakhovich, Yuri, 'Why Russia is Bailing Out Iceland', *Time*, 13 October 2008. • **15.** Dr Herbertsson cited a World Bank report in defence of this thesis: Easterly, William and Kraay, Aart, 'Small States, Small Problems?', *Policy Research Working Paper 2139*, World Bank, June 1999. • **16.** Traynor, Ian and Gunarsson, Valur, 'Iceland to be Fast-Tracked into the EU', the *Guardian*, 30 January 2009. • **17.** See US State Department, *A Report on the Resources of Iceland and Greenland, Washington D.C., Government Printing Office*, 1868. • **18.** At the time of writing, an offshore licensing round was underway for the Northern Dreki area with licences to be awarded by October 2009. • **19.** Glitnir Geothermal Research, *United States Geothermal Energy Market Report*, September 2007. • **20.** United States Geological Survey, *Assessment of Moderate- and High-Temperature Geothermal Resources of the United States*, USGS Fact Sheet 2008–3082, 2008. • **21.** National Energy Authority, *Energy in Iceland: Historical Perspective, Present Status, Future Outlook*, September 2006. • **22.** At the time of writing, the website of Saving Iceland (www.savingiceland.puscii.nl) included articles linking aluminium production to large-scale environmental damage, genocide and exploitation in developing countries. The website also suggested that several Icelandic politicians were involved in corrupt practices and provided information on the home addresses of Alcoa staff members and information about various direct actions taken during the course of 2007 and 2008 to disrupt energy-related projects in Iceland. The 'About SI' section of the website described the organisation in the following terms: 'We are a network of people of different nationalities, who do not intend to stand by passively and watch the Icelandic government in league with foreign corporations slowly kill the natural beauty of Iceland. Icelandic environmentalists desperately need outside help to drive away the corporate threat to their island.' • **23.** Public utilities take around 2,100 gigwatt hours per year (GWh/y) per year. The Alcoa Fjardaal aluminium smelter (on the east coast) alone uses some 4,700 GWh/y; the smaller Alcan plant (on the west coast) uses 2,800 GWh/y; the Century Aluminium plant uses 1,400 GWh/y. • **24.** Magnason, *Dreamland*, p. 206. • **25.** While the exact terms of the contract between Alcoa and Landsvirkjun are secret, the shape of the deal struck between the two companies is public knowledge: Alcoa buys electricity from Landsvirkjun on a 'take-or-pay' basis, the contract has a duration of forty years, and the price at which electricity will be sold to Alcoa depends on a formula linked, in part, to the world price of aluminium. The contract is denominated in US dollars – which means that the financial crisis may actually have increased Landsvirkjun's income in Icelandic currency terms. If Alcoa were to walk away from the project it would be sued: 'Alcoa Fjardaal is an Icelandic company and Icelandic law applies.' Hilmarsson seems confident that the dam itself will have paid itself back over twenty or twenty-five years, including a reasonable rate of return. Still, 'It's complicated,' he tells me. • **26.** Magnason, *Dreamland*, p. 157. • **27.** In 2000, coal-fired power accounted for 40.2 per cent of electricity used by Alcoa's plants, and hydroelectric power accounted for 45.8 per cent of electricity use. By 2007, the figure for coal had fallen to 27.7 per cent and the figure for hydroelectric power had risen to 50.7 per cent. Over the same time period Alcoa's reported direct emissions of greenhouse gases – measured in equivalent to metric tons of carbon dioxide – fell from 38.8 million to 31.1 million. • **28.** Smelting aluminium involves attracting the oxygen out of molten aluminium oxide (alumina) using carbon anodes. As the oxygen binds to the carbon anodes, carbon dioxide is produced. • **29.** More fancifully, perhaps, Hilmarsson argues that the Karahnjukar project itself – the dam, reservoir and hydroelectric plant which provides Alcoa with the electricity it needs for the smelter – is actually a net environmental positive. 'We are binding thirty times more [greenhouse gases] than we are emitting,' Hilmarsson tells me. The reason? Revegetation of areas around the flooded area. 'We didn't even think we were doing it,' he says. 'It just turned out that way.' Again, many environmentalists think this argument is just playing with words, abusing meaning, a travesty of common sense. Viewing the Karahnjukar project as a net environmental positive, they argue, depends on ascribing essentially no value to the highland area which will be flooded as a result, and suggesting that the revegetation could only have occurred as a result of the building of the dam. • **30.** In November 2008 *Frettabladid*, Iceland's most popular newspaper, published a series of articles by Magnason and Sigurdsson expressing their opposing views of Alcoa's contribution to the export revenues of Iceland. • **31.** Alcoa, *Álver Rís*, 2007 (English version entitled *Fjardaal Takes Off*, also produced in 2007). • **32.** Magnason, *Dreamland*, p. 269. • **33.** One exception to this overall picture is a man called Gudmundur Beck, a former resident of Reydarfjördur quoted by Saving Iceland as saying, 'When the authorities and my neighbours talked about construction I always corrected them and told them that the building of the smelter and the Karahnjukar dam was not construction but terrorism. I said that I felt sorry for them because they were bringing us seventy years back to the period of heavy industry.' • **34.** In the north-east Iceland area some 67.9 per cent of respondents supported the building of the Fjardaal smelter, compared to 56.5 per cent nationwide. The lowest support for the construction of the smelter was in north Reykjavik (at 43.9 per cent). There were wide disparities in the source of support for the plant, however. While 64.6 per cent of male Icelanders support it, only 48.1 per cent of women did. And while 67.3 per cent of respondents aged between fifty-five and seventy-five supported Fjardaal, only 45.5 per cent of those aged between sixteen and twenty-four were in favour of the smelter. Source: Gallup / Capacent, March 2007. • **35.** Helguvik had been expected to produce its first aluminium in 2010. • **36.** From Sigfusson's perspective, big infrastructure investments are as much an economic headache as a solution for Iceland, bringing too much money into the economy over too short a space of time. 'I am very concerned that the East – and this is my constituency – will not be as well-served as they [Alcoa] have claimed, unless they take a lot of side measures,' worries Sigfusson. For the region, as for the country, 'It's not wise to have all your eggs in one basket.' • **37.** This does not include smelters for which plans have already been submitted. • **38.** The main reasons for this are Iceland's low average temperatures (thus reducing the need for cooling equipment for the computers), its 'green' energy credentials, the country's educated workforce and its well-connected communications system. Invest in Iceland Agency, *Iceland – The Ultimate Location for Data Centers*, 2008. • **39.** Landsvirkjun has already signed a contract to supply electricity to a server farm near Keflavik, owned by Verne Holding. Another company, called Greenstone, has plans for a 50,000 square metre site in the new Thorlákshöfn industrial zone. • **40.** PriceWaterhouseCoopers, *Benchmarking Study on Iceland as a Location for Data Centre Activity: A Green Window To the World*, May 2007. • **41.** For an overview of the history behind the dam, and the beginnings of opposition to its construction, see De Muth, Susan, 'Power Driven', the *Guardian*, 29 November 2003. • **42.** He also makes a comparison with Canada. Newfoundland was a British colony up until 1949, with the prospect of independence, 'But they gave that up and became part of Canada, dependent on the federal government. Now some of them might say that they would like to be like Iceland.'

Selected Bibliography

Books

Applebaum, Anne, *Gulag: A History*, London, 2003.
Bamberg, James, *British Petroleum and Global Oil, 1950–1975*, London, 1982.
Bergesen, Helge Ole; Moe, Arild; Østreng, Willy, *Soviet Oil and Security Interests in the Barents Sea*, London, 1987.
Bothwell, Robert, *The Penguin History of Canada*, Toronto, 2006.
Brigham, Lawson W., ed., *The Soviet Maritime Arctic*, London, 1991.
Clarke, Duncan, *Empires of Oil: Corporate Oil in Barbarian Worlds*, London, 2007.
Cone, Martha, *Silent Snow: The Slow Poisoning of the Arctic*, New York, 2005.
Conquest, Robert, *Kolyma: The Arctic Death Camps*, London, 1978.
Diamond, Jared, *Collapse: How Societies Choose to Fail or Succeed*, New York, 2006.
Diubaldo, Richard, *Stefansson and the Canadian Arctic*, Montreal, 1998.
Eferink, Alex G. Oude and Rothwell, Donald R., eds., *The Law of the Sea and Polar Maritime Delimitation and Jurisdiction*, The Hague, 2001.
Gaddy, Clifford and Hill, Fiona, *The Siberian Curse: How Communist Planners Left Russia Out in the Cold*, Washington D.C., 2006.
Galbraith, John S., *The Hudson's Bay Company as an Imperial Factor, 1821–1869*, Berkeley, 1957.
Goldman, Marshall I., *Petrostate: Putin, Power and the New Russia*, Oxford, 2008.
Hanson, Earl Parker, *Stefansson: Prophet of the North*, New York, 1941.
Horensma, Pier, *The Soviet Arctic*, London, 1991.
Huntford, Roland, *Nansen: The Explorer as Hero*, London, 1997.

Jensen, Ronald J., *The Alaska Purchase and Russian–American Relations*, London, 1975.

Kavenna, Joanna, *The Ice Museum: In Search of the Lost Land of Thule*, London, 2005.

Kaye, Roger, *Last Great Wilderness: The Campaign to Establish the Arctic National Wildlife Refuge*, Fairbanks, 2006.

Magnason, Andri Snær, *Dreamland: A Self-Help Manual for a Frightened Nation*, London, 2008.

Mann, Chris and Jörgensen, Christer, *Hitler's Arctic War: The German Campaign in Norway, Finland and the USSR, 1940–1945*, London, 2002.

Maugeri, Leonardo, *The Age of Oil: The Mythology, History and Future of the World's Most Controversial Resource*, London, 2006.

McCannon, John, *Red Arctic: Polar Exploration and the Myth of the North in the Soviet Union, 1932–1939*, Oxford, 1998.

McGhee, Robert, *The Last Imaginary Place: A Human History of the Arctic World*, Oxford, 2006.

McLaren, Alfred S., *Unknown Waters: A First-Hand Account of the Historic Under-Ice Survey of the Siberian Continental Shelf by USS Queenfish (SSN-651)*, Tuscaloosa, AL, 2008.

Nansen, Fridtjof, *In Northern Mists*, London 1911.

—*Through Siberia the Land of the Future*, London, 1913.

Nuechterlein, Donald E., *Iceland: Reluctant Ally*, Ithaca, NY, 1961.

Paolino, Ernest N., *The Foundations of American Empire: William Henry Seward and U.S. Foreign Policy*, London, 1973.

Peterson, D. J., *Troubled Lands: The Legacy of Soviet Environmental Destruction*, Boulder, 1993.

Rothwell, Donald R., *The Polar Regions and the Development of International Law*, Cambridge, 1996.

Solzhenitsyn, Alexander, *The Gulag Archipelago: An Experiment in Literary Investigation*, London, 1985 (first published in Russian in 1973).

Sorensen, Axel Kjaer, *Denmark-Greenland in the Twentieth Century*, Copenhagen, 2006.

Stefansson, Viljhalmur, *The Friendly Arctic: The Story of Five Years in Polar Regions*, London, 1921.

—*The Northward Course of Empire*, London, 1922.

Stokke, Olav Schram and Honneland, Geir, eds., *International Cooperation and Arctic Governance: Regime Effectiveness and Northern Region Building*, London, 2007.

Tamnes, Rolf, *The United States and the Cold War in the High North*, Brookfield, VT, 1991.

Vaughan, Richard, *The Arctic: A History*, London, 1994.

Wohlforth, Charles, *The Whale and the Supercomputer: On the Northern Front of Climate Change,* New York, 2004.

Articles and reports

Balzer, Harley, 'Vladimir Putin's Academic Writings and Russia's Natural Resource Policy', *Problems of Post-Communism*, vol. 53, no. 1, January/February 2006, pp. 48–54.

Bøhmer, Nils; Nikitin, Aleksandr; Kudrik, Igor; Nilsen, Thomas; McGovern, Michael H.; Zolotkov, Andrey, *The Arctic Nuclear Challenge*, Bellona Foundation, 2001.

Campbell, Kurt M.; Gulledge, Jay; McNeill, J. R.; Podesta, John; Ogden, Peter; Fuerth, Leon; Woolsey, James R.; Lennon, Alexander T. J.; Smith, Julianne; Weitz, Richard and Mix, Derek, *The Age of Consequences: The Foreign Policy and National Security Implications of Global Climate Change*, Center for Strategic and International Studies and the Center for a New American Security, November 2007.

Corell, Robert, et al., *Arctic Climate Impact Assessment – Scientific Report*, Cambridge, 2005.

Durner, George M.; Douglas, David C.; Nielson, Ryan M.; Amstrup, Steven C., and McDonald, Trent L., *Predicting the Future Distribution of Polar Bear Habitat in the Polar Basin from Resource Selection Functions Applied to 21st Century General Circulation Model Projections of Sea Ice*, United States Geological Survey, 2007.

Ewing, Maurice and Donn, William L., 'A Theory of Ice Ages', *Science*, vol. 123, no. 3207, June 1956.

Fogelson, Nancy, 'Greenland: Strategic Base on a Northern Defense Line', *The Journal of Military History*, vol. 53, no. 1, January 1989, pp. 51–63.

Goodman, Sherri W.; Catarious, David M.; Filadelfo, Ronald; Gaffney, Henry; Maybee, Sean; and Morehouse, Thomas, *National Security and the Threat of Climate Change,* the CNA Corporation, 2007.

Gordon, Richard and Stenvoll, Thomas, *Statoil: A Study in Political Entrepreneurship*, James A. Baker III Institute for Public Policy of Rice University, March 2007.

Hegerl, G. C.; Zwiers, F. W.; Braconnot, P.; Gillett, N. P.; Luo, Y.; Marengo Orsini, J. A.; Nicholls, N.; Penner, J. E. and Stott, P. A., '2007: Understanding and Attributing Climate Change', *Climate Change 2007: The Physical Science Basis. Contribution of Working Group 1 to the Fourth Assessment Report of the Intergovernmental Panel on Climate Change*, Cambridge, 2007.

Holland, Marika M.; Bitz, Cecilia M., and Tremblay, Bruno, 'Future

Abrupt Reductions in the Summer Arctic Sea Ice', *Geophysical Research Letters*, vol. 33, 2006.
Huebert, Robert and Yeager, Brooks, *A New Sea: The Need for a Regional Agreement on Management and Conservation of the Arctic Marine Environment*, World Wildlife Fund, January 2008.
Ingimundarson, Valur, 'Buttressing the North: The Atlantic Alliance, Economic Warfare, and the Soviet Challenge in Iceland, 1956–1959', the *International History Review*, vol. XXI, no. 1, March 1999, pp. 80–103.
Johnston, V. Kenneth, 'Canada's Title to the Arctic Islands', *Canadian Historical Review*, vol. XIV, 1933, pp. 24–41.
Jones, B. M.; Arp, C. D.; Jorgenson, M. T.; Hinkel, K. M.; Schmutz, J. A.; and Flint, P. L., 'Increase in the Rate and Uniformity of Coastline Erosion in Arctic Alaska', *Geophysical Research Letters*, vol. 36, 14 February 2009.
Kolz, Arno W. F., 'British Economic Interests in Siberia during the Russian Civil War, 1918–1920', the *Journal of Modern History*, vol. 48, no. 3, September 1976, pp. 483–491.
Larsson, Robert, *Russia's Energy Policy: Security Dimensions and Russia's Reliability as an Energy Supplier*, Swedish Defence Research Agency, March 2006.
Lovins, Amory B. and Lovins, Hunter L., 'Fool's Gold in Alaska', *Foreign Affairs*, July/August 2001, pp. 72–85.
MacDonald, Brian, ed., *Defence Requirements for Canada's Arctic*, Vimy Paper, the Conference of Defence Associations Institute, 2007.
Mackinder, Halford, 'The Geographical Pivot of History', *Geographical Journal*, vol. 23, 1904, pp. 421–437.
—'The Round World and the Winning of the Peace', *Foreign Affairs*, July 1943, pp. 595–605.
McCannon, John, 'Positive Heroes at the Pole: Celebrity Status, Socialist-Realist Ideals and the Soviet Myth of the Arctic, 1932–39', *Russian Review*, vol. 56, no. 3, 1997, pp. 346–365.
Nansen, Fridtjof, 'Klimat-Vekslinger I Nordens Historie', *Avhandlinger Utgitt av Det Norske Videnskaps-Akademi I Oslo*, Oslo, 1926.
Nikitin, Aleksandr; Kudrik, Igor; Nilsen, Thomas, *The Russian Northern Fleet: Sources of Radioactive Contamination*, Bellona Foundation, 1996.
Noël, Pierre, *Beyond Dependence: How to Deal with Russian Gas*, European Council on Foreign Relations, November 2008.
Pearson, Lester B., 'Canada Looks "Down North"', *Foreign Affairs*, vol. 24, no. 4, July 1946.
Ries, Tomas and Skorve, Johnny, *Investigating Kola: A Study of Military Bases using Satellite Photography*, Norwegian Institute for International Affairs, 1987.

Riffenburgh, Bruce A., *The Anglo-American Press and the Sensationalization of the Arctic, 1855–1910*, D. Phil thesis, University of Cambridge, 1991.

Skorve, Johnny, *Megaton Nuclear Underground Tests and Catastrophic Events on Novaya Zemlya: A Satellite Study*, NUPI Paper 716, Norwegian Institute of International Affairs, 2007.

Skorve, Johnny and Skogan, John Kristen, *The NUPI Satellite Study of the Northern Underground Nuclear Test Area on Novaya Zemlya: A Summary Report of Preliminary Results,* Research Report no. 164, Norwegian Institute of International Affairs, December 1992.

Smith, Gordon W., 'The Transfer of Arctic Territories from Great Britain to Canada in 1880, and Some Related Matters, as Seen in Official Correspondence', *Arctic*, vol. 14, no. 1, 1961, pp. 53–73.

Stokke, Olav Schram, 'A Legal Regime for the Arctic? Interplay with the Law of the Sea Convention', *Marine Policy*, vol. 31., no. 4, 2007, pp. 402–408.

Timtchenko, Leonid, 'The Russian Arctic Sectoral Concept: Past And Present', *Arctic,* vol. 50, no. 1, March 1997, pp. 29–35.

Victor, David G. (project leader), *National Security Consequences of US Oil Dependency*, Independent Task Force Report no. 58, Council on Foreign Relations, New York, 2006.

Victor, David G. and Victor, Nadejda Makarova, 'Axis of Oil?', *Foreign Affairs*, March/April 2003.

Victor, Nadejda Makarova, *Gazprom: Gas Giant Under Strain*, Working Paper #71, Program on Energy and Sustainable Development, Stanford University, January 2008.

Webb, Melody, 'Arctic Saga: Vilhjalmur's Attempt to Colonize Wrangel Island', *The Pacific Historical Review*, vol. 61, no. 2, May 1992, pp. 215–239.

Whist, Bendik Solum, *Nord Stream, Not Just a Pipeline: An Analysis of the Political Debates in the Baltic Sea Region Regarding the Planned Gas Pipeline from Russia to Germany,* Fridtjof Nansen Institute, November 2008.

Young, Gordon, 'Norway's Strategic Arctic Islands', *National Geographic,* vol. 154, no. 2, August 1978, pp. 267–283.

Index